For the Betterment of the Race

For the Betterment of the Race

The Rise and Fall of the International Movement for Eugenics and Racial Hygiene

Stefan Kühl

Translated by
Lawrence Schofer

FOR THE BETTERMENT OF THE RACE
Copyright © Stefan Kühl, 2013.

All rights reserved.

First published in German in 1997 as *Die Internationale der Rassisten: Aufstieg und Niedergang der internationalen Bewegung für Eugenik und Rassenhygiene im 20. Jahrhundert* by Campus Verlag, Frankfurt, Germany.

First published in English in 2013 by
PALGRAVE MACMILLAN®
in the United States—a division of St. Martin's Press LLC,
175 Fifth Avenue, New York, NY 10010.

Where this book is distributed in the UK, Europe and the rest of the world, this is by Palgrave Macmillan, a division of Macmillan Publishers Limited, registered in England, company number 785998, of Houndmills, Basingstoke, Hampshire RG21 6XS.

Palgrave Macmillan is the global academic imprint of the above companies and has companies and representatives throughout the world.

Palgrave® and Macmillan® are registered trademarks in the United States, the United Kingdom, Europe and other countries.

ISBN: 978–1–137–28611–6

Library of Congress Cataloging-in-Publication Data

Kühl, Stefan.
   For the betterment of the race : the rise and fall of the international movement for eugenics and racial hygiene / Stefan Kühl ; translated by Lawrence Schofer.
      pages cm
      Includes bibliographical references.
      ISBN 978–1–137–28611–6 (alk. paper)
   1. Eugenics—History. 2. Racism—History. I. Title.

HQ751.K838 2013
305.8—dc23   2013005123

A catalogue record of the book is available from the British Library.

Design by Newgen Knowledge Works (P) Ltd., Chennai, India.

First edition: August 2013
10 9 8 7 6 5 4 3 2 1

Contents

Abbreviations		vii
Introduction		1
One	The Dream of the Genetic Improvement of Mankind—The Formation of the International Eugenics Movement	11
Two	The First World War and Its Effect on International Eugenics	29
Three	Racism, Internationalism, and Eugenics	47
Four	The Crisis of Orthodox Eugenics and the Rise of Human Genetics and Population Science	71
Five	National Socialist Germany and the National Eugenics Movement	91
Six	The Second World War and the Mass Murder of the Sick and Handicapped	121
Seven	On "Good" and "Bad" Eugenics: Refocusing on Human Genetic Counseling and the Struggle against "Overpopulation"	133
Eight	The Renaissance of Racist Eugenics	157
Nine	The Dissolution of the Eugenics Movement: Will There Be Eugenics without Eugenicists?	181
Afterword		187
Notes		191
Sources and Bibliography		237
Index of Persons		279
Organizations, Conferences, Journals and Newspapers		285

Abbreviations

The abbreviations for the archives are noted in the bibliography of sources and secondary literature.

AES	American Eugenics Society
FILDSE	Fédération Internationale Latine des Société d'Eugenénique
FHU	Foundation for Human Understanding
FREED	Foundation for Research and Education on Eugenics and Dysgenics
GbAEV	Gesellschaft für biologische Anthropologie und Verhaltensforschung
GRECE	Groupement de recherche et d'études pour la civilisation européene
IAAEE	International Association for the Advancement of Ethnology and Eugenics
IFEO	International Federation of Eugenic Organizations
IGRH	Internationale Gesellschaft für Rassenhygiene
IIA	Institut International d'Anthropologie
IIS	International Institute of Sociology
INED	Institut National d'Etudes Démographiques
IPPF	International Planned Parenthood Federation
IUSIPP	International Union for the Scientific Investigation of Population Problems
IUSSP	International Union for the Scientific Study of Population
NPD	Nationaldemokratische Partei Deutschlands
NSDAP	Nationalsozialistische Deutsche Arbeiterpartei
PAA	Population Association of America
PIC	Population Investigation Committee
PIEC	Permanent International Eugenics Commission
SD	Sicherheitsdienst
SS	Schutz-Staffel
T4	Tiergartenstraße 4
UN	United Nations
UNESCO	United Nations Educational, Scientific and Cultural Organization
UNFPA	United Nations Fund for Population Activities

Introduction

The attempt to improve the quality of the hereditary pool of a particular group of people was long connected directly with National Socialism. "Eugenics" and "race hygiene," aiming for the genetic "improvement" of the race, inevitably aroused thoughts about the Germany of 1933 to 1945. After all, National Socialists relied on these ideas of mass sterilization of the mentally and psychologically handicapped as a means of legitimating the ban on marriage between the handicapped and the non-handicapped and to justify the mass murder of the handicapped and the sick.[1] This led to the situation after the Second World War where the image of race hygiene and eugenics was frequently reduced to the inhumane reactionary and pseudoscientific practices of National Socialism.[2]

In the 1970s and 1980s, the immense advances of gene technology set off a great controversy about the possible risks of in utero diagnoses, fertilization outside the body of the mother, genetic germline therapy, and the creation of genetically identical human beings. The reduction of eugenics to a forerunner of National Socialist race policy played a central role in this discussion. Reference to the systematic application of eugenics and race hygiene in National Socialism served critics of human genetic measures as a convenient coup de grace against all attempts to bring the topic of eugenics back onto the political and scientific agenda.

By the beginning of the 1980s, however, the image of eugenics and race hygiene as the pathbreaker for National Socialism was crumbling. Studies of various national eugenics movements have shown that eugenics was in no way confined merely to Germany and some other European countries plus the United States, but rather it was also propagated in Brazil, China, the Soviet Union, and in Japan as well.[3] Clearly, a policy for improvement of the hereditary patrimony was in no way a monopoly of the National Socialists. Representatives of practically all political convictions used the idea of eugenics as a design for the genetic improvement of the human race. Neither socialists nor anarchists, neither conservatives nor liberals were immune to the enticements of the scientifically based social engineering of eugenics.[4] Even religious underpinnings, such as in Catholicism, Protestantism, or in Judaism, in no way prevented the propagation of a eugenic policy. On the contrary—Catholic, Protestant, and even Jewish varieties of eugenics succeeded in having an influence on legislation in various countries.[5] Historical case studies on individual eugenicists have shown that eugenics per se could not simply be reduced to a pseudoscientific ideology since a number of eugenicists of the twentieth century were numbered among the leading scientists of their time.[6]

The new, more finely tuned image of eugenics researchers has been marked by an emphasis on variety within the movement. More and more national, political, and scientific distinctive patterns of eugenics have been uncovered, and the resulting network of such special paths now constitutes the standard picture of eugenics. The original orthodox race-oriented eugenics, based on Mendelian laws of inheritance, in the meantime has come to be considered as only one among several variants of eugenics.[7]

Nonetheless, this picture of variation within the eugenics movement does not mean that the last word has been spoken regarding eugenics and race hygiene. In a way, more new questions have been raised than old ones have been answered by the excellent studies regarding socialist eugenics in Germany and the Soviet Union, on scientifically brilliant eugenicists in the United States and Great Britain, on the strongly social hygiene–oriented eugenics in France, on eugenics in South America, and on the German eugenicists who spoke out decisively against the killing of mentally handicapped and physically sick people. If eugenics was a worldwide phenomenon, what were the relations among the individual national movements? What conflicts were there between the various movements? If there were eugenicists who were scientifically prominent, what connections existed between eugenics as a scientific idea and eugenics as a political movement? By what means was the attempt made to make eugenics into a scientific enterprise, and what considerations led to a situation where this strategy was later abandoned? If eugenics was not a specific hallmark of National Socialism, what were the relationships between the various types of eugenics in the National Socialist race policies? If we understand the discussions about eugenics and euthanasia as two independently existing areas, how can we then explain the tolerance or even support by German eugenicists for the murder campaigns against the handicapped and the sick?

Toward a Revision of the Dominant Image of Eugenics

Based on the history of the international eugenics movement in the twentieth century, answers to these questions will be proposed in this book.[8] These answers are meant to provide a framework for doing justice to the varieties of eugenics and at the same time should correct certain assumptions that continue to be held, despite the change in history writing about eugenics and race hygiene.

First of all: A number of historians have shown close ties between the early variant of eugenics and National Socialism.[9] Under the impact of National Socialism, in which nationalist policy was merged with a program of eugenics, race distinctiveness, and destruction, racist tendencies have also been perceived in other countries and other periods as biological justification for nationalistic demands. My analysis of the international eugenics movement contradicts these case studies about single-nation-oriented eugenicists. To be sure, German, British, and French eugenicists were numbered among supporters of nationalist movements, but the orientation of the leading representatives of the early eugenics movement was decidedly international. Basing themselves on a eugenic perception, these thinkers did not place a specific nation in the foreground, but instead the white, Europoid race stood out in their thinking. The perception of an international of the "white race" was a driving force in establishing and expanding the international eugenics movement.

Second: Historians in Scandinavia, Germany, Great Britain, and the United States have created a picture of the racist, scientifically questionable eugenicists as the mainstream of the early eugenics movement. This mainstream is said to have been pushed aside in the 1930s by scientifically credible reform eugenicists.[10] The history of the international eugenics movement calls into question this implicit equivalence of racism and pseudoscientific thought. The subsequent discrediting of the racist notions of the early eugenicists as unscientific may be convincing from today's point of view, but the discussions from that period are not irrational. Eugenicists of race orientation and the supposedly antiracist ones each tried to discredit the other by making the accusation of being "unscientific." On the basis of the international eugenics movement, it may be shown that it was precisely the race-oriented eugenicists who stood for eugenics as an independent scientific discipline in the international arena. They were the ones who in the 1920s and 1930s attempted to save the claim of eugenics as true science by intensifying race research. In contrast, the supposedly scientifically oriented antiracist reform eugenicists represented the point of view that eugenics was not a scientific discipline.[11]

Third: For a long time, historians traced the roots of eugenics to the Social Darwinism of the end of the nineteenth century.[12] This approach implies that the eugenicists also adopted the outlook of the Social Darwinists in questions of war. Consequently, the "struggle for existence" in wartime clashes led to distinction by race, due to the fact that genetically inferior elements were wiped out. However, on the basis of the international eugenics movement, it can be shown that even though at the beginning of the twentieth century there were some eugenicists who maintained such a position, under the pressure of the First World War, the overwhelming majority of eugenicists became supporters of a eugenically motivated policy concern about the counterselective effect of modern war. This did not lead to an absolute rejection of war, but it did form the background for the radicalization of eugenic positions after the First World War.

Fourth: The standard works on the history of eugenics as a rule end with the Second World War. If treated at all, the period after 1945 is handled simply in the form of a brief review of the possibilities of a eugenic renaissance through new genetic techniques.[13] An impartial observer could get the impression that the racist variant of eugenics came to an end with the discrediting caused by National Socialism. On the contrary, however, the history of the international eugenics movement shows that though the First World War forms a giant hiatus for the race-based variant of eugenics, it did not lead to the complete disappearance of this variant from the picture. In the 1960s, an international eugenics organization with a racist outlook and with strong personal and institutional continuity was revived in the tradition of the early eugenics movement. Even if this association never achieved the scientific and political influence of the early international eugenics organizations, it is clear that a revival of race-oriented eugenics in the 1960s and particularly in the 1970s came about in the wake of the clashes over the ending of racial segregation in the United States.[14]

This book should make it clear that the "racist international," meaning the wing of eugenicists and race hygienists primarily concerned with the genetic improvement of a particular group defined by race was the dominant force forming international working relationships for the entire twentieth century. This trend within the eugenics movement concentrated on the "distinction" of the "Nordic" or "European" race and excluded alternative views from international meetings, views that included an explicitly antiracist orientation. For that reason, the history of the international eugenics movement is primarily the history of race-oriented eugenicists, not a comprehensive history of eugenics and race hygiene in the international context.[15]

When I use the concept of the eugenics "movement," I am adopting a label that was used by the eugenicists themselves. They wanted to express that they were concerned not only with differentiating a new scientific discipline, but also with the establishment of a political movement that on the national and international stage called for the implementation of a set of eugenic requirements. The identity of the movement was thereby not to be defined solely through academic research, but also through the pursuit of a positively defined end value—the genetic improvement of the race. This value was here so abstractly formulated that it could bind together persons of quite different political and professional origins, including those from opposing directions.[16] Thus, competitive wings came into being, and it became increasingly difficult to speak of a general unified eugenics movement.[17]

As opposed to formal organizations, there is no formal entry into a movement, and it is therefore more difficult to determine which people belong to a movement like the eugenics movement and which people do not belong. One cannot take as the sole indicator for belonging to a movement one's specific political position in private discussions, but instead it is important for being classified to a movement that a person becomes active for the corresponding political position. It is certainly through the level of activities that a person himself (or herself) determines whether he is to be numbered among the activists, who are the real participants, or simply numbered among the supporters or sympathizers of a movement.[18]

In movements, certain points crystallize when cooperation broadens; the collective ability to act as a movement results. In the case of the international eugenics movement, such nodes appeared at conferences and large-scale meetings, encouraging not only direct face-to-face communication but also allowing the creation of resolutions affecting public opinion. For eugenics, another important form of these crystallization points lay in the publication of journals in which the main lines could be defined by the editorial group. Above all, however, the central binding points lay in the overlapping cooperation of organizations. Unlike movements, organizations can decide who is a member and who is not, and therefore find it much simpler to define a specific program.

In addition to reconstructing the lines of international cooperation among persons, in this book, I use these crystallization points, expressed in international congresses and conferences as well as in journals and in organizations, in order to reconstruct the activities of the international movement.

On the Interplay of Racism, Internationalism, and Scientism

Expressed very simply, this book is intended to analyze more closely the areas of tension among racism, internationalism, and scientism. The approach using "isms" in

"Racism," "Internationalism 'and' Scientism" has the result that these concepts can be frequently used in political controversies to discriminate attacking the opposition or praising one's own positions. The challenge in this book is to use these "isms"— in particular the concept of racism and the counter-concept of egalitarianism—not as political campaign slogans, but rather to use them for completing an objective scientific description.

When in this book I speak of "racism," or its counterpart "egalitarianism," I am not trying to denigrate one position, but to give a label to a point of view from which political consequences can be drawn on the basis of scientifically accepted or rejected racial differences.[19] Racism stands for a worldview according to which human beings are divided into specific genetically determined races, whose differing properties permit an explicit or implicit ordering among them. Politically, an ethnic racism aims to "protect" the "native race" by preventing race mixing, while a eugenic racism functions to distinguish among the races by preventing supposedly inferior members of the "native race" from propagating.

The idea of "internationalism" is here described not as the obvious phenomenon of cooperation extending beyond national borders as one sees in scientific disciplines, but as the conviction that urgent political questions can be solved only through international cooperation. Under "scientism," I refer to the perception that all relevant questions can be answered only on the basis of scientific knowledge, including questions of politics, education, and economics.[20]

The case of eugenics shows that politically motivated racist perceptions do not automatically contradict internationalism or scientific objectivity. Rather, it is just the opposite—these three notions were closely interwoven in the various international eugenics and race hygiene societies over the course of the entire twentieth century.[21] This close connection is interesting because science and policy in modern society function according to different logics.[22] In science, the idea is to obtain new knowledge and to set up operations on that knowledge. It is to be tested whether new knowledge can be considered to be true or not, whereas in policy, the idea is to create collectively binding decisions that can be implemented through the use of force.[23] Not everything that is accepted by science as true must therefore be considered politically sensible as well, and not everything that is decided politically and that is implemented collectively must be based on scientific knowledge.[24]

Against this background of the differentiation between politics and science as two different societal realms, the case of eugenics stands out as interesting. At least at the beginning of the eugenics movement, there was barely any difference seen between the claims of a comprehensive scientific research program on one hand and the claims of an ambitious political program for race improvement on the other. Scientific knowledge by eugenicists was to result in direct political programs, and political programs by eugenicists were obviously to be supported by scientific research. To this extent, eugenics was practically a model of a scientism approach.

Because of this close connection between politics and science, the question arises as to what role the early markedly international orientation of eugenics played, in particular in comparison to other scientific fields. Based on a sociological perspective, one can assume that the early international orientation of eugenics was primarily to serve as support for its universal claims. The universality of science means that a statement is true, independent of where, when, and by whom it is proposed. Science demands that a statement about the supposed hereditary quality of intelligence must be valid, and it is unimportant whether the knowledge comes from a scientific institute in Washington, Nairobi, or Hong Kong, or whether this knowledge is already hundreds of years old or first been made known just two months ago, or whether the person making the discovery is a white-skinned German, a Chinese American, a European, or an Inuit living in Greenland.[25]

The situation in eugenics was basically different. The striving for the internationalization of eugenics was not primarily intended to assure the worldwide validity of scientific knowledge, but—and this is one thesis of this book—it was the international political interest of eugenicists to prevent the genetic deterioration of a supposedly well-defined "white race." Since the "white race" in the eyes of eugenicists was not limited to a national state, the eugenicists' political efforts from the very beginning had a significant international component.[26]

The fuzzy contours of eugenics as a scientific discipline made this politically motivated internationalism the primary driving force of the international eugenics movement. In the field of eugenics, established academics from such various fields as anthropology, psychiatry, biology, psychology, agricultural science, and sociology were active in striving for eugenics as applied science. They were joined by nonprofessional researchers who were not established in universities or research institutes, but who felt themselves drawn in by the political program of eugenics and who wanted to see this program supported by scientific discoveries.

Despite various isolated successful attempts at founding professorial chairs in eugenics, in the first half of the twentieth century, there was little success for eugenic research institutes and courses of study to establish eugenics as an accepted scientific discipline. It is one of the arguments of this study that eugenics, with its borrowings from biology, anthropology, psychology, and sociology, was too widespread academically to be able to be accepted as a scientific discipline. It was never able to set up a concrete disciplinary set of scientific principles, and at the same time, claimed too much on the basis of this insecure scientific basis. In the meantime, research into eugenics took on a broader meaning, and finally became only a "modern" path to speak about social problems using biological terminology.[27] While this extremely broad orientation of eugenics made it increasingly uninteresting for academic researchers to participate in these discussions, the narrower professional orientation of human genetics and demography as scientific disciplines allowed them to become established. At the same time that human genetics and demography served practically as the scientific pace car of eugenics, the connection weakened between eugenics and these fields, with their professorial chairs, research institutes, and courses of study. In the face of the politicized notion of eugenics in its initial period, this differentiation into relatively strictly defined scientific disciplines seems to have made the critical contribution that human genetics and demography became strongly oriented to a scientific code, and the ties to the political demands of eugenics grew ever looser.[28]

This distinction came to light as early as the 1920s and 1930s between relatively narrowly defined scientific disciplines like human genetics and demography, which made a distinction between science and politics, and eugenics, oriented less to "scientism" than to policy making. It is true, Nazi Germany—and some imitators like Fascist Italy under Mussolini—stood as an exception to this general trend, because here there was no strict separation between science and policy. Race policy was presented as applied science, and both science and politics profited from their symbiotic relationship. Even if science under the National Socialist regime had not played the role of a "precursor" or had not been "misused," the relationship was in the end a symbiotic one, from which both sides drew advantages.[29]

In the final analysis, the integration between Nazi race policy, the German race hygienists, and their foreign supporters broadened the divide between eugenics as a political program and the distinct scientific disciplines of human genetics and demography. Aside from the fact that eugenics by its broad interdisciplinary orientation was unable to fulfill its claim to scientific status, the close connection between eugenic science and policy discredited any serious attempts after the Second World War to establish eugenics as applied science.

The only exception in comparison to the first half of the twentieth century came in the very small international movement of racist eugenicists founded as an international organization for eugenics in the 1960s and financed mainly by a foundation in the United States. In addition to increased legitimation by being set up as a supranational organization, the international orientation aimed at a thoroughgoing political program to spread beyond national borders for "race improvement." Very much in the keeping of the tradition of the early eugenics movement, the boundaries between science and policy melted away here.

This very small, politically driven international eugenics movement stood and continues to stand in marked contrast to the scientific disciplines of human genetics and demography. After 1945, many scientifically respected human geneticists and demographers not only had hidden sympathies for a eugenics policy, but they also actively supported it. The connection between scientific research projects and political programs, however, became looser and looser. This is seen particularly in human genetics, where internationalization has been driven primarily by the desire to establish scientific contact networks across national boundaries and thus to permit the creation of a worldwide uniform standard of human genetics research.

On the Presentation of International Eugenics in the Twentieth Century

Chapter 1 shows that even at the moment of the formation of the eugenic and race hygiene movement at the beginning of the twentieth century, the national societies in Germany, Great Britain, and the United States took special notice of each other; they promoted the internationalization of eugenics through the founding of other national eugenics societies. Even if this internationalization was primarily politically motivated, a central goal was still the differentiation of eugenics as an independent scientific discipline.

Chapter 2 illustrates how the First World War interrupted the work of the international eugenic umbrella organizations. However, the general concern of eugenicists about the supposedly counterselective effect of the war due to the losses of the best "germ plasm" at the front led to a situation of popular support for many eugenics measures, such as sterilization of the mentally handicapped, a ban on marriages, and the furthering of genetically "high-value" couples. The coupling of the counterselective effects of the war with a policy opposed to marginally social groups, seen as supposed profiteers from the war, accelerated the reestablishment of the international eugenics movement after 1918.

In chapter 3, it is shown that the focus of the internationally active eugenicists on the "white race" moved across national boundaries. The increase in race research in the 1920s was also used by eugenicists to firm up the scientific claims of eugenics. It was here that the efforts of the eugenicists to distinguish a particular racially defined group increasingly made common cause with the claim to be representatives of an internationally organized and accepted academic discipline.

The topic of chapter 4 is the pressure to which eugenics was subjected as a freestanding scientific subject. The crisis that embraced many eugenics societies claiming to be scientific specialty societies at the end of the 1920s originated in the rise of the influential scientific branches of human genetics and demography, as well as by a boom in eugenic policies. Considered from an international perspective, it is clear that the "descientization" of the eugenics movement—renunciation of the claim that eugenics is an independent scientific discipline—was connected, particularly in the United States and Great Britain, with the transfer of eugenic research approaches to human genetics and population science.

Chapter 5 describes how after the takeover of power by the National Socialists, the German state developed into a "eugenic model state," one that was supported by an appreciable number of non-German eugenicists as well. The leaders of the international eugenics umbrella federation willingly allowed themselves to be used to legitimate German race policies.

In chapter 6, it is shown that the beginning of the Second World War meant the temporary end of the international eugenics movement, but at the same time, the war opened up the possibility in Germany for the Nazis to realize eugenics measures in their most radical form. It was against the background of the death of supposedly eugenically "high-value" German soldiers on the front that the systematic murder of mentally handicapped and psychologically ill persons could be carried out in Germany. The collapse of an international "eugenic peace ideology" was the matrix through which the racist and eugenically motivated mass murder of socially marginal groups was carried out.

After the Second World War, when the public learned of the crimes carried out in the name of "improvement of the race," the racist variant of eugenics was internationally discredited. Chapter 7 shows that a modernized eugenics policy, however, continued to exist in many countries. Particularly in the United States and in Great Britain, the eugenics movement maintained political and scientific influence through the rejection of open racism and the switch of its focus to the supposed "overpopulation" in the Third World; it also gained influence through the "individualized eugenics" that had become possible through genetic consultation.

In the 1960s, as indicated in chapter 8, a renaissance in racist eugenics occurred outside still existing eugenics societies. Racially oriented eugenicists created the basis for the popularization of race research in the 1970s by means of the founding of an international eugenics organization and the publication of an international journal. Even today, ongoing attempts to show scientifically a genetic determinism and the "inferiority" of specific "races" are being made theoretically, personally, and organizationally on the basis of this race-oriented international eugenics movement.

CHAPTER ONE

THE DREAM OF THE GENETIC IMPROVEMENT
OF MANKIND—THE FORMATION OF THE
INTERNATIONAL EUGENICS MOVEMENT

> *What we need is a world-wide movement that is engaged in these important topics since it is the basis of our existence... What we need is a joint effort of all the civilized nations of the world to improve the race.*[1]
> —The American eugenicist Robbins Gilman at the National Conference for Race Betterment in Battle Creek, 1914

When Francis Galton, the researcher into natural science and the founder of eugenics, told the members of the British Sociological Society of London on May 16, 1904, about the goals of his hitherto quite unknown theory about "natural inheritance," he was emphasizing the national character of his concept. According to Galton, the overriding goal of eugenicists must be to make clear the "national significance" of an intentional intervention into human evolution. As a first step, the British eugenicists would have to establish their ideas as scientific theory in order then to introduce them as the basis for a social movement and ultimately as a "new religion" in the "national consciousness."[2]

Francis Galton was the typical case of an amateur scientist interested in a whole variety of topics, such as geography, meteorology, statistics, heredity, and psychology.[3] In the eighteenth and nineteenth century, scientific research did not take place as part of a classical professional occupation, but was carried out by amateur scientists and was therefore the expression of a class-connected style of life of the British upper class.[4] Galton, like many other amateur scientists, was a wealthy man, and he did not find it necessary to be paid for his scientific research. It was just the opposite—he could finance his research out of his own means.

Born in 1822, Galton was so much a product of Victorian Great Britain that he originally wanted to restrict the activist circle of eugenics to Great Britain. Someone whose accomplishments included being among those who discovered the uniqueness of fingerprints, he also made a name for himself as the inventor of correlation coefficients in statistics, the originator of systematic meteorology, and the founder of research on twins. He could hardly have imagined initially that his theories would find a great reception outside of Great Britain. For a long time, he spoke of "national eugenics" to describe the theory of the factors that might make it possible to improve the "racial qualities of future generations and to develop them to their highest level of perfection."[5]

In opposition to the prevailing scientific notions of his time, Galton developed his theory in the 1860s that talent and character are primarily inherited. The effect of the environment was said to play only a secondary role in the mental development of human beings.[6] Galton used the index *Dictionary of Men of the Time*, a kind of nineteenth-century *Who's Who*, to show that the overwhelming majority of well-known British scientists, poets, writers, lawyers, musicians, politicians, and generals were related by blood. It followed that famous families on the average would bear a more talented younger generation than did the normal British population.[7]

From this observation, it was a relatively small step to his conviction that one could and even should improve the human race genetically. Galton promoted the idea that people who were especially talented should have an above average number of children (positive eugenics).[8] The "unfit ones" should to a large degree be excluded from propagation (negative eugenics). It was Galton's hope that by building on a better understanding of human inheritance, mankind could intervene in evolution. Galton's eugenics was directed against the fatalism found in the Darwinian theory of natural selection. His approach envisaged that human beings would use their intelligence to change the external "influences" so that only the most gifted people would reproduce.[9]

This idea of "national eugenics" makes obvious the question of how an ideology conceived originally in national bounds could at the beginning of the twentieth century become an international movement. In what follows, it will be shown how and why the originally nationally oriented eugenics movements in Great Britain, Germany, and the United States became international in scope and how three distinct and partially competing motivations in the incipient international movement came together for international cooperation. The British eugenicists, organized in the Eugenics Education Society, argued in the original sense of Galton for including eugenics in the national social reform movements. They considered an international eugenics union merely as a forum for the exchange of ideas and experiences. On the other hand, the German racial hygienists in particular called for an international eugenics organization as the racist amalgamation of white peoples. Starting with the first International Eugenics Congress in 1912, the efforts to establish eugenics as a worldwide accepted science led to a situation where the internationalization of eugenics more and more became a tool for forging an instrument of scientific stature.

Between Racist Internationalism and the "Knightly Combat" of Nations in the Art of "Distinguishing Individual Races"

Although Galton's eugenic ideas for a long period received little attention, at the beginning of the twentieth century, national eugenics societies in Great Britain, Germany, and the United States arose almost simultaneously. The background of the sudden popularization of eugenics lay in the shift of evolutionary Social Darwinism, dominant at the end of the nineteenth century, which simply legitimized existing social relationships as a given in nature. The new emphasis lay on the propagation of the idea of targeted interventions in the evolution of mankind.

Up until the end of the nineteenth century, the "winners" in the industrial revolution were able to present a simplistic version of the Darwinian theory of evolution in order to legitimize the dark side of industrialization—the pauperization of broad segments of the urban population, catastrophic living conditions, terrible health conditions, and increasing susceptibility to infectious diseases. Industrialists like the oil magnate John D. Rockefeller and the American "steel king" Andrew Carnegie used Darwin's theory of the "struggle for existence" to elevate biologically the laissez faire principle of Manchester liberalism, according to which the free play of forces would further the common good. In this way, they could justify growing social problems in industrial societies as "natural."[10]

At the end of the nineteenth century, however, the more crises in industrial capitalism appeared in the form of economic instability and growing social tensions, the less did Darwin's theory of the "struggle for existence" provide a conclusive mode of explanation.[11] The "Lumpenproletariat"—the dregs of society—to be sure, suffered under catastrophic living conditions, but, despite the expectations of the representatives of an evolutionary Social Darwinism, this group was not "eradicated" through the process of natural selection. The free play of the forces of selection in industrial society clearly did not lead to biological progress.

In this situation, Galton's eugenics came to the fore. Galton's basic premise was that in the industrial states, the principle of natural selection would be set aside and an over-proportional increase of the supposedly "inferior" population groups would result in the degeneration of the human race. The weak ones, unable to adjust, would no longer be "weeded out" as in preindustrial societies. Instead of that, they could hold on to life because of "pseudo-humanism" and reproduce themselves to an even greater extent. Hygiene, medicine, and social policies from this perspective are not blessings, but rather dangerous enemies of human progress.[12] From the point of view of eugenics, civilization was only a "chain of Pyrrhic victories," which in the end would see the decline of Western states just as the Roman Empire had declined earlier.[13]

Building on the generally pessimistic atmosphere at the end of the nineteenth century, the eugenicists gave a biological explanation for the problems of industrial society and thus offered what seemed to be a scientifically grounded solution. Instead of going back to the classic Darwinian natural selection in the sense of "survival of the fittest," they argued for removing those things that worked against positive selection and called for a state-managed reproduction policy based on rational criteria.[14] At the same time, in Western Europe and the United States, organized capitalism with its industrial bureaucracies and industrial conglomerates began to take over from laissez faire capitalists. Programs of state intervention in politics were gaining in popularity, and the eugenicists presented their demand for a "more rational" organization of human reproduction and selection.[15]

The proposals of the American engineer Fredrick Winslow Taylor to use exact measurements, scientific improvements in industrial production, and wages based on production in order to make the production process more rational sprang from the same *Zeitgeist* as the eugenicists' demand to replace natural selection with a more rational, more efficient, more humane form of selection.[16] The "rationalization of sexual life" promised to humanity increases in efficiency similar to those of the rationalization of the production process in the economy—at least from the perspective of the eugenicists in Western Europe and the United States.[17]

Against the background of this general trend toward rationalization in the industrial states, the eugenics movements in Germany, Great Britain, and the United States took on special national orientations in each country. In Germany, the striving of the racial hygienists for general high-level breeding of the white race linked up with a strongly medical-oriented concept of inferiority, focusing in particular on combating those who were psychologically ill and on epileptics. In the sharply outlined class society of Great Britain, the pauperization of broad segments of the population was the main concern of the eugenicists. The supposed genetic differences between the middle class and the lower classes prompted the British eugenicists to join the British social reform movement. Psychological diseases, epilepsy, and alcoholism were regarded simply as one part of the genetically related problem of pauperization.[18] In the United States, the focus of attention of the eugenics movement lay on what were called the "feebleminded." The struggle of the American eugenicists against the "psychologically inferior population" joined up with a racism against immigrants who could not satisfy the standards of the WASPs (White Anglo-Saxon Protestants).

These divergent centers of interest resulted in a situation where the various national eugenics societies gave varying degrees of emphasis to the idea of international cooperation. Based on their focus on the distinction of the white race, the German race hygienists were the driving force in the internationalization of the eugenics movement. The German physician Alfred Ploetz founded the first eugenics society in the world in 1905, the Society for Race Hygiene (*Gesellschaft für Rassenhygiene*). In contrast to the other early thinker of German eugenics, the physician Wilhelm Schallmayer, who conceived of eugenics as a social technology to strengthen national efficiency in an "international struggle for existence," Ploetz turned outward internationally with the race hygiene society that he dominated.[19]

As early as 1895, in a book on the "efficiency" of the "race" and the "protection of the weak," Ploetz formulated approaches to a systematic distinction of the race. In this book, in accord with the sense of the general feeling of decline at the end of the nineteenth century, he portrays a horrifying picture of rule by the weak and the infirm. The "protection of the weak" has come to dominate modern society instead of selection, a development that will lead to a decline in the quality of the race. As with Galton, he was concerned here with an acceleration of the conflict between the selection mechanisms first presented by Darwin versus the humanistic ideals that had influenced him and many eugenicists of his age. Instead of the gruesome struggle for existence between men, he had in mind a selection even in the days before insemination. In his utopian vision, only the couples with the best germplasm were to have children, and thus would determine the genetic future of the race. "Inferior elements" would no longer be "eradicated" through a bloody battle, but would be prevented from coming into existence in the first place. The more strongly one could prevent the production of "inferior variants," the less one would need the "struggle for existence."[20]

By "racial hygiene," Ploetz understood the same thing that Galton meant by "eugenics." In all the discussions within the international eugenics movement about the possibilities of setting clear boundaries for the field, the two concepts were generally used synonymously. For Ploetz, race was the designation for "an entity of human beings living over generations with a set of physical and mental characteristics."[21] This definition was so vague that depending on one's preference, a small ethnic group, a psychological race, or all of humankind could be understood by it.[22] Encouraged by this inexact determination of the concept of race, many German race hygienists focused their interests on a general distinction of humankind favoring the white or Nordic race. Ploetz made no secret of the fact that the "direct building ground" for his race hygienic ideal was the "Nordic race." In his opinion, this "race, tall, light-skinned, blue-eyed, blonde-haired, with a long and thin face, with a long and large head" was marked by intellectual qualities such as "joyful interior being, path-breaking intelligence and artistic creative powers, persistence, loyalty, and courage." For Ploetz, the Nordic race could still to some extent claim to be dominant in Germany, Denmark, Sweden, Great Britain, and The Netherlands, and also in Ireland, Russia, Finland, northern Italy, and northern France. However, in North America, South Africa, and Australia, the race was involved in a "difficult struggle for existence against the other races, a struggle seen by many others as hopeless." A "Nordic-Germanic race hygiene" would be necessary to assure "its survival for all time."[23]

It was this focus on the distinctiveness of the Nordic or white race that led the members of the race hygiene society, initially limited to German-speaking areas, to found the International Society for Race Hygiene in January 1907. According to a memo by Ploetz, it is necessary that the white race, "which rolls through time like a great organism in separate countries and other social entities," should have created for it a "spiritual center, consciousness, a conscience, and an organ of desire." The members should be recruited from "the top quarter of the population of today's European cultural humanity or from those whose origins lie in Europe."[24]

Information from the American physician Robert Stone to the effect that a number of people in the United States had become active for the cause of racial hygiene gave the German race hygienists the immediate stimulus to expand their membership recruitment to Europe and the United States. The United States at the end of 1906 stood on the verge of passing the world's first sterilization law in the state of Indiana. Moreover, the American Breeders Association in 1906 established a section for eugenics, which supported the choice of marriage partners according to eugenics perspectives and called for the prevention of the reproduction of "defective portions of the population."[25] On top of that, the construction of the Laboratory for Experimental Evolution Research in Cold Spring Harbor by the zoologist and geneticist Charles Davenport in 1904 laid the foundation for the construction of a eugenic research establishment.[26]

The forefront of the work of the International Society for Race Hygiene lay in the recruitment of new members in Scandinavia, a place that Ploetz considered the motherland of the Nordic race. In March 1907, Ernst Rüdin, a young Swiss physician and brother of Alfred Ploetz's first wife, made his way to Denmark, Sweden, and Norway as the official representative of the International Society for Race Hygiene. Despite the reticence of some of his Scandinavian contacts of too public a discussion of questions of reproduction, Rüdin nevertheless succeeded in winning over as members to the International Society for Race Hygiene two future key figures of the eugenics movement, the Danish heredity researcher Wilhelm Johannsen and the Norwegian chemist Jon Alfred Mjöen.[27] In 1903, Johannsen succeeded in showing that in the selection of variants from specific races, no new types arise, but rather only hereditary subraces, the so-called pure lines.[28] Stimulated in part by an early acquaintance with Ploetz, Mjöen had by the end of the nineteenth century developed an intense interest in race hygiene. In 1906, he founded a private eugenics laboratory, in which he dedicated himself primarily to research on the heredity of musical abilities and on the effects of "race mixing" between Norwegians and Lapps.[29]

Impressed by the founding of an autonomous British eugenics society, in April 1908, the general meeting of the International Society for Race Hygiene discussed "easing membership entry by non-German countries into the society."[30] The German race hygienists decided to intensify their impact in Scandinavia and to expand their recruiting activity to France and Great Britain.[31] In a trip through France in 1909, Rüdin did succeed in winning over three well-known French academic figures as members of the society, the pediatrician Eugene Apert, the Paris professor of anthropology Leonce Manouvrier, and Jacques Bertillon, the head statistician of the prefecture of Paris.[32] He was even more successful in his attempt to gain other Swedish members. In a trip to Sweden in May 1909 more than 20 new members joined. Stimulated by Rüdin, the Swedish race hygienists together with the Svenska Sällskapet för Rashygien founded a national branch of the International Society for Race Hygiene.[33]

Although the German leadership of the International Society for Race Hygiene was able to integrate the Scandinavian and French eugenicists into the original structure of the society relatively seamlessly, it was clear that it would have to make some compromises in the integration of the British Eugenics Society, founded in 1907. In a letter to the British statistician and Galton student Karl Pearson, Ploetz played down the role of the Germans in the International Society for Race Hygiene. Ploetz explained that the society "was conceived with a strong international bent" and that the individual country groups would maintain a large degree of independence. The "central office" really had only the "role of a facilitator." The headquarters of the society, temporarily still in Munich, would soon be moved to Switzerland or to Holland. It was the intention to make the International Society for Race Hygiene "as international as possible" by bringing onto the board of directors people from Great Britain, France, and Scandinavia. Ploetz suggested to Pearson that the previously sporadic contacts between the British eugenicists and the German racial hygienists be intensified and that the British Eugenics Education Society be integrated as a national grouping within the International Society for Race Hygiene.[34]

While the British eugenicists noted many similarities between the program for the Society for Race Hygiene and the goals of the Eugenics Education Society, for them it was out of the question to downgrade the Eugenics Education Society into a national grouping of the International Society for Race Hygiene.[35] In addition to a fundamental mistrust of German dominance in an international eugenics movement, a basically different concept of international cooperation was responsible for the refusal of the German offer. The Society for Race Hygiene in Germany worked on a single issue, while the British Eugenics Society was tied into a network of social reform initiatives and organizations, all of which were perceived as an instrument for the strengthening of British "national efficiency" in the struggle between the various imperialist powers.[36] Thus, the Eugenics Education Society cooperated with such groups as the Central Association for Mental Welfare, the Moral Education League, the National Council for Mental Hygiene, and the New Health Society. Many members of the Eugenics Society were at the same time active in other British reform societies.[37]

Due to this tight integration into a network of British welfare organizations, the members of the Eugenics Education Society represented a strongly national focus on the concept of eugenics.[38] Quite in keeping with Galton's original ideas, the founding members of the society defined their goal to be the establishment of "the national significance of eugenics" and a shift in public opinion so that "all affairs related to human parenthood" would be considered from the eugenics points of view.[39] In the eyes of the British eugenicists, the various national eugenic societies should compete with one another as though in a "knightly tournament" in the art of "distinguishing the races."[40] International cooperation was initially important for the British eugenicists only insofar as they could get information about developments in other countries that would allow them to strengthen their political influence in Great Britain by pointing to progress in other countries and to the international character of eugenics.

Their basically different concepts of international cooperation became clear at a meeting between the leadership of the International Society for Race Hygiene and the Eugenics Education Society in London in 1910 and again on the margins of the International Hygiene exposition in 1911 in Dresden. It is true that the British eugenicists agreed with the German race hygienists in general that it was of primary importance for the threatened "white race" genetically to be "set apart." They accepted the program of the International Society for Race Hygiene with their notion of promoting the furtherance of race hygiene among the white peoples as the basis for international cooperation, but they fended off any integration of the Eugenics Education Society into the International Society for Race Hygiene.[41] While Ploetz strove for "inclusion of the British" in his primarily racist-oriented, transnational organization, the British eugenicists argued for cooperation among equals within the framework of an international organization. This international organization was meant to concern itself only with the exchange of information and the organization of large meetings.[42]

The First International Eugenics Conference as a Reflection of Research on Heredity at the Beginning of the Twentieth Century

Fear of a dramatic degeneration of the peoples of the Western culture area was the theme of the first International Eugenics Congress held in London from July 24 to July 30, 1912, arranged by the Eugenics Education Society in consultation with their German and American colleagues. The organizers from the British Eugenics Society justified the necessity of an international meeting of eugenicists, race hygienists, and heredity researchers by stating that in all "culture peoples a selection in favor of the inferiors" had set in, a phenomenon that presented "a growing danger for the future of the entire human race." "Weakness and uselessness" would not have the result of "direct eradication" either of individuals or of particular races. Instead, practically all societies would now be so organized that "the physically and psychologically incapable" people would have children whose existence would be favored at the "cost of the better suited." In light of the imminent "racial degeneration," Galton's eugenics was of "extreme importance."[43]

At the London Congress, it was clear how much eugenics at the beginning of the twentieth century was a mixture of emerging science, a social reforming political movement, and a close confederation of eugenically motivated men and women. Since the Eugenics Education Society had invited to London the various eugenic, race hygiene, and genetic societies as well as state and municipal institutions from such various areas as administration, religion, education, medicine, and law, the attending public was quite diverse.[44] The more than 700 participants included physicians, biologists, statisticians, sociologists, and anthropologists, as well as genealogy researchers, military people, politicians, church leaders, and representatives of the feminist movement and social reformers.[45] A wide variety of personalities acted as vice presidents of the congress, such as Sir Thomas Barlow, the president of the Royal College of Physicians; the Lord Bishop of Oxford; the later British Prime Minister Winston Churchill; the Swiss psychiatrist Auguste Forel; the Italian criminal anthropologist Alfredo Niceforo; the inventor Alexander Graham Bell; and the American social reform politician Gifford Pinchot.

The extensive reporting on the congress by the international press reflected how unclear it was in 1912 what eugenics was exactly and who represented it. The reporters in attendance could not agree as to whether they were present at the birth of a new innovative science, the founding of a political movement marked by class and race prejudices, or simply an international meeting of dreamers and visionaries. Some newspapers did not refrain from making fun of the scientific dilettantism of many a participants. So the German *Vossische Zeitung* mocked the Turin professor Roberto Michels, who had claimed in one lecture that the most handsome people were also the best politicians. According to a reporter for the *Vossiche Zeitung*, this surely did not include "feminine beauty," but only the handsome male, who radiated "desire, superior knowledge, and self-confidence."[46] The *Berliner Tageblatt* complained that "the female element" was too broadly spread in London. The editors printed a drawing of street beauties discussing among themselves whether their attractions were in demand at the "Congress on Racial Improvement" in London.[47] The Paris newspaper *Le Matin* presented caricatures to its readers of how the congress participants would most likely imagine the ideal cricket player, boxer, businessman, Oxford professor, pickpocket, or public speaker.[48]

The uncertainty over the status of eugenics was connected to the call for the tight interconnecting of science and policy, an idea that emerged with Galton. The eugenicists were not only attempting to have a scientific monopoly on the research into human heredity and population, but they also claimed for themselves the status of a political movement. As the British newspaper *Public Opinion* wrote regarding the congress, the eugenicists were concerned with a "marriage between science and practice." The research program and a political program were to be merged into one unit.[49]

This merger of science and politics at the beginning of the twentieth century essentially established eugenics. From one perspective, the reference to the threatening signs of degeneration in modern society could help mobilize the necessary resources for researching the laws of human heredity and selection. Reference to a politically critical social situation helped just such a young research area as that of eugenics to participate in the distribution of research money from the state and from foundations. On the other hand, the reference to the scientific character of eugenics served the eugenicists in legitimating their political demand for a genetic "distinction" of race. Political decision makers could utilize the scientific nature of eugenics to make the basis of eugenics policies seem objective.

As the basis for their theory of the distinction of the races, the eugenicists looked to the science of genetics for support for its statements about the inheritance of human physical and psychological characteristics. In particular, a whole series of supposed hereditary diseases, such as syphilis, tuberculosis, alcoholism, and manic-depressive insanity, could, in the view of the eugenicists, be explained and combated with the aid of genetics. Building on this genetic biological foundation, the eugenicists sought to determine the various laws of selection that are dominant in modern societies by studying the regional, race, and class-specific development of populations. It was from these insights from natural and social science that they filtered out a scientifically grounded political program.

Despite the worry mentioned by some observers that the eugenics movement could overreach with its scientific and political claims, among the London Congress-goers, the confidence ruled that new scientific insights from research on human heredity and on populations would lead to legitimation of the eugenicists' political program. This optimism was stimulated by the feeling of breakthrough that reigned in heredity research on the basis of the rediscovery of the Mendelian laws of inheritance and the increasing spread of the Weismann theory of the continuity of the germplasm.

The Austrian Augustinian monk Gregor Johann Mendel had observed in his experiments with garden peas that crossing plants with green and yellow seed shells in the first generation would result in all peas being green-seeded in the first generation, but in the second generation for every three green-seeded plants, there would be one yellow one. The predisposition for the quality of yellow seeds thus remained, even if it did not appear in the first generation. From here, Mendel concluded that in every parent plant, two factors must exist, each of which is passed on to the daughter plant. If the two factors that come together in the daughter plants are different, so Mendel reasoned, then the dominant factor would rule over the recessive factor. Only the dominant factor would be visible, such as the green seed of the pea plant. However, if two recessive factors were to come together in the second daughter plant, the suppressed material would be apparent again in the next generation.[50]

August Weismann, professor of zoology in Freiburg, honorary member of the International Society for Race Hygiene and one of the vice presidents of the London Congress, in 1892 presented the theory that only the hereditary material in the germ cells or seed cells is passed on to the next generation. The body cells would of course come out of these germ cells; an alteration of the body cells would not lead to a change in the hereditary material. Like Galton before him, he therefore completely ruled out heredity of characteristics acquired during one's life.[51] Going on from here, he concluded that it was not the organism shaped by the environment that was important for maintenance of the species, but the hereditary material that was determined even before birth.

Even though at the beginning of the twentieth century both Mendel's and Weisman's theories were still questioned by many heredity researchers and eugenicists, at the London Congress, American and British eugenicists in particular applied the findings from the experimental studies on heredity directly to humans. The congress gave short shrift to the neo-Lamarckians, who continued to believe in the theory of Jean-Baptiste de Lamarck regarding the inheritance of acquired characteristics and who engaged in vigorous arguments with the representatives of the Weisman germ plasm theory. Neither did the biometricians, who in contrast to the experimentally oriented Mendelians, quantified the heredity factors in biological populations and made prognoses about their further development.[52]

Hence, the American eugenicist Raymond Pearl essentially proclaimed without opposition in London that the results of his studies on the fertility of hens could be transferred to human beings. He felt that he had proved that high fertility was strictly hereditary under Mendel's laws, and he speculated that the declining fertility among "highly civilized races" was connected to the loss of one or several "fertility genes."[53] Reginald Punnett, professor of biology at the University of Cambridge and member of the Eugenics Education Society, claimed that "mental weaknesses" are inherited based on the rules of Mendel's laws.[54] David Weeks, medical director of the New Jersey State Village for Epileptics, declared in his presentation that epilepsy alone was not inherited according to Mendel's laws, but that in the Mendelian schema, mental weakness should be considered as part of recessive inheritance.[55]

At the congress exhibitions as well, the eugenicists attempted to prove the genetic foundation of various diseases, handicaps, and social oddities on the basis of Mendel's rules of heredity and the Weismannian theory of the continuity of the germplasm.[56] Davenport and colleagues from the Eugenics Record Office that he set up in 1910 presented genealogical trees that purported to prove that most mental illnesses were inherited recessively. Every social oddity, from alcoholism through mental retardation down to epilepsy, was forced into a Mendelian schema.[57] German race hygienists, using the form of the traditional ancestor or clan tables, presented studies on the degeneration of the Habsburg clan. The peculiarities extended from the protruding edges of the mouth—the so-called Habsburg lip—through mental abnormalities down to declining fertility.[58] The British eugenicists were represented at the exhibits by the research studies of Ernest Lidbetter. Lidbetter as a young social worker in the poverty areas of the London East End had determined that a large proportion of his "clients" were related to one another. He concluded from this that there was a hereditary propensity to poverty that was passed on from generation to generation.[59] At the London Congress, he presented family trees that were supposed to prove that poverty is inherited and is accompanied by mental and physical defects. Unlike Davenport, he did not attempt to force every mental or social deficiency into a Mendelian schema, but declared himself satisfied that he had proved through his family trees that there was something like a genetically determined social underclass.[60]

In London, it was clear that both the Mendelian Laws and the Weismann theory of the "continuity of the germ plasm" could lead to a radicalization of eugenic positions. The theses that Mendel developed on plants, that specific recessive properties are not expressed in the first-order generation, could in the opinion of many eugenicists explain why mental diseases did not appear automatically in each successive generation. Since human beings could carry hereditary traits or mental diseases without themselves being sick, the eugenicists assumed that a creeping "degeneration" was an imminent danger.

Weismann's germ plasm theory assigned a meaning to environmental influences only in the formation of the individual. In the opinion of the geneticists who adopted this theory, an improvement of the environment would consequently have no positive effects on the race. The "racial degeneration" could be halted only through interventions in the area of reproduction, not through an improvement of the medical, hygienic, and economic conditions of human life.[61]

Against the background of the fact that the birth rate had been declining since the end of the nineteenth century in the industrial states, in particular among the middle and upper classes, several speakers in London warned of a disproportional increase of the racially or socially "inferior" population groups. The "differential reproduction" between "superior and inferior" portions of the population was built up into a threat for the civilized peoples.[62] For example, the American statistician and eugenicist Frederick L. Hofman reported on a change in the composition of the immigrants to the United States. Of the approximately 1,000,000 immigrants annually, the greatest number no longer came from northern Europe, as they had done in the past, but from the southern countries of the continent. According to Hoffmann, the drastic reduction of the share of Germans and the sharp increase in the share of Italians would make clear the gravity of the "race problem" in America. The disproportionate number of children of immigrants from southern and eastern Europe was an unambiguous indicator of "degeneration."[63] The British eugenicist Murray Leslie agreed with Hoffmann's analysis and declared that the situation in the London East End was comparable to that in the United States. There too, the share of British in comparison to the residents born outside of Great Britain was declining radically because of the higher fertility of the immigrants.[64]

Bleecker van Wagenen, chairman of the sterilization committee of the eugenics section of the American Breeders Association, warned in London that the number of "defectives" in America and Europe was growing rapidly and that the "asocial and defective character traits" were creeping even into the "normal population groups." Wagenen went on that the "mental retarded," the "pauper," the criminals, the epileptics, the insane, the constitutionally weak, the "deformed," or "those having defective sense organs, as the blind and deaf, or the kakaistetic class" must be considered "socially unfit" and should, if possible, "be eliminated from the human stock."[65]

Wagenen informed his European colleagues in London about the introduction of sterilization laws in the states of Indiana, Washington, California, and Connecticut. Despite a certain doubt about the practice in the United States and the desire for more knowledge about the genetic bases of the mental diseases, for him, sterilization was the long-term proper alternative to the very expensive prevention of reproduction through strict isolation of the sexes in institutions.[66]

Only rarely was any criticism at the London Congress heard about the demand for eradication of socially marginal groups. Prince Pëtr Alekseevič Kropotkin, the Russian anarchist living in exile in Great Britain, protested against sterilization of the "so-called unfit." He rhetorically asked who was useless for mankind—the presumably inferior pauperized women of the industrial proletariat, who courageously nursed their many children, or the presumably genetically superior women of the upper class, who were increasingly renouncing pregnancy.[67] Samuel G. Smith, professor of sociology at the University of Minnesota, in his speech rigorously opposed Galton's supposition that talent is inherited—Luther, Napoleon, and Lincoln were as much "biological surprises" as were Beethoven, Mozart, or Wagner. The impact of the environment was many times more important than hereditary traits. Thus, the British in an exemplary way had successfully turned criminals into prime ministers when they sent their lawbreakers to Australia.[68]

International Cooperation as an Instrument for Establishing Eugenics as a Scientific Enterprise—The Permanent International Eugenics Committee

The dawn of the twentieth century saw a boom in the founding of international organizations, which sprang from the earth in the areas of medicine, science, technology, industry, labor, religion, sports, education, and politics. While the last two decades of the nineteenth century had seen the founding of a handful of international organizations each year, in the period from 1900 to the outbreak of the First World War, up to 50 international organizations were founded each year.[69]

From this perspective, it is understandable that there was also a trend toward the formation of an international eugenics organization. In the case of eugenics, it was the first International Eugenics Congress that gave the impetus to the establishment of an international eugenics organization. In London, it also became clear to the eugenicists that international cooperation could increase their influence in politics and in science. However, just as in the preceding years, there were controversies about the setup of such an international eugenics organization. In the run-up to the conference, Ploetz spoke against the founding of such an organization and for the integration of the various national movements into the International Society for Race Hygiene. A change in the bylaws of the International Society could lead to the building up of a "confederation of all serious societies for race hygiene that are based on scientific principles." Since Ploetz had been aware of the opposition of the British eugenicists to integration of the Eugenics Education Society into his society at least since their joint meetings in 1910 and 1911, he first wanted to come to some agreement with the American and Swedish eugenicists before informing the British eugenicists of his suggestion. He hoped that by negotiating his plan ahead of time with the Swedish, German, and American representatives, he would be able to get the British to give in. Then an international network dominated by the Germans could arise out of the German section of the International Society for Race Hygiene, the Svenska Sällskapet för Rashygien, the American Breeders Association, and the Eugenics Education Society. This group could then decide on the acceptance of eugenicists from other parts of Europe and North America.[70]

Ploetz's plan did not succeed. The Swedish race hygienists could not come to London. The attempt to make contact with the American eugenicists before the congress failed. In addition, the background of the growing political tensions in Europe made the British, French, and Italian eugenicists strongly distrust the claim for leadership by the German race hygienists.[71] A disappointed Ploetz had to admit that despite the broad agreement of the eugenicists on matters of race, his idea of a race internationalism under the International Society for Race Hygiene did not find any particular response. The British eugenicists pushed through their idea that a newly founded Permanent International Eugenics Committee should coordinate cooperation among the eugenicists of the various countries and support the national movements by establishing eugenics as a science and as political practice. Ploetz's idea of an international eugenics merger of the white race was indeed considered during the founding of the Permanent International Eugenics Committee, but it did not stand in center stage as he wished.[72]

When the Permanent International Eugenics Committee met for the first time on August 4, 1913, in the offices of the French Central Statistical Office in Paris, it was clear in the meantime that a broad, international eugenics movement had grown up through the founding of other national eugenics societies from the original troika of British, German, and American societies.[73] The impulse toward institutionalization for national eugenics initiatives in France, Italy, Denmark, Norway, The Netherlands, and Belgium came directly from the International Eugenics Congress. At the forefront, the British Eugenics Education Society had stimulated the founding of the national "consultation committee," which was to coordinate the participation of academic figures from the various countries. While the American and German committees were identical with their national eugenics organizations, the French, Belgian, Italian, and Norwegian committees formed a very informal group of people who simply had a diffuse interest in eugenics and who were drawn together by the concrete goal of participating in the congress. Stimulated by further discussions and the examples of British, American, and German anarchist societies, various national eugenics initiatives arose out of this "consultation committee."

The 18-person French delegation in London called for the founding of the French eugenics society by referring to the already existing societies in Germany, Sweden, the United States, and Great Britain.[74] In 1912, in Italy, two Italian vice presidents of the London Congress, the criminal anthropologist Alfredo Niceforo and the statistician and demographer Corrado Gini, founded the Italian Eugenics Society.[75] From the Belgian consultation committee, a eugenics department in the Solvay Institute emerged, one of the most significant centers for sociological research in Europe.[76] In The Netherlands, spurred on by the congress, a eugenics committee was formed that called for a medical examination based on eugenics criteria before every marriage.[77] Jon Alfred Mjöen took advantage of the Norwegian eugenics consultation committee, founded on the occasion of the first eugenics congress, to achieve influence on Norwegian social policy up until the 1930s.[78] The Danish delegate in London, physician and anthropologist Sören Hansen, under the influence of the international meeting founded a small eugenics section in the Danish anthropological Society.[79]

As delegates of the various national societies, many of the eugenicists present in Paris represented those who would be the hard core of the international eugenics movement in the decades to come. The statistician Lucien March represented the French eugenicists, along with the later French state President Paul Doumer, the publisher of the *Revue Politique et Parlementaire* Fernand Faure, and the biologist Frederic Houssay. Leonard Darwin, who was to become the first president of the Permanent International Eugenics Committee, and Sybil Gotto, who took over the role of general secretary, represented the Eugenics Education Society. From Scandinavia there came Jon Alfred Mjöen as representative of the Vinderen Laboratory and Sören Hansen for the Danish Anthropological Society. Louis Querton and Louis Caty of the Eugenics section of the Belgian Solvay Institute came to Paris. The Italian movement was represented by Corrado Gini, who later became chairman of the Italian Eugenics Society. Ploetz, as representative of the International Society for Race Hygiene, took part in the meeting. Since the leader of the American eugenics movement Charles Davenport was unable to attend, Adam Woods represented the eugenics section of the American Breeders Association.[80]

Even though the direct impulse for institutionalization for the various eugenics initiatives in Scandinavia, France, Italy, Belgium, and Holland had originated in the International Eugenics Congress and in the promotional activity of the International Society for Race Hygiene, the existing societies also developed their own national and regional-specific eugenics programs. In France, with its sharp decline in births at the end of the nineteenth century, the dominant worry was that the French people might be outnumbered by their German neighbors, and the eugenicists emphasized measures to promote the birthrate of genetically superior couples. In Italy as well the Catholic tradition led to the highlighting of a pronatal eugenics policy. Negative eugenics measures such as sterilization and targeted contraception were regarded very skeptically by the Italian eugenicists.[81] On the other hand, in Protestant Scandinavia, demand for negative eugenics in the form of isolation of the handicapped in institutions and their sterilization did not essentially conflict with the social policy in place there.

In light of the nationally specific direction of the various eugenics societies, it was an important step that the assembled eugenicists from Great Britain, the United States, Italy, France, Belgium, Denmark, Germany, and Norway could agree on a general programmatic platform.[82] As the basis for the platform, the eugenicists used the Norwegian program for race hygiene, which Mjöen had presented for the first time at a meeting of the Norwegian physicians organization in 1908.[83] Mjöen's program could serve as the basis for a minimal consensus among the various eugenics societies because it did not clearly acknowledge either Mendelianism or the Weismann germ plasm theory. Instead, it placed in the foreground the eugenics enlightenment and its battle against the so-called racial poisons of alcohol, tobacco, and sex-related diseases, and it did not make any distinctions within the white race.[84]

Mjöen's basic idea was to distinguish between the "right to life" and the "right to give life." While the former was a basic right to which every person was entitled, the latter was a privilege belonging only to selected couples.[85] The "valuable race elements" were to be encouraged by reduced taxes and subsidies for healthy families with lots of children and special insurance for mothers to create more progeny. In the eyes of the eugenicists, subjects like "Biology (renewal of the family),"

"Chemistry (feeding the family)," and "Hygiene (protection of the family)" ought to be taught as compulsory subjects from preschool to university. In addition, state institutes for race hygiene were to spread the possibilities of eugenics far and wide among the population.

As an add-on to Mjöen's catalog of requirements, the eugenicists present in Paris favored a system for preventing the reproduction of "the mentally ill, epileptics, and similarly physically and mentally crippled individuals" as well as "alcoholics, habitual criminals, professional beggars, and those who refuse to work" by isolating them in institutions and work colonies. Eugenic sterilization, as demanded in particular by some American eugenicists, was to be used only in exceptional cases. Certain criminals were to be given the possibility of being sterilized as an alternative to being imprisoned.

The eugenicists took over from Mjöen as a "preventive" race hygiene measure the demand for campaigns against "racial poisons" like alcohol, tobacco, and sexual diseases. They made the argument that the entire population should be recorded biologically and that a national genetic and health register should be set up. Immigration from other countries was to be organized according to biological standards. In their view, every country experiencing immigration should have the possibility of turning away "generally inferior" persons. They felt that before marriage, a eugenics couple should undergo a health examination, at which the physician was to counsel the pair against "marriage between widely separated races."

The program decided on in Paris by the eugenicists included almost the entire palette of eugenics measures that were technically possible at the beginning of the twentieth century. Selection at the level of the nucleus, as demanded by Galton and Ploetz, was still not possible because of the lack of technical knowledge. It therefore seemed to the eugenicists that managing reproduction through the financial and ideological promotion of "racially superior" population groups and through the reduction of reproduction of supposedly "racially inferior" population groups was a humane alternative to the otherwise threatening social Darwinian "struggle for survival." Despite the humanistic concern in the early eugenics movement, it was still common among all eugenics program points that human beings were reduced to being objects of a scientifically based policy of race distinction. Rights applicable to all humans were subordinated to the supposedly superior goal of genetic distinction.

The eugenicists who gathered in Paris were unanimous in expressing that strengthening the scientific basis of eugenics was necessary to realize their political demand. A political eugenics program, so they reasoned, could be successfully introduced only if legitimized by eugenics science. For that reason, the debates at the London Congress included a major role for a discussion within the international movement of strategies for establishing the science of eugenics.

How did the geneticists plan to establish eugenics as a science? How could it be distinguished from neighboring disciplines like genetics, medicine, sociology, and psychiatry and developed into a uniform eugenics research program? How did they wish to justify the premature announcements by the French eugenicist Eugene Apert of eugenics as a "new science"?[86]

The establishment of research institutes, professorships, study areas, and professional journals all had a decidedly national approach, but in the opinion of the eugenicists, international cooperation was to play a central role in the establishment of eugenics as a science. As early as the London Congress the one-time Conservative prime minister and honorary vice president of the Eugenics Education Society, Arthur James Balfour, had written to the participants in the official visitors book that eugenics, this "splendid applied science," need know no bounds. The geneticists grappled with problems that concerned not individual peoples, but everyone.[87] International congresses, organizations, research initiatives, and international uniform standards of methodology would be a precondition for eugenics to be accepted as a scientific discipline.[88]

Although the internationalization of eugenics was principally politically motivated, considerations of setting up eugenics as a science also played a role. The eugenicists coupled their efforts to differentiate eugenics as an independent scientific discipline to the experiences of the second half of the nineteenth century, when many scientists had raised the status of their research discipline by internationalizing their mode of work. The research results were to be considered independent of the national, religious, political, and social background of the researcher in a kind of scientific universalism, and the growing possibilities of international scientific cooperation through the improvement of modes of transportation and communication in the nineteenth century led to a situation where international scientific joint projects mushroomed everywhere.[89]

Since eugenics had a twofold orientation, both as research into human hereditary processes and as the study of population change, international cooperation was essential for the establishment of eugenics as a science. Since genetics was propounded as the basis of eugenics in natural science, experiments and measurements of hereditary changes formed the central scientific methods of the eugenicists. The use of experiments and measurements were things that since the middle of the nineteenth century had increasingly moved into the center of scientific activity in biology, chemistry, and physics, and they demanded international standardization of measurement and experimental techniques. A transnational culture in scientific laboratories was promoted, leading to a strengthening of scientific communication among scholars of various nations.[90] On top of this, the focus of eugenicists on population change suggested both the international standardization of research methods and the transnational comparison of research data.

Following this vein of thought, the eugenicists present in Paris agreed to exchange bibliographic and anthropological data, information on research projects, and updates on initiatives for eugenics laws. It was acknowledged by all delegates that a eugenics science would have to be systematized internationally.[91] The French eugenicist Lucien March took the opportunity of a meeting of the Permanent International Eugenics committee to ask for the development of an international eugenics classification schema built on preliminary work of the Eugenics Record Office in Cold Spring Harbor, the Eugenics Education Society, and the International Society for Race Hygiene.[92] March's schema defined the scientific methodology of the eugenicists as consisting of experiments, observation, statistics, biographies, and family histories. He noted the relationship of eugenics to other sciences such as genetics, evolution research, natural history, anthropology, ethnography, demography, and economics. Moreover, March noted the areas in which the eugenicists wanted to develop new knowledge: human heredity research, the determination of environmental influences on the development of mankind, the mode of functioning of the selection processes in subgroups, and research into the possibilities of how social conditions could be altered in a eugenic sense.[93]

Subsequent to the first International Eugenics Congress, the beginning of international cooperation as an instrument for making eugenics into a science supplemented and even replaced the idea of a racist international, so strongly promoted by the German race hygienists. An international umbrella organization to organize congresses, create guidelines for a common program, and set uniform standards for research appeared in particular to the British and French eugenicists as an alternative to what Ploetz favored, an international confederation of European race hygienists and eugenicists.[94]

With agreement on a common eugenics program and the rudimentary exchange of information on literature, legislative proposals, and research projects, the international eugenics movement seemed to be on the high road to establish itself as a well-rounded player in the world of science. In fewer than ten years after the founding of the first national society and even before the setup of eugenics organizations in most countries, the eugenicists had made available a stable international network. They began planning for a large international eugenics congress in the United States.[95]

However, the outbreak of the First World War caused a break in these plans. Looking back from 1930, Eugen Fischer, Germany's most important anthropologist in the 1920s and 1930s, expressed regret that the World War had broken off this development before "true international cooperation" of eugenicists could be made apparent to the outside world.[96]

Chapter Two
The First World War and Its Effect on International Eugenics

Then this war broke out, having myself studied eugenics, it nearly broke my heart.[1]
—Irving Fisher, Economist at Yale University

Shortly after the United States entered the war in 1917, the American economist and eugenicist Irving Fisher, at a lecture in Portland, Oregon, told the story of a dinner with a young, hale, and hearty student at the University of California. "As a great admirer of health" I thought, "What a wonderful example this man is, what a wonderful physique, what an alert mind, what a fine character." That is the real tragedy of war, that such fine examples of humanity are to be sacrificed on the European front.

It did not disturb Fisher in the least that so much money was being wasted in the war, since at the most in one generation, prewar prosperity would be regained. He was not bothered by the fact that valuable works of art would be destroyed by the war. They could be restored. Nor even the fact that during the war, unprecedented numbers of human lives would be wiped out would cause him sleepless nights. Everyone has to die sooner or later. What caused his greatest fears was not the quantitative destruction of human life, but the elimination of quality, the killing of such highly valuable people such as the one whom he had met at that dinner.[2]

This chapter will show that Fisher's open concern about the dysgenic effect of modern wars was shared by many eugenicists in Europe and America. The First World War reinforced the change in thinking that had set in after the turn of the twentieth century—away from sanctioning war as the proper instrument of selection to damning war as one of the greatest dangers to distinction by race. Even if most of the national eugenics societies during the First World War saw it as their patriotic duty to support the war efforts of their own governments, and even if the international relations among eugenicists remained unbroken, still the common worry about the counterselection consequences of the war formed the basis for an intensification of eugenics policy after the war.

From Social Darwinian Warmongering to a Eugenics Peace Policy

In the first two decades of the twentieth century, the attitude of eugenicists toward modern war-making underwent an astounding change. At the turn of the century, one group of eugenicists regarded war as an effective means to filter out the inferior parts of the race or even to eradicate whole races. In the tradition of evolutionary Social Darwinism, they used Darwin's ideas of the struggle for existence in an extremely simplified version in order to emphasize the positive influence of war in the selection process.[3]

In Germany, anthropologists like Otto Ammon represented the view that wars would prevent social and moral degeneration. Ammon called Social Democratic pacifism a great danger, since the weak military defensive power of other nations would be reversed, and those who still stood well below the Germans in natural talents and culture would come out on top.[4] In Great Britain, Karl Pearson in particular stood out as a spokesman for eugenically based war propaganda. At the climax of the war between the Boers and the British in South Africa in November 1900, he claimed that the struggle for existence in a particular case might mean suffering and misery, but still this selection process was the cornerstone of any biological progress. In Pearson's eyes, humanity would stop developing further if "higher and lower races" were to be at peace with one another. If peace were to continue, there would no longer be any instruments to reduce the "fertility of inferior elements." The "relentless law of heredity" would no longer depend on the natural selection process.[5] Colonel Charles H. Melville, professor of Hygiene at the Royal Army Medical College, claimed that military service eugenically made sense since it would transmit the ideals of physical ability, efficiency, courage, and patriotism. The "occasional war" could definitely be helpful since only in times of danger could a nation develop its "virility."[6] In the United States, it was Ronald Campbell Macfie who made the claim in many magazines that wars absolutely had a positive effect on the collective heredity pool. The eugenic consequences of war would be "a winnowing down of men." This group of men would then have a greater choice among the surplus of women and would presumably choose among them only the eugenically most valuable females.[7] In this way, war in the final analysis would lead to an improvement of the "health and beauty of the fighting races."[8]

The views of Ammon in Germany, of Pearson and Melville in Great Britain, and Macfie in the United States, all shared by some Italian and French eugenicists, were the direct and simplistic transfer of Herbert Spencer's adaptation of Darwin's idea of "survival of the fittest" to international relations. A historic development of peoples was in the end reduced to the history of the war of the "biologically superior races" against the less-well-equipped races. As expressed in Spencer's rather crude interpretation, Darwin's theory of the struggle of species to survive provided a welcome biological justification for the imperialism of the various great powers at the turn of the twentieth century. It formed a well-fitting binding element for specific eugenicists of the nationalistic and imperialistic movements in various countries.[9] As a program for state intervention came to displace an evolutionary Social Darwinism that legitimized existing relationships, the American eugenicists in particular increasingly called into question the position of their militaristic and imperialistic colleagues. In the tradition of Galton's eugenics, which saw selection in industrialized states as counterproductive, they also criticized modern wars as the expression of dysgenic selection processes in industrial society.[10]

Vernon Kellogg, a professor at Stanford University and a leading American eugenicist, questioned the claim of eugenics militarists that a high mortality rate was an indication of the improvement of the race through the medium of war. Kellogg emphasized that "military selection" had in general nothing to do with "natural selection." Quite contrarily, modern wars are deep down "unnatural." There could hardly be a greater obstacle to "progress in human evolution" than wars. Militarism could do nothing more than bring forth "a rotten brute."[11] Kellogg's Stanford colleague, David Starr Jordan, director of the World Peace Foundation and president of the International School, similarly explained that modern wars lead to "unavoidable deterioration of heredity material." The "strongest investment" would be killed or wounded and leave behind no or few children. On the other hand, the "weak survivors" could remain at home and visibly reproduce.[12]

In the United States, where the growing tensions among the European great powers were observed by many eugenicists with great dismay, Frank Smith, a member of the House of Representatives, initiated a eugenics peace initiative shortly before the outbreak of the First World War. Supported by eugenicists like the sociologists Edward A. Ross and the economist Irving Fisher, he promoted a coalition of Great Britain, France, Germany, and the United States to "spread the superior human element further."[13] "The union of Britain, France, Germany, and the United States would constitute an international executive power strong enough to ensure universal peace." Since the interest of the purest and most gifted races was considered to be the interest of mankind as well, such a league would assure a permanent race hygienic peace. In a speech to the House of Representatives, he noted ways in which the superiority of the "white race" could be safeguarded against the threat from the East. Smith specified the danger that the "Asiatics" would soon be able to rule the world militarily. If this were possible—and this was apparently his chief concern—then unlimited migration to the United States and Western Europe would continue unabated and would thus endanger the dominance of the white race in these countries. In contrast to Alfred Ploetz, Smith emphasized that the "Asiatics" still did not have military superiority. Nonetheless, the eugenicists and race hygienists in the United States, Great Britain, France, and Germany should prepare for this danger. To Smith, only "a union of the white race" could mediate all conflicts in the "interest of the culture-bearing race of Western Europe" and could "insure universal peace and the union and supremacy of the white race."[14]

This negative attitude of the American eugenicists toward war was shared by the British eugenicists Edgar Schuster and William R. Inge, as well as by the German race hygienists Alfred Ploetz and Wilhelm Schallmeyer, but one should not thereby conclude that all eugenicists shared a basic condemnation of war.[15] The eugenicist opponents of war could not be defined strictly by a morally grounded pacifism.[16] Their first concern was not a moral condemnation of the killing activities by soldiers, but rather involved the prevention of the counterselective effect of wars. Under certain conditions, the eugenicists who had warned about the dysgenic effects of modern war could also imagine a eugenic war.

Ploetz, as one of the first German race hygienists to point to the counterselective effects of war, put forth the notion that the mentally and physically weak should be drafted for wartime service. If war were to come, then these "especially gathered poor varieties" could be used as "cannon fodder."[17] Paul Popenoe and Roswell H. Johnson, authors of the standard American work on eugenics, considered it possible to shape a war in a way that it would be positive for the composition of a people. Such a "eugenics war" would be fought "with elderly men as officers and with mental defectives in the ranks."[18] Even Fisher made the point that he would welcome a war if the warring powers could agree not to send the "best young men" to the front, but rather the "worst"—the "idiots." Such a eugenic war could present the possibility of quickly and simply getting rid of the "degenerates."[19] Kellogg himself agreed with the equation of "dysgenic war = bad war, eugenic war = good war." In the journal *Social Hygiene*, he asserted that "military selection" could be biologically advantageous if the entire population were equally exposed to the war.[20]

Based on this logic, many eugenicists adopted the view that war between "primitive tribes" would have a positive selection effect, while more modern wars would be dysgenic. The German race hygienist Fritz Lenz wrote that war between "primitive peoples" would lead to the "physical and psychological advantage" of the superior groups, while modern wars present a danger to the race hygienic health of a people. In modern wars, the defeated peoples would not be rooted out, but would instead maintain their ability to reproduce. He gave as an example African Americans as the way in which an "inferior race" could convert a "wartime" defeat into a biological victory. According to Lenz, it was only through their enslavement that the "Negro race" had succeeded in spreading over a large part of America.[21] A very similar line of argument was given by the British eugenicist and biologist J. Arthur Thomson, who emphasized that in earlier periods, wars have led to the elimination of the weakest parties on both sides of the warring powers. "Times and wars" had unfortunately changed, thought Thomson. The victorious people would no longer completely "eradicate" the defeated one. It was his point of view that in modern wars, the elimination of race elements was in the best case eugenically neutral. As a rule though, selection would move in the "wrong direction"—the best squads would have to perform the most daring operations. The especially brave soldiers would be the most likely to meet death on the battlefield.[22]

At the beginning of the twentieth century, the discussion about the dysgenic or eugenic effects of modern war became quite heated. Internally, the eugenics movement reflected the general political controversy of "war opponents" versus "militarists" that was raging in view of the growing international tensions in Europe. In contrast to the overwhelmingly moral tone of the general political discussion between "pacifists" and "militarists," both sides within the eugenics movement shared the belief that in the end, history and politics could be reduced to a subset of biology. The laws described by a social scientist or a historian in a study of the causes of the rise and fall of empires in the end were the same that the biologist was attempting to uncover in the study of human types.[23] It was only over the question as to which laws of selection would come into play were there opposing opinions between the two groups of eugenicists.

These two factions within the eugenicist movement clashed directly for the first time at the international eugenics congress in London. Kellogg, a skilled debater, agreed in a speech with the militarists like Otto Ammon regarding the dysgenic effects of war, whereby higher mortality in wars in itself was not a racial catastrophe. Nonetheless—and here the basic assumptions of the eugenic pacifists and eugenic militarists were not at all at odds with each other—the principle of selection would be affected in the face of an "overproduction" of individuals. The question in Kellogg's mind, however, related to the notion of which population groups would be particularly affected by the war. Soldiers were unfortunately chosen with reference to sex, age, and physical strength, and came from that portion of the population that was manly, young, strong, and free of disease.[24] He pointed to the fact that the conscript armies in Germany and France and the professional army in Britain turned away as unsuitable almost half of all those called for duty or who volunteered. It was precisely these who were turned away who would not be subject to the principle of selection during a war.[25]

Kellogg's speech provoked vehement protests in particular among several German and British eugenicists who were convinced of the health promoting, race hygienic positive effects of war. General Carl von Barderleben, president of the Herold Association and one of the founding members of the congress, proclaimed that wartime service is not bad for the body, but rather is healthy and that the human spirit would receive important inspiration and stimulation through war.[26] Arnold White, representing the British National Service League at the congress, noted "the eugenic effect of discipline, of training, of obedience, and of learning the secret of willingness to die for a principle" is undoubtedly eugenically positive.[27] Colonel Melville and Colonel Warden, representatives of the British army at the congress, agreed with Baderleben and White in resolutely rejecting Kellogg's theses. They admitted with Kellogg that during the war, certain eugenic drawbacks would appear, but it was precisely in peacetime that they claimed that military service undoubtedly brought physical, mental, and moral advantages.[28]

The clash between Kellogg and his critics was marked by the lack of agreement on the question of the heritability of acquired characteristics. Baderleben, White, Melville, and Warden were those at the congress who most clearly agreed with the neo-Lamarckian notion that environmental influences could improve the genetic structure of humans. The American eugenicists had more unreservedly accepted Weisman's thesis of the continuity of the germplasm than had their German and British colleagues, and to them the "ignorance" of such laypeople was absolutely incomprehensible. In criticizing the views expressed at the London Congress, the American eugenicist Roswell H. Johnson wrote that only through "a strange reversal of cause and effect" could eugenicists claim that a "the waste of virility could be repaired by universal military drill."[29]

Despite the marked decline of Lamarckian theory after the turn of the century, and despite the replacement of evolutionary Social Darwinism by state intervention in society, it was only the destructive consequences of the First World War that led eugenicists in various countries to develop a generally negative attitude toward war. Although internal controversies over the eugenic and dysgenic effects of war prevented the eugenics societies in Germany, Great Britain, France, and the United States from formulating a single position prior to the First World War, with the outbreak of the war, a broad coalition emerged among eugenicists to prevent modern wars.[30]

In Great Britain, where the question of war and eugenics was extensively discussed, the leadership of the Eugenics Education Society condemned the war as a biological catastrophe.[31] Leonard Darwin and the Oxford eugenicist and professor of biology Edward Poulton agreed in the *Eugenics Review* that war undoubtedly kills off the "better types and is therefore highly dysgenic."[32] Poulton was especially concerned that the young men who were voluntarily leaving Oxford and Cambridge to serve their country and to defend freedom in the world would die in the trenches. "Their courage is intellectual and moral rather than physical, so they were precisely the men we most need in the great social reconstruction that is coming."[33] Caleb W. Saleeby, an activist in the British Eugenics Education Society and member of the Royal Society of Edinburgh, proclaimed that "war is dysgenic" and will kill many of the best possible fathers.[34] Saleeby's Edinburgh colleague, the geneticist and eugenicist Francis Albert Eley Crew, maintained in a lecture before the Eugenics Education Society that he had been brought up by Great Britain's scientific "militarists" such as Galton and Pearson; he had been influenced in his thinking about war by them. However, in light of his own experiences at the front, he could now reject war as counterselective.[35]

In the United States, Johnson summarized the fears of many of his colleagues in the *Journal of Heredity*, where he noted that the "in-dwelling quality" of the human race would decline faster during war than ever before.[36] Together with Paul Popenoe, Johnson complained that the First World War had therefore been "especially destructive," since the hostilities had been marked by the high quality of the fighters on both sides.[37] Irving Fisher stressed that the "loss of 7 million germ plasms of the fittest men" on the Entente side represented unbelievably great damage to the race. Future generations would have to bear this loss in their heredity for many years.[38]

In Italy, the eugenicists also adopted an increasingly critical attitude toward modern wars. In what was a typical Italian joining of eugenics and social hygiene views, the demographer and eugenicist Marcello Boldrini labeled three decisive grounds for the "waste destructive" effect of the war—the dropping out of the dead soldiers from the selection process, the unhindered continuing reproduction of the physically and mentally infirm who had been exempt from wartime service, and the wartime favorable environment for the spread of tuberculosis, malaria, and mental diseases.[39] The Italian anthropologist and eugenicist Giuseppe Sergi promptly transferred his theory of the decline of the Roman Empire to the World War. He emphasized that the wartime "scarcity of food" and "mental unrest" were additional reasons for the dysgenic effects of war.[40]

In Germany, the race hygienist Ernst Haeckel, who before 1914 had been a firm proponent of selection through war, argued under the impact of heavy war losses that one must prevent an unavoidable competition between states from sliding into a bloody and murderous battle for survival.[41] In agreeing with Haeckl, Geza von Hoffmann, the Austro-Hungarian vice consul in the United States and an important go-between for the European and US eugenicists, bemoaned the fact that in the First World War a "significant proportion of the best, the bravest, and the healthiest had been eliminated for all time."[42] At the end of the war, the *Archiv fur Rassen- und Gesellschaftsbiologie* marked a basic shift in the thinking of the German race hygienists. Even the race hygienists, who before 1914 had emphasized the positive effects of war, could no longer deny the destructive counterselective results of modern war. The euphoria for war as a heightened form of the "struggle for existence" was completely gone. The eugenic "mass experiment of war" had unequivocally collapsed.[43]

The Effects of the First World War on the Development of Eugenics

The First World War struck the international eugenics movement at a particularly unpropitious moment. On June 28, 1914, at the moment when the shots at the Austrian heir to the throne Franz Ferdinand unleashed the First World War, the next meeting of the Permanent International Eugenics Committee was about to be convened. The planning for a second international eugenics congress had already begun in the United States. The national enthusiasms for the war in the early months did not spare the various eugenics societies. Despite the growing worry about the destructive dysgenic effects of the war, the eugenics societies saw it as their patriotic duty to support the war efforts of their respective governments.

It was the German race hygienists who were most caught up in patriotic passions and who cut off contact with eugenicists in enemy nations.[44] In the light of the rebuff that the German race hygienists had received at the London Eugenics Congress, since 1912, a certain sympathy had spread within the International Society for Race Hygiene for a change in orientation to a pan-German organization for race hygiene.[45] Due to the intensification of the war situation, the International Society for Race Hygiene in July 1916 decided to convert to a national organization. As outlined in the Congress report by Fritz Lenz, the organization should remove the "appearance of an international character," for since the start of the war, this had proved to be "a great obstacle in recruiting members."[46]

In place of an international of white people, the Munich race hygienists around Ploetz, Rüdin, and Lenz in particular during the war called for a German-Austro-Hungarian alliance of race hygienists. The race-oriented Munich race hygienists, who were carrying on a growing conflict with the more technocratic and welfare state–oriented Berlin eugenicists, strove for close cooperation with Austrian and Hungarian organizations of the ultra-right. They began contact with the Hungarian Society for Race Hygiene and Population Policy directed by Count Paul Teleki, and began planning for a general meeting in September 1918 of German, Austrian, and Hungarian race hygienists in conjunction with a meeting of the physicians section of the association of their comrades in arms.[47]

In the first ten years, the Society for Race Hygiene had had no noticeable political influence in Germany, but this transition to a national organization opened access for the race hygienists to government officials. The government was interested in a healthy and growing population as a precondition for a military victory and as a way to assure the role of Germany as a world power. In 1916, Emperor Wilhelm had noted the necessity of a systematic German population policy. At the initiative of the Center Party, the Reichstag established a "Committee for Population Policy," which numbered important ministerial officials and high-ranking physicians among its members.[48]

Even when German population policy in the First World War merely promoted a purely quantitative buildup of the population, the race hygienists, with their call for a qualitative population policy, could also join in. Their plan for "exploiting the interest in population policy aroused by the war to spread racial hygiene propaganda" was at least partially met.[49]

The British eugenicists were bemused by the resonance achieved by the race hygienists in Germany. In the annual report of the Eugenics Education Society of 1917, it was noted that "in Germany there is very considerable eugenic activity, and the War has brought about the formation of various new societies and directed the attention of several existing organisations to eugenics."[50] Still, in Great Britain as well eugenics was the object of heightened attention. In addition to the increasing number of war dead, since 1915, there had been an increase in child mortality, and a report of the Royal Commission for Sexual Diseases had concluded that the British public was increasingly interested in questions of eugenics.[51] The Eugenics Education Society attempted to react to the challenges of the war by founding the Professional Classes War Relief Council and by participating in a national baby week in July 1917. In the view of the board of directors of the Eugenics Education Society, the returning war wounded still capable of fathering children would have their way made easier in seeking out a wife and starting a reproductive career.[52] In Leonard Darwin's eyes, only through a quick marriage and founding of a family of the returning soldiers could their "manly qualities" be assured for future generations.[53]

The French Eugenics Society was hit the hardest of all by the First World War. Lucien March and the other leading French eugenicists were drafted into military service. As early as August 1914, the eugenicists were forced to suspend the activities of the French eugenics society until the end of the war.[54] On the other hand, the eugenics movement in the United States was affected but little by the war. The American eugenicists continued their research and publication activity unhindered, and in August 1915, the National Conference for Race Betterment met for the second time.[55]

Even if the war interrupted the international cooperation of the eugenicists and race hygienists, and even if the development of eugenics societies was impeded in several European countries, the supposedly destructive dysgenic effects of the war led the eugenicists to put forth their political demands even more vigorously. It is only on the background of the disappointment of the eugenicists over the "racial consequences" of the World War that the rising popularity of eugenics as a political and scientific movement may be understood. The World War for the first time allowed the transition from a purely intellectual debate over the possibilities of race distinction into a concrete implementation of eugenic concepts in political practice. Only under the pressure of war did various eugenic measures receive greater attention, such as financial aid to especially "valuable" parents, the required exchange certificates of good health before marriage, the sterilization of "inferior beings," abortion on eugenic grounds, and the killing of "beings unworthy of life."[56]

For the eugenicists in Europe and in the United States, intensifying eugenics efforts were the only adequate answer to the destructive dysgenic effects of the war. The Swedish race hygienist Hjalmar Anderson, on the occasion of the imminent opening of the Swedish State Institute for Race Biology in 1921, declared that the "great War, with all its horrors and pitiable consequences," had been the occasion for many to "put their hope in race biology and eugenics as the possible saviors of suffering Europe."[57] In the United States, Paul Popenoe stated that despite all the worrisome consequences of the war, it at least had convinced many people to think "about the value of race and artificial selection" in a way that had never been done before. To Popenoe, the war was one of the major reasons that Galton's ideas had gained more ground in America than even the greatest optimists of ten years before had dared to hope.[58]

Under the motto of a quotation from Shakespeare, "There is some soul of goodness in things evil, would men observingly distil it out," the Northern Irish physician and eugenicist James Alexander Lindsay claimed that the British nation had to accept the war as the occasion for a comprehensive eugenic new order.[59] Leonard Darwin played from the same score by demanding that reconstruction in Great Britain be tied up with a broad eugenic reform. The "slaughter" of the "best types in this horrible war" was the major reason for the growing demand to treat the question of racial progress more forcefully.[60]

For the European and American eugenicists, the "problem of inferior beings" took on a whole new dimension because of the war. The sex researcher and temporary vice president of the Eugenics Education Society, Havelock Ellis, wrote that "It may now appear clear that the problems in eugenics which we have to face as a result of the war are not new problems. They are the same old problems, only they have acquired a new urgency."[61] Irving Fisher called on the eugenicists to draw lessons from the World War that were directly entwined with one another—first, the establishment of a League of Nations to prevent another war, and second, the prevention of reproduction by the mentally handicapped and psychologically ill.[62]

How much the eugenicists linked the eugenic peace policy that they demanded with the "problem of inferior beings" is shown by the manner in which the race hygienists balanced the loss of "genetically valuable" soldiers with the eugenic gain of the death by hunger of tens of thousands of mentally handicapped and socially weak persons in Germany.[63] The Munich social hygienist and eugenicist Ignaz Kaup maintained that the hunger blockade against Germany fortunately had "biologically done away with mainly inferior defectives," but on the whole, the "flower of the physical virility of the German people" had not been affected by the tremendous war losses.[64] Alfred Ploetz as well considered the wartime widespread hunger in Germany during the war as eugenically positive. In total, however, "the counter selection" outweighed the positive "eradication" of the handicapped and of social weaklings. "Terribly many of the most beautiful branches and boughs on the flowering tree of the German race have been cut off."[65]

To create a theoretical foundation for their effort to tie in eugenics peace policy the problem of "inferior beings," the eugenicists displaced the selection process from the level of the group or the state to the level of individual reproduction.[66] In their view, the struggle for existence did not stop with a eugenically motivated peace policy, but had merely become more rational. The trick of displacing the selection process from the level of the group to the level of the individual had already existed in the nineteenth century through Francis Galton and Alfred Ploetz. Galton in his book *Hereditary Genius* claimed that the struggle between various races or groups for evolutionary progress is not necessary. Instead, the progress of humanity might be achieved by a rational organization of the selection process on the individual level.[67] Ploetz too even before the turn of the twentieth century thought to replace the "struggle for existence with all its wretchedness" with selection on the level of the nucleus. For him, the solution of the conflict between the nonselective goals of a war and those of race hygiene lay in the "selection and weeding out of human beings at the cellular level."[68]

In light of this close bonding of war policy with eugenic policy on the handicapped, the eugenicists increasingly supported their "struggle" against the mentally handicapped and socially frail with military analogies. The German eugenicist Alfred Grotjahn saw the isolation of "the army of hoboes, alcoholics, criminals, and prostitutes, of psychopaths, epileptics, mentally ill and mentally defective, oddballs, and cripples" as a necessary condition for the revitalization of the German people.[69] The American professor of zoology Edwin G. Conklin justified the "elimination of the worst" with a reference to the "army of the defective and criminal persons" who populated the hospitals, mental institutions, and penal institutions, and with a reference to the enormous costs for the care of these "human wrecks."[70] The leading Swedish race hygienist Herman Lundborg differentiated between "two types of enemies," who threatened "every culture nation"—the external enemies, by which he meant rapacious neighbors, and the internal enemies, the "inferior" citizens of a state. The internal enemies were just as strongly responsible as the external ones for the "deracination of a people," and he called on physicians and sociologists to lead the way in the battle against these "internal enemies."[71]

The Revival of the International Eugenics Movement and the Second International Eugenics Congress

The enormous concern regarding the dysgenic facts of the World War was the major reason that the Permanent International Eugenics Committee set itself up again shortly after the Versailles peace agreement of June 28, 1919. While many other international scientific organizations did not at first resume activity due to the embittered front lines between the scientists, the worry about the imminent degeneration in the "civilized" countries made the eugenicists move with greater speed. At a meeting of the Permanent International Eugenics Committee on October 18, 1919, in London, eugenicists from Great Britain, France, Italy, the United States, Belgium, Denmark, and Norway agreed to hold another international eugenics congress as soon as possible. Due to this increased cooperation, it was decided to discuss which social policy measures could reduce the dysgenic effects of the war, how one could maintain the eugenic value of wounded soldiers for the race, and what consequences the increased appearance of "racial mixtures" would have on the hereditary structure in Europe and America.[72]

The director of the American Museum of Natural History, Henry Fairfield Osborn, on September 22, 1921, opened the Second International Eugenics Congress in the presence of almost 400 eugenicists from the United States, Great Britain, Italy, France, Belgium, Norway, Sweden, Denmark, Japan, Mexico, Venezuela, India, Australia, New Zealand, San Salvador, Siam, and Uruguay. For them, the question of the eugenic consequences of the war stood in the foreground.[73] In his opening speech, Osborn emphasized that he doubted if there had ever been a moment in the world's history when an "international conference on race character and betterment has been more important than the present." Through "patriotic self-sacrifice," Europe had lost a large portion of the "heritage of centuries of civilization." In some parts of Europe, the "worst elements of society have gained ascendancy."[74] For Osborn, the International Congress for Eugenics should commit itself to save those states that had been struck hard by the war by suggesting doable methods to improve the race. The tendency of valuable parents driven by individualism and egoism to one- or even zero-child families must be stopped and reversed.

Osborn emphasized that in the United States unfortunately the assumption that "all people are born with equal same rights and duties" had become confused by "political sophistry" with the notion that "all people are born with equal character and ability to govern themselves and others." The right of the state "to safeguard the character and integrity of the race or races on which its future depends is...as incontestable as the right of the state to defend the health and morals of its people." Just as science had informed governments about the spread and prevention of diseases, it must "enlighten government in the prevention of the spread of worthless members of society."[75]

The Second International Congress for Eugenics clearly showed that the American eugenicists had taken over the leading role in the international eugenics movement.[76] In addition to a general foreign policy expansion by the United States after the World War, this change was particularly tied to the enormous progress of eugenics in the United States during and immediately after the war. Several American states had passed sterilization laws, and at the beginning of the 1920s' sterilization of the mentally handicapped was possible in 15 states.[77] The American Congress passed a new immigration law to make difficult immigration by Slavs, Southern Europeans, Asians, Russian Jews, and the mentally handicapped. The national prohibition on alcohol was celebrated by many eugenicists as another important race hygiene step. Paul Popenoe saw in these measures clear proof that the United States was developing into a model of a eugenics country. To be sure, thought Popenoe, the "thousand year empire of race hygiene" was still far away for the United States, but the society there would strive to rebuild along eugenic lines.[78]

In contrast to their European colleagues, the American eugenicists at the beginning of the 1920s had direct access to the most important political institutions on the federal and state levels. The international congress was supported by such influential politicians as the later governor of Pennsylvania Gifford Pinchot and the later US President Herbert Hoover.[79] The American eugenicists also had available a number of eugenics organizations that were more or less closely connected with one another. The American Genetic Association, a successor to the American Breeders Association, had the influential eugenicist and sociologist David Fairchild as chairman. He cooperated closely with the Eugenics Research Association, which at the beginning of the 1920s was led by Baltimore medical professor Llewellyn F. Barker. In New York, the Galton Society was formed, a club of eugenicists restricted to Americans of European heritage.[80] Research institutions included Davenport's Eugenics Record Office, the Eugenics Registry of the Race Betterment Foundation, and the Genealogical Record Office created by Bell.[81]

At the congress in New York, the American organizers were at great pains to present eugenics as a serious science with political implications. These organizers were haunted by the nightmare that the second congress, like its London predecessor, would be descended upon by scientific dilettantes, and that representatives of the press would take the occasion to ridicule eugenics. For Clarence C. Little, general secretary of the congress, the eugenicists had to go all out to work on the image of eugenics so that it would not seem like a mere nervous tic of fanatics and delusional souls.[82] A serious science was unworthy of being robbed of its seriousness by caricatures and jokes, such as that which occurred at the first congress.

In order not to appear to be a congress of laypeople as had occurred in London in 1912, leading American scientists were represented on the central organization committee—Charles B. Davenport, Raymond Pearl, the geneticist Thomas H. Morgan, the biologist Herbert S. Jennings, and the head psychologist of the American army, Robert M. Yerkes.[83] This group was to monitor the professionalism and scientific character of the congress. Lectures had to be submitted ahead of time so that any possible sensationalism and cheap effects could be removed in time. The organizers signed up an experienced press spokesman especially for the conference, the race theoretician Lothrop Stoddard. The domestic and foreign presses were seriously asked to dispense with "little mockeries."[84]

The organizers began with the assumption that a strong emphasis on genetic research would best represent the scientific character of eugenics. Little hoped that by including the latest results of genetic research one could counter the criticism that eugenicists knew too little about the foundations of their own field and that they had constructed a "superstructure" without sufficient basis.[85] The linkage that the organizers wanted between genetics and eugenics was successful. Lucien Cuenot, Thomas H. Morgan, Hermann Joseph Muller, Edwin G. Conklin, Ronald A. Fisher, and Ruggles Gates, leading American, French, and British heredity researchers, participated in the congress.[86]

In his address as president of the Permanent International Eugenics Committee, Leonard Darwin emphasized that heredity research had to be the "lodestar" of eugenics. Genetics was the "pure science" of heredity, and therefore the basis on which the "superstructure of eugenics" would have to be built.[87] The same theses were put forth by the conference organizers in the foreword to the first volume of the conference proceedings. They claimed that the "germ plasm" was the foundation for all social progress. The chromosome was said to have brought about all the progress of the organic world. It had produced the forerunners of human beings and would certainly have an effect on evolution, an influence which would far surpass all attempts of humans to manipulate it.[88]

For the eugenicists, genetics was more than a support science; it was inextricably tied up with eugenics as a partner science. Morgan's goal was to bind the Mendelian principles together with the structure and behavior of chromosomes, and a new eugenicist interest in plant and animal genetics would emerge from the subsequent merger of cytology genetics. If genes, chromosomes, and cells were the basic units of plants, animals, and humans, then knowledge gained in plant and animal genetics would also give an important push forward for research into human heredity.[89]

Davenport was firmly convinced that ongoing development in human, animal, and plant genetics would confirm eugenic principles. In his address to the congress, he claimed that "an imbecile is an imbecile for the same reason that a blue-eyed person is blue-eyed." The core significance of heredity for the formation of psychological characteristics would be unambiguously confirmed by modern genetics.[90] George Adam, vice chancellor of the University of Liverpool, argued to the congress participants that genetics students were always also eugenicists. Their studies would force them to recognize that humans are not equal. Geneticists would comprehend that humans come into the world with unequal possibilities and capabilities, and that this capability, or lack thereof, was inherited from their ancestors.[91]

The dominant genetic themes at the congress were chromosomes and mutations. The geneticists were especially interested in the question of whether all genetic changes could simply be traced to recombination of already existing genes, or whether completely new genes could be formed through mutation. Hermann Joseph Muller, through his work on drosophila (fruit flies), had determined that X-rays increase the number of mutations. Before the eugenics congress, he maintained that the great majority of mutations are harmful, possibly even fatal. The formation of "bad" mutations would illustrate that selection was indispensable for life. Without selection, these undesired genes would increase unhindered for so long that the entire germplasm would be completely infiltrated by defective genes. Muller was convinced that this theory applied not only to his fruit flies, but also to human beings.[92]

The high-ranking geneticists who were represented at this International Congress for Eugenics were a clear sign of how closely the direct tie between genetics and eugenics had become after the First World War.[93] The only question critical of eugenics appeared from a non-geneticist in 1921. The American psychiatrist Abraham Meyerson came out in opposition to the claim that most psychological diseases were not linked to heredity. According to Myerson, heredity research was still too far away from being able to draw definite conclusions about the relationship between genetics and psychological diseases. Insane people might have "normal" progeny, and the "normal population" might have psychologically disturbed children in a baffling and inexplicable manner. Referring to the results of the X-ray experiments on the fruit fly, Myerson wrote in the conference proceedings to the heredity researchers that it was more important to look for the causes of damage to the germ cells than it was to explain all psychological diseases with Mendelian laws or other biological principles.[94]

The section on heredity research was the biological backbone of the other sections at the congress, which debated the "questions of social and legal controls of fertility," the differences between the races, and the relationship of eugenics to the state, society, and education.[95] Optimism reigned in all the sections. People hoped to spread even further the influence that genetics had already achieved in science and politics. The *Journal of Heredity* celebrated the fact that the lectures given at the congress were of supreme importance, and could be compared in significance only to Galton's publications.[96] Darwin joined in with the shouts of enthusiasm, and after his return to Great Britain reported that the congress had been the starting point for new "eugenics efforts in America" and had given a mighty push forward to the international eugenics movement.[97]

Alfred Mjöen, as representative of the tiny Norwegian delegation, presented a resolution that a central eugenics organization should be founded in every country.

Mjöen gave the reason for his initiative that the large majority of people in the Western states were still ignorant about eugenics, and that "mentally inferior and abnormal individuals" could still reproduce at will; consequently, "odd, defect-riven germ cells" could "infect" the collective heredity pool of still relatively high-value people. In light of this situation, coordinated action on the national and international levels was necessary.[98]

The suggestion to found central national organizations met favor among the US eugenicists in particular. In the United States, there were various groups, some of them quite successful, but a coordinating center was still lacking where the various initiatives could be gathered together. The congress participants supported a suggestion from Irving Fisher to set up a working committee to prepare an American eugenics society. The task of the society would be to inform the American public about the goals of eugenics and to represent the eugenics movement to the government in the United States.[99] With this proposal at the international congress, Mjöen and his Norwegian colleagues laid the cornerstone for perhaps the most successful national eugenics organization. Within a few years, the American Eugenics Society had developed into a very powerful lobbying organization. Scarcely any other eugenics organization within or outside the United States was as active.[100]

To Mjöen's mind, the national eugenics organizations could become much more effective if an international institution or organization for eugenics were available for support. He felt that the chances of success for national eugenics legal initiatives were quite high among the various countries that had voted for the proposal.[101] Cyril E. A. Bedwell, director of King's College Hospital in London, in his lecture read at the congress had listed the various stages that a movement based on "scientism"—jointly representing both politics and science—had to climb through. Initially, there was a need for scientists to meet their colleagues at international congresses. Starting from that point, one had to attempt to intensify the relationships. First, a permanent organizational office or a standing council had to be set up, which after some time would be able to move into its own building. Such an international center could coordinate research, gather statistical and biographical information, and publish materials. With growing professionalization, an attempt could be made to convince governments to name an official representative. After this official recognition, international agreements and then their implementation in national laws could follow.[102]

The Second International Congress allowed the assembled eugenicists to achieve one of the crucial stages of institutionalization—the permanent international organization, a goal ever since 1912. The Permanent International Eugenics Committee was established on stable footing at the 1921 congress. During a meeting among Darwin, March, Mjöen, Pearl, Davenport, Osborn, and Irving Fisher—officially the third and last meeting of the Permanent International Eugenics Committee—a resolution was prepared that was approved by the general assembly of the international congress on its last day. This resolution stated that the "progress of eugenics science and education makes necessary a permanent international organization."[103]

In addition to the preparation and administration of further international congresses, the international organization in the eyes of its founding fathers was also to function as an interim committee. All tasks that occurred between international congresses, such as the publication of an international journal, the coordination of research projects, or the internationally agreed upon exertion of influence on governments or international institutions, were to be undertaken by the international organization of eugenicists. It was little emphasized in his discussion of the resolution that the expansion of the interests of eugenicists to questions of geographic distribution and race mixture made international coordination urgently necessary. Migratory movements would have to be studied on an international plane, and the question of race mixing in particular took on growing significance.[104] To make this expansion of the tasks of the international eugenics organization meaningful, the Permanent International Eugenics Committee was transformed into a Permanent International Eugenics Commission.[105]

The founding countries included Belgium, Czechoslovakia, Denmark, France, Great Britain, Italy, Norway, and Sweden in Europe; Argentina, Canada, Columbia, Cuba, Mexico, Brazil, Venezuela, and the United States in the Western Hemisphere; and New Zealand and Australia. Despite the protests of several Scandinavian and American eugenicists, membership was initially denied to Germany, Hungary, and Austria.[106]

The Reintegration of Germany into the International Movement

A brisk argument broke out at the beginning of the Second International Eugenics Congress regarding a suggestion of the American eugenicists to invite the race hygienists from Germany, Austria, and Hungary to the congress. The French and Belgian eugenicists in particular refused to have any participation of their German, Austrian, or Hungarian colleagues at the preparatory meeting in London in 1919; they threatened to boycott any international congress with German participation.[107]

Quite early on, it was the Scandinavian and American eugenicists who brought up this exclusion of the German race hygienists from the international movement. While in many other international societies a reintegration of German scientists was never even considered until the mid-1920s, some eugenicists pleaded for a participation of German race hygienists at the Second International Congress for Eugenics, referring to the international character of eugenic science. Herbert S. Jennings, only recently having become a member of the planning committee for the Second International Congress for Eugenics, protested to Davenport regarding the agreement by the American eugenicists at the Permanent International Eugenics Committee to exclude the German race hygienists. Supported by his colleague Thomas H. Morgan, he worked against the attempts of the French and Belgian eugenicists to make science a national affair and called on his American colleagues to acknowledge the international character of eugenic science.[108]

Mjöen, well known to both congress president Osborn and to Ploetz, also made the case that the German race hygienists should be admitted unconditionally to the international congress.[109] His Swedish colleague Herman Lundborg, who like many Scandinavian scientists had strong sympathies for Germany, protested to Davenport in an outraged letter at the exclusion of the German race hygienists from the congress. An international eugenics congress, he said, which was not open to all "civilized nations," could not be international, but would instead damage the eugenics cause. In light of the necessity for states to be united in their eugenics measures, it was deeply disappointing that even scientists could not escape from their political prejudices.[110]

In view of the broad criticism of the exclusion of German race hygienists from the international eugenics movement, the leading American and British eugenicists in particular got into difficulties. While they did not want to ignore the national resentment expressed by the French and Belgian eugenicists, they also did not want to vigorously rebuff the pro-German eugenicists from Scandinavia and the United States.[111] While Darwin and Davenport actually voted for the exclusion of Germany from the Second International Eugenics Congress and from the Permanent International Eugenics Commission, they gave a clear indication to the German colleagues that they were interested in a rapid acceptance of the German race hygienists. In this vein, Davenport lamented to Ploetz's close associate Agnes Bluhm that "international complications" had prevented a "formal invitation" to the German hygienists for the international congress in New York. He gave expression to his hope that by the time of the next international congress, such "complications" would rapidly drop out of sight.[112]

While the American and British eugenicists pushed for a rapid integration of their German colleagues in the 1920s as part of a general internationalization of the political, cultural, and scientific life, the Germans were increasingly isolated. The German race hygienists at the beginning of the twentieth century had still stood out as opposed to the British eugenicists as eugenic internationalists, but now in the wake of the isolation of Germany, they became nationalistic lone wolves. While in other countries the growing number of international organizations—one need mention here only the founding of the League of Nations—also had effects on the attitude of eugenicists for the international eugenics movement, the German race hygienists refused their international cooperation. For example, Ploetz refused to agree to an exchange of the *Archiv für Rassen- und Gesellschaftsbiologie* for publications of the Bureau of Ethnology of the Smithsonian Institution. Since the "Americans" had attributed sole guilt for the war to the "Germans," Ploetz in an ironically worded letter to the leadership of the Smithsonian Institution called it logical to exclude the "Germans as filthy barbarians" from the international eugenics congress in New York. Since the *Archiv für Rassen- und Gesellschaftsbiologie* was an undertaking by people who "consciously belonged to the German nation," they could not maintain any contact with the Americans.[113]

By December 1921, Darwin was considering ways in which the Germans could best be integrated into the work of the Permanent International Eugenics Commission. Darwin feared that any further delay in cooperation with the Germans would harden the front lines, and he suggested that the topic be taken up at the very next meeting of the international eugenics organization.[114] When the Permanent International Eugenics Commission met in October 1922 in the Maison des Médecins in Brussels, it unanimously invited the German Society for Race Hygiene to cooperate in the international eugenics movement.[115] Davenport personally urged the German race hygienists to accept the invitation. He wrote to Erwin Baur and

Fritz Lenz that despite his fears, even the French eugenicists in Brussels had voted for the invitation to the German rice hygienists. Davenport gave expression to the hope that German and French scientists could come together in the same room as eugenicists, and not as representatives of their countries. International meetings of eugenicists could "heal the wounds that the war had inflicted on science and on the progress of humanity."[116]

The invasion of the Ruhr Valley by French and Belgian troops, however, stiffened the German race hygienists in their drive to national self-isolation. Baur thanked Davenport for his energetic action in favor of resuming international relations in eugenics, but he noted that in light of the "state of war along the Rhine and the Ruhr," any participation by Germany would be impossible. As long as the "French white and colored soldiers" occupied "in a most shameful way this highly cultured area of Germany," it would be impossible for German delegates to sit at the same table with Frenchmen and Belgians. Although Baur personally wanted the war wounds to be closed up and wanted scientific relations to be resumed, in light of the situation in the Ruhr, it was inconceivable that German race hygienists would participate in meetings of the Permanent International Eugenics Commission.[117] Lenz also maintained that "now is not the time for international congresses." France had placed itself "outside civilized humanity" by oppressing Germany, and it was impossible for him to cooperate with Frenchmen. For Lenz, only a new ordering of the international situation through a regrettable but unavoidable Second World War would again make international cooperation possible.[118] It was from the point of view of Baur and Lenz that the board of directors of the German Society for Race Hygiene in principle declared itself ready to cooperate with the Commission, but it emphasized that it could not cooperate with representatives of specific states in the existing situation in Europe.[119]

Despite the offers of cooperation from the Permanent International Eugenics Commission at its meetings in 1923 in Lund, Sweden, and in 1924 in Milan, and despite Darwin's persistent requests to Alfred Ploetz to participate in the international meetings, the German race hygienists continued to stand back.[120] At their general meeting in October 1924, the German Society for Race Hygiene chose its chairman Otto Krohne and Ploetz as delegates to the International Commission, but they continued to refuse to cooperate with the French and Belgian eugenicists. They made their participation in the meetings of the commission contingent on German being admitted as an official language and that neither Brussels nor Paris would be chosen as a site for meetings.[121] The leading members of the international eugenics organization were unwilling to submit to the ultimatum of the German race hygienists. As long planned, in 1926, they met in the French capital. It was only in 1927 that the German race hygienists resumed participation at an international meeting.[122]

CHAPTER THREE
RACISM, INTERNATIONALISM, AND EUGENICS

Just as we isolate bacterial invasions, and starve out the bacteria by limiting the area and amount of their food supply, so we can compel an inferior race to remain in its native habitat, where its own multiplication in a limited area will, as with all organisms, eventually limit its numbers and therefore its influence.[1]

—The American eugenicist Prescott F. Hall

At the meeting of the international eugenics organization in 1928 in Munich, Charles Davenport announced a very well-financed scientific competition over what he considered to be the central question in eugenics. The so-called Draper Prize was to be given to that scientist who most convincingly could record the differences in the "raw birth numbers," the "fertility numbers," and the "vital index" between the "Nordic" and "non-Nordic peoples" in Europe and America. Nordic peoples were defined here as coming from all Scandinavian countries north of the sixty-third degree of latitude, The Netherlands, Scotland, Northern Ireland, and the German regions of Schleswig-Holstein, Mecklenburg, Hanover, Westphalia, plus all the emigrants out of these countries. All other European areas, Asia, and Africa north of the Zambezi were labeled as non-Nordic.

The prize was named for its initiator and financier, the New York textile manufacturer and millionaire Colonel Wickliffe P. Draper. Draper wanted to safeguard the maintenance of the "Nordic qualities" in the American population through a selection of the immigrants based on criteria of race and through influence on the marital behavior of genetically superior "whites."[2] He was anxious to return his black American co-citizens to Africa, and attempted everything that he could to have scientific legitimization of his political demands, which were based on racial segregation.[3] Over several decades, he financed various eugenics societies interested in questions of race through large contributions to the Pioneer Fund, which he had founded.

Using the international prize competition that he funded, one of his first services for the eugenics movement, he wished to draw attention to the declining birthrate of the "Nordic peoples" as opposed to the "non-Nordic peoples." To Davenport's great disappointment, the winner of the prize competition, the Dutch eugenicist and statistician Jacob Sanders, was unable to find any significant difference in the birthrate between the Nordic and non-Nordic peoples in Europe, but his contribution did leave open the possibility that the birthrate in European countries was dropping in comparison to that in Africa and Asia.[4]

With a little effort, Draper could make a case from the submissions that the European peoples with a high Nordic blood component were numerically declining in comparison to Asian and African peoples.[5] This prompted him to finance a second competition, in which the causes for the decline in the "civilized countries" were to be examined. Particular attention here was to be paid to the appearance of peoples of Nordic and predominately Nordic ancestry in all parts of the world.[6] Almost 100 eugenicists in Europe and North America took part in the prize competition, which had a prize of US$3,500.[7] The winner this time was the grand old man of German population science, Roderich von Ungern-Sternberg. The study that he handed in for the Draper Prize, *The Causes of the Decline in Births in the Western European Cultural Area*, was an internationally accepted standard work of demography.[8]

The Draper Prize was part of a broad-based campaign with which the eugenicists in the 1920s developed race research in an international context. In what follows, it will be shown that the idea of European and American eugenicists to have a racist international was the basis for intensifying race research under the auspices of the international eugenics movement. Efforts to promote the "white race" independent of national borders became increasingly tied up with the efforts to make eugenics a scientific endeavor. The eugenicists' claim of eugenics to be a science and the close connection between eugenics and race research were distinguished from alternative strains of development within the international eugenics organization. By intentionally squeezing out Socialist and Lamarckian eugenicists from the international eugenics movement, the international organization increasingly developed into a bulwark of racist eugenicists against various other reform efforts within the movement.

From the "Blonde International" to the "Race Confederation of European Peoples"

The premier scientific and political point of reference for the eugenicists engaged in the international eugenics organization was not individual nations, but a social construct, something they called the white, Nordic, European, or Europoid race.[9] Even though the eugenicists differed on exactly where to draw the boundaries of their own race, the campaign for the collective hegemony of the peoples of European culture ever more strongly colored the thinking of the European and American eugenicists. In their view, the danger of "race suicide" threatened not merely individual nations, but the "white race," the "culture peoples," "Western civilization," the "Occident" in its entirety.

In this vein, Leonard Darwin warned that the Western world as a whole was threatened by biological collapse. If a broad eugenics reform were not to be undertaken in the coming century, Darwin claimed, "Western civilization" would be condemned to the same inexorable "decline" as the civilizations of Rome and Athens in antiquity.[10] In a publication with the pretentious title "The Decline of the Culture Peoples in the Light of Biology," Darwin's German colleague Erwin Baur voiced the fear that today's culture peoples—meaning the white race—were threatened in the same way as the ruling class in ancient Rome. For him, the major cause of the "degeneration of cultures and culture peoples" was clearly "of a biological nature."[11] From one angle, the "inferior" elements of the white race were reproducing faster than the carriers of the "superior hereditary material," and on the other. there was the threatening change in the "race makeup."[12]

The American, German, and Scandinavian eugenicists who were especially interested in close cooperation among eugenicists interested in the "white race" were the ones who had developed the eugenic "world concepts." By biologizing international relations, these proposals went far beyond the prevailing nation-state concepts. Albert E. Wiggam, one of the most successful science-popularizing writers in the United States and in the 1930s. the director of the American Eugenics Society, denied a "world state" in his bestseller *The New Decalogue of Science*, but he nonetheless spoke out especially for eugenic internationalism. National borders would have to remain, since race mixing would undermine the "homogeneous national mind," the "common racial outlook," and the common culture. Without national borders "peoples on a lower level of development" would in great numbers "pour their mongrel blood into richer radical streams." "Free wandering and migrating en masse...would shortly plunge the word back into savagery" and result in the "lowering of the blood of the enterprising pioneers who discovered and developed any country." Only an international authority could guarantee biological diversity and in so doing guarantee the "purity" of the various races.[13]

Roswell H. Johnson carried on Wiggam's intellectual mélange of internationalism and racism. Johnson, Davenport's student from 1905 to 1907 in Cold Spring Harbor and later a professor in Pittsburgh, gave up a lucrative job in the American oil industry in order to devote himself completely to international eugenics. From 1926 in 1927, he was president of the American Eugenics Society.[14] In a study on the differences in the evolutionary situation of various countries, Johnson claimed that a gradual integration of nations into a world society would be eugenically advantageous. The elimination of wars, an unambiguous determination of the optimal level of population for each nation, and the expansion of the authority of international organizations would be eugenically much more advantageous then the continuation of international conflicts. A large portion of Johnson's eugenics standards in promoting "the gradual integration of the nations into world society" was comprised of an international migration policy, which would allow only people with "superior hereditary material" to migrate to another country. Immigrants whose intelligence lay under the average of the population in the target country or who suffered from serious health effects were to be blocked from entering.[15]

Harry H. Laughlin, director of the Eugenics Record Office, emphatically wanted to set up a eugenically oriented world constitution in the 1920s. He thought that without a "democratic world government," states would fall back into the anarchy characteristic of the World War.[16] Laughlin's democratic-international ideal was however strongly colored by racism. Instead of "one man-one vote," he called for a "world parliament" in which the vote of a European citizen of the world would count for approximately ten times as much as the voice of an African co-citizen. Countries with a high rate of literacy and high exports and imports—that is to say, the European and North American states—would choose the majority of the 600 world senators. Africa, Asia, and South America in Laughlin's view would altogether send somewhat more than 150 delegates.[17] The goal of Laughlin's fantasy was a eugenic world order. The tasks on the list of his world government included "the establishment of race ideals," the "maintenance of an optimal population number," and the "improvement of the quality of physical, intellectual, and spiritual hereditary material."[18]

In Germany, it was the racially attuned Bavarian race hygienists Fritz Lenz and Alfred Ploetz who in particular wanted a merger of the "white race." Their extensive isolation during the war had caused even the German race hygienists to give up their propagandizing of the Society for Race Hygiene as the embodiment of such a white race merger, but racist internationalism continued to play an important role in their theoretical observations. The Munich professor for race hygiene Fritz Lenz expressed his deep sorrow in the standard textbook of German eugenics that the "people of the modern culture lands...were permeated with sick hereditary deposits."[19] He hoped to hold off the direct threat to the "best racial elements" through eugenic cooperation of "all nationals of European race and civilization."[20] Lenz desired the creation of a "blonde international," a racially defined confederation of peoples that would represent the "common interests of the Nordic race." This confederation would not draw members solely from the peoples of Scandinavia, Germany, Switzerland, and Austria. In Lenz's opinion, "all people of European culture" were "together a specific part of the Nordic race," and southern or eastern European states could also hope for acceptance into his "blonde international." Leadership should fall to the United States. Lenz expected that the United States could best pull together a policy of systematic race improvement with an ambitious internationalism of the "white race." Lenz understood that the time had not yet come for his "blonde international," but he felt that it could serve as an ideal orientation point for greater cooperation of peoples of European culture.[21]

Ploetz too, harking back to prewar considerations, called for a "confederation of states of linguistically and racially related peoples."[22] Ploetz's racist internationalism was strengthened by his conviction that "war is one of the most meaningful race hygiene factors." He thought that the "counter-selection of war" had debased the "intellectual and physical constitutional strength" and "the share of blood" of the "Nordic race." Wars involved an "ongoing destruction of the good race effects of extermination." Ploetz appealed to adventurous "young hotheads" not to put their "race hygienic" struggle to the test by war, but rather to create new "virile human material" for the German people.[23]

In Scandinavia, it was Alfred Mjöen in particular who propagandized for the merger of the "Nordic race." He desired the establishment of an "All-Nordic Institute" to "protect the interests of the Nordic peoples in a racial biological sense and according to uniform international rules." The French race researcher Georges Vacher de Lapouge and the American eugenicist Madison Grant took up Mjöen's suggestion and from their own point of view called for an international convocation to maintain the "Nordic race."[24]

Wiggam's eugenic internationalism, Laughlin's eugenically oriented world constitution, Johnson's "eugenic standards for a planned world order," Lenz's "blonde international," Ploetz's "confederation of states of linguistically and racially related peoples," and Mjöen's merger for the maintenance of the "Nordic race" were not part of a systematic campaign for the internationalization of eugenics. They were more or less thought experiments thrown out for discussion—thought experiments that, however, point to the fact that influential eugenicists after the First World War were internationally oriented in their thinking and were looking beyond the bounds of the individual state. These mind games formed the background of a *Weltanschauung* that looked to the concrete international cooperation of the European and American eugenicists in questions of race.

The eugenic world models à la Johnson, Laughlin, and Lenz show that the intellectual, scientific, and political "boundary drawing" of orthodox eugenics was not a national affair, but rather one of race. Eugenicists were clear that nations were political and cultural constructs, not race constructs. In this, they consciously turned away from the race theory of Arthur de Gobineau, who in an essay on the "Inequality of the Human Races," had claimed that a people's cultural assets and its ability to develop historically were determined by a people's "race substance." According to Gobineau, every "nation" is therefore the result of racially determined abilities and lack of abilities.[25]

Again and again eugenicists in Great Britain, America, France, and Germany warned against mixing up "race" and "nation." All peoples, including nation states, were in the end nothing more than the result of "race mixtures." Erwin Baur emphasized in a memo for the Council for Race Hygiene of the Prussian Ministry of Welfare that "all present-day culture peoples are...mixed peoples." They are the result of the "sexual intermingling of a very large number of exceedingly different initial races." Thus, the German people included not only components of the "Nordic race," of the "Alpine race," and the "Mediterranean race," but also "Mongols" and even "Negroes."[26] André Siegfried, French eugenicist and sociologist at the Paris École des Sciences Politiques, referred in a lecture at an international eugenics conference in 1926 in Paris to the mixture of Germans, Scots, Irish, Spanish, Jews, Danes, and Swedes in the United States in order to make clear the complex racial composition of one nation.[27]

On the basis of this understanding of peoples and nations as mixtures of races, the members of the international eugenics organization set themselves off from the race theorists like Arthur de Gobineau, Houston Stewart Chamberlain, and Ludwig Woltmann. Genetically educated eugenicists modernized the race theories that had emerged in the nineteenth century—geneticists like Charles Davenport, Fritz Lenz, Erwin Baur, Hermann Muckermann, Ruggles Gates, and R. A. Fisher. In place of a rigid race typology based on nations and languages, they used population genetics and the Mendelian theory of heredity for a modern variant of scientific race research. It was now no longer necessary to establish an identity among race, language, and nation; it was sufficient to identify various race proportions in the individual peoples and cultures. Thus, Baur relativized his thesis about nations being formed from various races by commenting that these mixed peoples arose out of a constant, "dappled characteristic." The "percentage of blondes and dark-haired people, the short-nosed people and the long-noses, the diabetics, paranoids, the stupid ones, the smart ones, those with great talent, and the congenital criminals" would without selection remain approximately the same from generation to generation in a people.[28] Lenz also maintained that a "white race" in the ordinary sense of the term no longer existed. Instead, the "whites" had a larger "share of Nordic blood" and a smaller share of "Negro blood."[29]

By limiting the definition in this way, the internationally organized eugenicists could maintain the idea of race-related superiority and inferiority, and at the same time, they could replace the biologically outmoded notion of nation with a European-centered and race-centered internationalism. The eugenicists thereby intentionally destroyed national boundaries, and from then on introduced a scientifically legitimated construct that served to distinguish humans from each other. The nation-state distribution of the global population into various groups with distinguishable cultures and collective abilities was superseded by dividing the total population by race. The Polish–British sociologist Zygmunt Bauman correctly therefore speaks of "non-national nations."[30] Races, in the words of the French social scientist Étienne Balibar, are "supranational." While the eugenicists dissolved the classical boundaries of nations, they adopted the exclusive character of the races in their racist "supernationalism."[31]

Race Research and Making Eugenics Scientific: The International Federation of Eugenic Organizations

The International Federation of Eugenic Organizations (IFEO), successor to the Permanent International Eugenics Commission in 1925, was a melting pot for eugenicists. It bound together their scientific interests in internationally organized research on race with political propaganda for the distinctiveness of the "white race." Like most of their contemporaries of European origin, they were convinced that the "white race" formed the *crème de la crème*, and its hegemony had to be protected against the tendencies to degeneration and racial mixing. The American representatives in the IFEO, Charles B. Davenport, Irving Fisher, and Harry H. Laughlin, considered it as scientifically proven that the "white race is the best race" in intellectual accomplishment. Davenport was concerned that the Americans of European origin would not succeed in preventing the "feebler races" from penetrating into the United States and that in the long term, the country would have to be turned over to the "blacks, browns, and yellows."[32] At the beginning of the 1920s, Laughlin had great success in influencing American legislation that reduced the quota for immigrants from the countries of Eastern and Southern Europe almost to zero. Fisher, one of the most important economists of the twentieth century, was of the view that the "white race is the best race" and that it therefore absolutely must take a stand against the "other races."[33]

The delegates to the IFEO from Great Britain, along with Leonard Darwin and Ernst W. MacBride, were the geneticists Ronald A. Fisher and Ruggles Gates.[34] Fisher, Darwin, and Gates were known for the fact that not only were they convinced of the existence of three separate races, but also that these could be ranked in a hierarchy. In intellect and spirit, the "white race" and the "yellow race" were particularly superior to the "black race." Gates in his textbook *Heredity and Eugenics* unequivocally stated that "race mixtures between white and African races" were undesirable both "from a eugenic" and from "any other reasonable point of view."[35] Cora Hudson, the British director of the IFEO, also spoke out against race mixing and gave as an example the sterility of couples because of the racial incompatibility of the marriage partners.[36] In 1929, Captain George H. L. F. Pitt-Rivers became one of the British delegates to the International Federation.[37] Pitt-Rivers later was one of the driving forces of the British Fascists and actively supported Hitler's expansion policy into the East.[38]

The German members of the IFEO, Eugen Fischer, Ernst Rüdin, and Alfred Ploetz, supported a scientifically grounded concept of race that differed only in the details. It is true that at the beginning of the 1930s, Fischer attempted to reduce any association with the racial anthropological concepts of the German race theoretician Hans F. K. Günther by emphasizing the concept of "eugenics" in place of "race hygiene," but he left no doubt that there were genetically determined psychological differences among the three large races.[39] Ploetz, supported by Rüdin, defended the concept of an "anthropological system of races" in the sense of "race hygiene." The North American eugenicists had shown "in their conflicts with the block of 11 million Negroes who are a thorn in the flesh of the people of the United States" that one cannot separate eugenics from questions of race.[40]

The Scandinavian members of the IFEO—Herman Lundborg, Herman Nilsson-Ehle, and Torsten Sjögren from Sweden; Sören Hansen and August Wimmer from Denmark; Harry Federley from Finland; and Alfred Mjöen and Wilhelm Keilhau from Norway—emphasized the Nordic character of the Scandinavians.[41] In this, they were very close to the position of the "Nordic wing" of the German race hygienists around Ploetz, Rüdin, and Lenz. The Swiss delegate Auguste Forel proclaimed the central task of eugenics to be to protect the "higher races" against the "profiteering by inferiors" such as the "Negroes." "Mixed marriages" between Europeans and "Negroes" had to be avoided as much as possible.[42] The long-time Swiss delegate, the Zurich anthropologist Otto Schlaginhaufen, very firmly rejected race mixing.[43] From Italy, Corrado Gini, an outspoken opponent of "race mixing" between Europeans and Africans, came regularly to the meetings of the International Federation.[44]

Harold B. Fantham, the South African delegate to the IFEO, regarded mixing between different, unrelated races as "an unnecessary irritant." Once the chromosomes of Bantu origin get mingled in white families, they cannot be bred out...but will exhibit themselves in unfortunate ways and unfortunate times throughout the ages."[45] Even the French eugenicist Georges Schreiber, who in many questions was on a confrontation course with the leadership of the IFEO, spoke out firmly against marriage between "individuals of different colors."[46]

Given the broad consensus within the IFEO on questions of race, it is not surprising that down through the 1930s, the European and American eugenicists had a silent agreement to accept only eugenicists of European origin into the international eugenics organization. Mjöen in particular, who even before the Second International Eugenics Congress had demanded a restriction of eugenicists to the "white race," insisted that the international eugenics organization be reserved for eugenicists of European origin.[47]

Race-related restrictions on membership in the IFEO resulted in a long period where Asian and African eugenicists were denied entry into the international eugenics organization. Active eugenics movements had been formed in the 1920s in particular in India, China, and Japan. The Indian Eugenics Society had been founded in June 1921 under the impact of the imminent Second International Eugenics Congress. It aimed to combine the Western idea of eugenics with elements of Eastern philosophies into a typical "Indian point of view" in eugenics.[48] In the 1920s, the Chinese eugenicists were strongly influenced by their American colleagues. Pan Guangdan, director of the first Chinese eugenics institute, in his eugenics pamphlet painted a picture of the eugenics societies in Europe and America as models worth copying.[49] The Japanese Society for Race Hygiene was founded in 1930 at the initiative of Hisomu Nagai, a professor of physiology at the Imperial University in Tokyo. The members of the society exercised great influence on the Imperial Commission for Population and Family and directly affected Japanese social welfare legislation.[50]

Despite the racist exclusion of the Indian, Chinese, and Japanese eugenicists, the leading circles of the IFEO were interested in building up their organization into a worldwide federation, and they courted the "white"-dominated eugenics societies from South America, South Africa, Kenya, and The Netherlands East Indies. Under the impact of the Second International Eugenics Congress in New York, a pan-American organization for eugenics was founded by eugenicists from Argentina, Bolivia, Chile, Columbia, Costa Rica, Cuba, the Dominican Republic, Guatemala, San Salvador, Mexico, Panama, Peru, Uruguay, Venezuela, and the United States. Their central coordinating office, the Oficina Central Panamericana de Eugenesia y Homicultura, became a member of the IFEO.[51] The South African Eugenics Society, dominated by Boers and Englishmen, was strongly inspired by the British and American examples, and directly after its founding in the mid-1920s, became a member of the IFEO.[52] The Eugenetsche Vereeniging in Nederlandsch-Indië, in which eugenicists from Holland were predominant, became a member of the international eugenics organization in 1930, followed in 1935 by the British colonialist-dominated Kenya Society for the Study of Race Improvement.[53]

It was only at the beginning of the 1930s that some members of the International Federation came around to the view that because of the desired scientific status of the organization, Asian and possibly African eugenicists should be admitted. The attempt to establish eugenics as an internationally recognized science required concessions to scientific universalism, according to which research results would be judged independently of nation, race, sex, religion, or social origin of the researcher. This was the view of the British members of the International Federation, who were interested in strongly developing eugenics in their Indian colonies, and were therefore more interested in the scientific status of the eugenics societies and less so in their racial origin.[54] Because he was impressed by the progress of Japanese eugenics in the 1930s, Davenport too supported accepting the Japanese Society for Race Hygiene into the International Federation.[55] Ernst Rüdin himself, elected as chairman of the International Federation in 1932, argued in a discussion with his mentor Alfred Ploetz for accepting African and Asian eugenicists.[56]

Even if in the end no expansion of the IFEO took place for African or Asian eugenicists, the discussion regarding their acceptance was a sign of disregarding the idea favored by Ploetz and Mjöen of an international merger of European eugenicists into an umbrella organization of eugenic scientists based on scientific procedures. This sidelining, however, did not mean that racist positions within the International Federation were given up; rather, it meant that the promotion of international race research became a basic scientific pillar of eugenics.

As the American eugenicists and German race hygienists moved into the leadership of the international eugenics movement, the intensification of international cooperation in race research became an important building block in the efforts to make eugenics more scientific. With the aid of genetically funded race research in an international context, such varied fields as anthropology, psychiatry, genetics, and sociology were fused together into a unified scientific construct. The members of the IFEO who were active in race research made it clear that the orthodox racist variant of eugenics in the 1920s could not allow itself to be reduced to a group of scientifically questionable individuals.[57] It is true that the scientific reputation of eugenicists like Darwin, Laughlin, Ploetz, and Mjöen was quite modest. Even eugenic colleagues doubted their academic abilities. More generally, however, the orthodox eugenicists who were in charge in the 1920s were numbered among the leading lights of European and American academic study. Ruggles Gates, Herman Lundborg, and Charles Davenport had laid the foundation for the study of human heredity in their studies at the beginning of the twentieth century. Ronald A. Fisher, Herman Nilsson-Ehle, and Harry Federley were listed among the leading general geneticists of the first half of the twentieth century. The scientific reputations of Auguste Forel, Ernst Rüdin, August Wimmer, und Torsten Sjögren in the realm of psychiatry were sterling, while Eugen Fischer and Otto Schlaginhaufen were oracles of physical anthropology as it related to human genetics.

From the point of view of many orthodox eugenicists, the migratory movements unleashed by the First World War and the increase in world trade had sharpened the "race problem," and research on the race problem had been pushed forward formally onto the agenda of eugenic science. In their eyes, it was a shame that at a time when more "race research" appeared necessary due to developments in world politics, the consensus among European and North American scientists regarding the superiority of the white race was increasingly falling apart. Eugenicists, anthropologists, and physicians, who at the Second International Eugenics Congress in 1921 had faced no opposition to their theses regarding the superiority of the white race over the Africans and Asians, now were the objects of increasing criticism in the scientific community. Studies by the anthropologist Melville Herskovits on racial crossings in the United States, the antiracist reinterpretation of the American army intelligence tests in the First World War by the New York anthropologist Otto Klineberg, and the methodological critique by the geneticist William Castle of his one-time teacher Davenport threatened to destroy the ground on which racist research was based.[58]

To most eugenicists it was clear that they had to adapt their proposals to new scientific knowledge and to the ever sharper criticism. Many studies of racial anthropology of the nineteenth century were methodically too questionable to stand up to a critical review. Now the eugenicists saw the need to adapt the study of race, which for a long time had been dominated by race theoreticians and race ideologues educated in the humanities, to new knowledge in genetics.

During his term in office as chairman of the IFEO between 1927 and 1932, Davenport turned the central focus of the international organization to international research on the spiritual, intellectual, and physical differences between the various races and on the "crossings" between the "major races."[59] In his view, a precondition for research into "race differences" and "race crossings" was making uniform the anthropological and psychological measurement techniques used by eugenicists. Despite progress made by eugenics research, in the 1920s, there was still an extensive lack of uniformity regarding important methodological questions and the interpretation of scientific results. Eugenicists in various countries used varying forms of phylogenetic trees and measurement techniques. The anthropological measurements by American eugenicists were often not comparable with the results achieved by their European colleagues, and varying standards and assumptions underlay the intelligence tests that were becoming ever more popular among psychologists and eugenicists.[60]

For Davenport, it was clear that only through international standardization of the measurements could "the differences of races" be determined and the "reactions of hybrids" be studied scientifically.[61] Davenport's goal was to show that through international standardization of anthropological measurements and of aptitude tests, there was a genetic basis for the varied results for the two races in intelligence tests. In that way, his prior research results on the "mental and temperamental reaction of races" could be confirmed.[62]

The foundation's efforts to meet international standards efforts began at a meeting of the IFEO in 1930 in the British town of Farnham. It became clear here how much interest in making eugenics scientific by international standardization of anthropological and psychological measurements of humans was tied up with the motivation to prove that there were physical and mental differences between the races. Miriam L. Tildesley of the British Royal College of Surgeons explained to the assembled eugenicists that standardized anthropological techniques could clarify race characteristics, the laws of heredity, and the influence of the environment on human development.[63] Barbara Schieffelin of the American Eugenics Research Association in her lecture emphasized that not only physical, but also mental characteristics of humans must be measured. The intelligence tests of black and white recruits in the American army had shown clearly that there were intellectual characteristics typical of particular races.[64]

Interest in intensifying research on race also played a role in the establishment of two committees at Farnham that were to set international standards for human measurement.[65] The first committee, which included as members besides Tildesley, the Swedish race researcher Herman Lundborg and the Swiss anthropologist Otto Schlaginhaufen, concerned itself with standardizing the measurement of the physical characteristics of the various races. Since no international anthropological congress had taken place since 1907, this committee was to set standards for the anthropological measurement of people that later could be adopted by an international congress of anthropologists.[66] The second committee, which included the British eugenicists and psychologists Charles Spearman and W. R. S. Stephenson as well as the American researcher on race Morris Steggerda was tasked with making uniform the intelligence tests developed in Europe and America and revising these tests in a way that they could be used for the "psychological examination" of "non-European and primitive peoples."[67] In that way, not only could a direct evaluation of each individual be made possible, but the presumably intellectual abilities and characteristics of the races could be shown quantitatively.

Research into "race mixing" in the entire world constituted the focus of the race research coordinated by the IFEO. The migration of the various European and Asiatic "sub-races" to North America, the mixing of the original American inhabitants with "whites" and "Negroes," the formation of "Eurasians" in India, and the "Mongolization" on various oceanic islands had, according to the eugenicists, made "race mixing" a significant problem of humanity.[68] Despite the fact that scientists "almost everywhere came into contact with the problems of race mixing," said Davenport, there reigned "a lamentable paucity of exact data, whether physical or psychological."[69]

Due to the presumably "great importance of race crossing and the international nature of this problem," the International Federation in 1927 set up a committee to research "bastardization" and "race mixing."[70] The immediate initiative to establish the committee came from Davenport. He had determined that since 1917 "crossings" between two different races had led to physical and mental "disharmonies." As proof, he used crossings between two different types of hens. If one would cross leghorns, hens with a high egg production but without good brooding ability, with brahmas, hens with lower egg production but with good brooding ability, the result would be extremely poor hybrids. The resulting hybrid hens would have small egg laying capacity and insufficient brooding ability. Davenport transferred this research to humans.

Although he had to admit that there were scarcely any sufficient data about "race mixing" among humans, he was of the opinion that he had observed "half-breeds" with disproportional body structure (long legs, short arms), inadequate blood circulation, and psychological problems.[71] To shore up his thesis about the disharmony involved in race crossings, between 1926 and 1929, he worked intensively with his colleague Morris Steggerda on research on "whites," "blacks," and "racial half-breeds" in Jamaica. This research project of the two academics of the Cold Spring Harbor Laboratory, financed by Wickliffe Draper, used intelligence tests and anthropological measurements to illustrate basic, genetically related differences between the "whites" and the "blacks."[72] In this study of half-breeds, they observed several physical and intellectual "disharmonies." Such "disharmonies" were for Davenport and Steggerda an example of the mixture of ambition, desire to work, and focus on power of the whites with the laziness, instability, stupidity, and lack of self-control of the blacks.

Davenport, in presenting the results of his "bastard researches" for the first time in 1928 in Munich to an international public, claimed that the results of the tests showed that "half-breeds" as a rule were better than the "blacks." To his surprise, some "half-breeds" even produced outstanding results. He concluded from this that race mixing would be eugenically desirable if one would allow only these superior "half-breeds" to reproduce, such as one did when raising horses, cows, and pigs. Davenport realized that this would be politically and morally impossible to implement, and it would be more sensible to take the precautionary measure of making difficult any form of race mixing.[73]

The German anthropologist Eugen Fischer, a well-known expert on "race crossings," took over the chairmanship of the international committee for research into "bastardization" and "race mixing." In 1908, Fischer had anthropologically measured over 300 so-called Rehoboth bastards. These "half-breeds," at that time living in German Southwest Africa, descended from the Dutch colonial masters mixed with the African native population. In his study published in 1913, Fischer, as Davenport before him, expressed the view that Mendelian laws are applicable to humans.[74] True, Fischer could not find any increased occurrence of disease among the "Rehoboth bastards," but nevertheless he argued for preventing race mixing as a precautionary measure. He was absolutely adamant in opposing the notion that "half-breeds" should be considered part of the "white race," and argued that they should be treated as "native born."[75]

Fischer, whose scientific prestige rested mainly on his study of the "Rehoboth bastards," did not hesitate to draw political conclusions from his studies. He feared that the 100 "Negroes" who had settled in Germany after the Berlin colonial exposition would bring race mixtures to German soil. Even the so-called Rhine bastards were a mote in his eye. He perceived as a serious threat this group of about 1,000 children, the children of German mothers and African soldiers of the French army who had been stationed in Germany between 1920 and 1927.[76]

Fischer's concern here was that because of the increasing "race mixing," it would soon be impossible to find any "pure races." As a result, the "F1-bastards"—half-breeds with two different pure-race parents—who were especially valuable for race research, would become increasingly rare.[77] At a meeting of the IFEO in Munich, he pointed out that fortunately there were still a sufficient number of "Jewish-Christian half-breeds" of the first generation who were available for anthropological observations. Anthropologists would then not have to take such difficult trips into the African bush for their "bastard researches."[78]

Fischer won over one colleague, the tropical medicine physician Ernst Rodenwaldt, for the Commission of the IFEO. In 1922, Rodenwaldt published his study on the "mestizos on Kisar"—"half-breeds" of Europeans and Indonesians. In this study, he found no physical disharmonies, and refrained from any judgment on possible intellectual degeneration.[79] Rodenwaldt's moderate position later changed drastically. In 1934, he published a study on the "conflict over the soul of the half-breed," in which he chastised "race crossings" as "irresponsible." Any people that "unreservedly mixed with a foreign race" would "reduce its percentage of self-assured leader personalities" and would register a loss that no people could bear in the long run.[80]

The geneticist Hermann Nilsson-Ehle, one of the two Swedish representatives on the Commission on Race Crossing, maintained that "hybrids" of racially distant parents would often inherit no or very few "valuable" qualities from one or the other race. A large group of the "half-breeds" would inherit some of the "valuable" characteristics, and many would inherit the "less desirable" characteristics. As evidence, he pointed at his research on hybrid plants. Particularly in the second cross-generation a large percentage of "defective and sick examples" would appear. That half-breeds would inherit the positive characteristics of both races was something that Nilsson-Ehle considered to be quite unlikely for far-apart races. On the other hand, with very similar "parent types," both with plants and humans, one could count on a "superior" set of hereditary materials.[81]

Herman Lundborg, his Swedish colleague on the Commission, had himself completed research on race crossings between Lapps in northern Sweden and the "Northern Swedish population." Lundborg was convinced that the "mixing" of two distant "races" as a rule resulted in "inferior" children. He feared that particularly in the large cities of America and Europe, the many forms of "race mixing" would give rise to a great potential for the intellectually or physically infirm.[82] At a meeting of the IFEO in Rome in 1929, he called for comprehensive studies on race mixes. They should preferably be carried out in those areas where races with clearly different characteristics had mixed together. In this way, the "great race biological institutes" in the United States and Great Britain would coordinate the studies on race by scientists from the various European and American countries.[83]

The Norwegian Alfred Mjöen was the most industrious member of the Commission. His initial hypothesis was that there were "astounding similarities" between race mixing among humans and among animals. The mixing of two widely separate races would reduce the physical and intellectual level both of animals and of humans. This was made clear by the fact that prostitutes and malingerers are more often to be found among the "bastards" than among "relatively pure types." Mjöen concluded that one would do better to prevent the mixture of "widely separated races" and should promote a "healthy race instinct" of the various races until there was more information about the results of race mixing.[84]

The British representative on the Commission on Race Crossing was Ruggles Gates, qualified by his research on "hybrids" between "whites" and "Indians" in Canada. Gates, who even after 1945 made no secret of his preference for racial segregation, warned that "race half-breeds" would have a "chaotic constitution." He was convinced that the differences between human races was often greater than the differences between many types of animals, and for that reason, he thought that "race mixing" constituted an extremely risky human experiment.[85]

For the eugenicists organized in the IFEO, race mixing was one of the areas in which international cooperation among scientists could bear fruit. In light of the explosive increase of race mixing and its significance for humanity, Lundborg desired the establishment of an international institute for research on race crossing.[86] Davenport supported the demand for a "World Institute for Race Crossings," and declared that the time was right for "concerted international action" to take in hand the solution for this pressing "world problem."[87] At the meeting of the IFEO in Farnham, Gates used the worldwide phenomenon of "race mixing" as the reason for establishing a "world eugenics"—the study of the eugenic consequences of "race mixing" on the international level.[88]

Following through on Davenport's desire for "concerted international action," the IFEO began a campaign for the systematic survey of the "half-breed races." Davenport asked eugenicists from around the world to inform him of the areas in which "dissimilar human races" come into contact with one another. In this regard, he was interested not only in "primary races, like white, negro, Indian and orientals, but also very dissimilar European races."[89] With these data, Davenport planned to supplement his "world map of mixed race areas," which he had presented at the IFEO meeting in Munich in 1928. In these "mixed race areas," the emphasis of this study would be on the "inheritance of physical and mental characteristics" of the various races.[90]

Based on this localizing of "bastardization areas," Fischer and Davenport agreed on a worldwide survey of anthropological data on race crossings. Davenport and Fischer at this time were directors of the two most important research institutes for eugenics and human genetics—the Cold Spring Harbor Laboratory on Long Island and the Kaiser Wilhelm Institute for Anthropology, Human Genetics, and Eugenics in Berlin. In an article that appeared in German, British, and American magazines, Fischer and Davenport maintained that despite the "biological significance" of race mixing, "knowledge regarding the bastardization among humans was still extraordinarily weak." There are only a few pioneer studies on the "Rehoboth bastards," the mestizos on "Kisar," the "race crossings" in Jamaica, the "race mixtures" in South Africa, and the "hybrids in Hawaii." The contrast between these "minimal scientific activities in the area of bastard research" and the "extent of race crossing" in all areas of the earth were for these two eugenicists "monstrous." "Systematic research into the many forms of combinations of individual races" was felt to be an urgent need. Before trained researchers could gather the "anthropological descriptive and metric data," the eugenicists would have to make an exact picture of the "scope and extent of human bastardization."[91]

For this purpose, Fischer and Davenport prepared a questionnaire in British, German, French, Spanish, and Dutch, which was to be sent at the cost of the IFEO to reliable physicians, missionaries, officials, merchants, farmers, and travelers in the overseas areas. The same questionnaire would later be used for surveying the "bastardizations" in European and North American large cities.[92] The observers were to report on the questionnaire whether there were "crossings between pure races" in a specific area and how frequently they occur. Fischer and Davenport asked whether the "bastard population" lived separately from Europeans, whether they came from stable marriages, whether public or private prostitution prevailed, whether the "half-breeds" made cultural contributions, and whether there was a separate "bastard culture." In addition, the observers were to supply information as to whether any remarkable psychological behavior of the "bastards" could be determined, and whether the "bastard population" was "morally inferior" to the two parent races. Furthermore, the observers were to supply information about "intelligence, energy, industriousness, stamina, planning for the future, temperament, violent temper, pride, kindness, cruelty, musicality, technical aptitude, mental diseases, alcoholism, criminality, prostitution, and vagabondage of the bastard population."[93]

The basic assumption of all of the "bastard research" was that the various races differed not only physically, but also mentally and intellectually. As Lenz had correctly observed, "The entire race question would be meaningless if there were only physical differences of race."[94] The "mental disharmonies of half-breeds" uncovered by "bastard researchers" could appear only if the various races were different not only physically but also from one another in their hereditary material for psychological characteristics.

The IFEO Commission for Race Psychiatry that was set up in 1930 was assigned the task of bringing out these differences in the mental and psychological constitution of races. The topic of "race psychiatry" included the "psychopathological differences of the various races," meaning the manner in which the "different races and race mixes of the world were affected by mental disturbances and abnormalities."

The Chairman of the IFEO Commission Ernst Rüdin saw the work on the "race pathologies" simply as a first step to a "race psychology," which was to involve systematic worldwide research into the various mental and intellectual tendencies of "races and race mixtures."[95] Rüdin felt that in the initial stage, the IFEO Commission should restrict itself to examining and evaluating the tendency of the individual races to "mental deficiencies and intellectual diseases and defects." This would require the "close collaboration of all researchers in this area...and careful comparison of the research results on as uniform a basis as possible."[96]

The eugenicists organized in the IFEO politicized the race questions that they discussed primarily from medical and anthropological points of view. For them, race anthropology, race psychiatry, and bastard research were primarily scientific research programs, but research programs with clear policy implications. So Davenport explained that he would prefer to limit international eugenic work to human genetic anthropological studies, but unfortunately the problems with "criminals and degenerates" made it necessary for the international eugenics movement to take a position on political questions.[97]

A close connection between science and politics was called for within the IFEO. By making their political demands flow as direct results of the scientific results, the eugenicists in the International Federation could systematically shield themselves from any criticism of their political demands. Any critical discussion of their eugenics program could be disqualified as being unscientific.

The dangers of race mixing that the eugenicists had confirmed made them feel obligated to be active politically. As long as science had not delivered absolute proof that the crossing of distant races was not harmful, this crossing should absolutely be prevented. Since they thought that races are intellectually and physically different, and since race crossing quite likely led to psychological disharmonies, global migration should be managed eugenically.[98]

Eugenicists saw the purity and health of the races threatened by race mixing brought on by migration. In a letter to Madison Grant, Davenport maintained that selectively accepting immigrants might quiet the need of the "capitalists" for cheap labor, but it would at the same time require later generations to turn over America to the "blacks, browns, and yellows" and even force him to seek asylum in New Zealand.[99] Mjöen declared at a meeting of the International Federation that in the three years following the World War, "race elements from a foreign source" had "fallen upon" the large cities of Western Europe and North America. He went on: "Nobody who with open eyes has observed the masses in the great modern cities, Paris, Berlin, New York, or Chicago, will have failed to be struck in which the racial physiognomy of the population is in process of changing...All unity of form is dissolved, and a hideous confusion of all possible colors and shapes from all the races of the earth has taken its place."[100]

It was Mjöen who in 1925 had first put the topic of immigration onto the agenda of the IFEO. In the middle of the 1920s, he had forwarded to the Norwegian Parliament a legislative proposal for "biological immigration controls." Since Norway had not yet become a target area for "increased immigration of foreign race elements," Mjöen felt that measures had to be undertaken for protection against immigrants "with a lower culture but with a higher birth rate." Immigration to

Norway was to be reserved to immigrants from countries with predominantly "Nordic population." Entry was absolutely to be refused to "idiots, mental defectives, the feeble-minded, alcoholics, and persons without proper employment."[101] To the eugenicists assembled in London, he proposed that immigration be made a central topic for international cooperation and that international guidelines be worked out for "biological monitoring" of migratory movements.[102]

The eugenicists went further into this topic at the 1926 conference in Paris. At the center of their considerations were the experiences of the immigration law of 1924 in the United States. This eugenic law had definitely brought to a close the political concept of the United States as the country of a melting pot of immigrants from the various parts of the world. As eugenicists in Europe and North America confidently noted, it effectively ended a policy that had devoted itself to the "ideal of general citizen equality." The American politicians had turned to the goal of "race purity and uniformity."[103] André Siegfried pointed out to the scientists gathered in Paris that the immigration law was an excellent example of how the penetration of eugenic knowledge into everyday thinking had exploded the "old theory of human equality, racial equality, and the possibility of educating a man of any race."[104]

In addition to excluding people from migrating to the United States with mental and physical ailments, the US immigration law also excluded Southern and Eastern Europeans, Asians, and Africans. In order to permit migration by people from countries with a higher percentage of "Nordic" hereditary material, a quota was set based on the 1890 census. In contrast to the censuses of 1910 and 1920, which had included the mass immigration from Southern and Eastern Europe, the earlier census had shown the overwhelming majority of Americans from abroad to come from Scandinavia, racial Germany, Austria, Ireland, and Great Britain.[105] In this way, by setting the immigration quotas based on the census of 1890 and not that of 1920, the overwhelming Nordic racial character of the United States was to be maintained.[106]

The immigration law of 1924 came into being heavily influenced by the American eugenicists and the Immigration Restriction League, which they dominated. Prescott F. Hall, a Boston eugenicist and cofounder of the Immigration Restriction League, in 1919 demanded that limitations on immigration be used as a "method of world eugenics." Just as the eugenicists tried to encourage virile people to reproduce and to limit reproduction by the non-virile, "world eugenics" had served for the protection and interests of the superior race. For Hall, restrictions on immigration were race separation writ large. Carriers of "inferior stocks" would thereby be hindered from "diluting and supplanting good stocks."[107]

A committee of seven eugenicists under the chairmanship of Harvard professor Robert DeCourcy Ward took the arguments of the American Eugenics Society right to the point—instead of being a great place of asylum for the poor and infirm of the world, the United States had to implement an immigration policy for "improvement of the race." Poor results on the Army intelligence test by immigrants coming from Southern and Eastern Europe and the high percentage of Americans from abroad in mental institutions were said to show that the United States was considered a place for the European countries to dump their "inferior" individuals.[108] The committee, which included Madison Grant, Harry H. Laughlin, and Albert Johnson, the chairman of the congressional committee for immigration matters, considered immigration to be a political question of the first order. Since eugenic management of immigration policy would be of decisive importance for the improvement of the race in the United States, it must be left to each state to determine how many and which immigrants it wished to accept.[109]

The American representatives to the IFEO aggressively supported the view that immigration policy is a matter for each individual nation, and Davenport and Laughlin pushed for an international regulation according to which each nation would have the right to decide for itself who could leave and who could enter.[110] Their views ran into opposition from eugenicists from emigration countries. The background of the argument lay in the economic difficulties of the countries of Southern and Eastern Europe that were reinforced by the rigid immigration policy of the United States in the mid-1920s. The oversupply of workers in countries like Italy, Spain, and Poland, which had been reduced by the streams of emigrants at the beginning of the century, grew significantly.[111] That is why the Italian eugenicists around Corrado Gini in particular, but also supported by some Scandinavian eugenicists, supported an internationally agreed-upon management of the streams of emigrants.

A resolution from the Dutchman G. P. Frets at the meeting of the IFEO in 1928 in Munich, which gave every state the right to do research on the origin of each immigrant and to decide on acceptance or rejection, was barely voted in with votes from the Dutch, Austrian, Swiss, and German eugenicists. The representatives of Italy, Great Britain, and Norway voted against.[112] At the Third International Congress of Eugenics, a self-satisfied Davenport noted that the self-determination rights of each state in questions of immigration were generally acknowledged. Henceforth, no state could free itself of its "inadequate individuals" by shoving them off to other states. In the future, every state would have to bear the burden alone of its own "unfit."[113]

The Exclusion of Lamarckian, Socialist, and Feminist Eugenicists

By the early 1930s, over 30 eugenic societies and scientific institutes from 22 countries belonged to the IFEO,[114] including two additional transnational organizations for eugenics, the Oficina Central Panamericana de Eugenesia y Homicultura and the International Union for the Scientific Investigation of Population Problems.[115] It is true that the leadership of the most important eugenics confederations and institutes in the IFEO came from Europe and America. One should not overlook the fact, however, that though the members of this international organization agreed on a broad rejection of the influence of the environment, the idea of the propagation of eugenics as a science with direct political consequences and of integrating race research into eugenics represented only one specific orthodox variant of eugenics.

The leadership group of the international eugenics organization—Davenport, Darwin, Rüdin, Gini, Lundborg, and Mjöen—defended this orthodox position against alternative strands of eugenics development, which came especially from within Social Democracy, the feminist movement, the neo-Malthusian movement, and the social hygiene reform movement. In France, the members of the eugenics society used their Lamarckian outlook to take strong social hygiene positions, and in Britain, the society for a long time included socialist and feminist eugenicists. Nonetheless, the policies of most membership organizations of the international eugenics organizations until the mid-1930s followed the orthodox eugenics position represented by Darwin, Davenport, Rüdin, Mjöen, and Lundborg. There were of course continued attempts by eugenics organizations from Catholic countries to reform the course of the international federation from within, but in the end, these efforts proved in vain.[116]

In ways strikingly similar to developments in eugenics societies in Europe and the United States, alternative eugenic discussion threads emerged within the socialist movement, the feminist movement, and the neo-Malthusian movement, coming from early eugenic thinkers like Wilhelm Schallmayer in Germany, Havelock Ellis in Great Britain, and Edward Bellamy in America. The socialist and feminist eugenicists usually did not get involved in the established eugenics organizations, but rather were involved in political groups using eugenic assumptions; as a result, the orthodox eugenicists in the international eugenics organization essentially remained among themselves. Even though critics of orthodox eugenics, particularly in Scandinavia and France, increasingly took over the leadership of the eugenics societies, the leaders of the IFEO maintained control over the composition of the membership and thus over the scientific and political orientation of their organizations. They did this by using bylaws that stated that new groups could be accepted only upon vote of a large majority of the current member organizations.

The leadership of the IFEO firmly took a stand against the program of the neo-Malthusian movement. The neo-Malthusians, based on the writings of the British economist Thomas Robert Malthus, saw the world threatened by an exponential growth in population and thus demanded a drastic reduction in population growth. In the early twentieth century, they achieved international significance in Europe and in the United States.[117] For a long time, the movement was led by the American birth-control activist Margaret Sanger, who in the magazines she founded, *Woman Rebel* and *Birth Control Review*, and through the creation of several birth-control clinics in the United States had unleashed a vigorous controversy over the use of contraceptives.

Sanger at first had maintained close contact with the workers' movement and saw birth control overwhelmingly as a means to emancipate women from their duties as mothers, but in the 1920s, she increasingly spoke of contraceptive means as an instrument of the social control of reproduction. Like other leading birth-control activists—one might mention here Marie Stopes in Great Britain, Thit Jensen in Denmark, and Helene Stöcker and Henriette Fürth in Germany—she made no secret of her sympathy for eugenics, and hence her proposition of "more children from the able-bodied, fewer from the inferior." Her notion that a eugenic policy could be successful only through the implementation of the right of women to obtain contraceptives was typical of the ideas of leading feminist family planners in Europe and in the United States in the 1920s.[118]

The spokespersons for birth control shared the view of the orthodox eugenicists that differential birthrates should be in favor of the genetically superior classes. In contrast to eugenicists like Davenport, Darwin, or Ploetz, who wished to achieve increased reproduction of the more viable classes by doing away with contraceptives, these birth-control activists thought that this trend could be overturned if one made contraception available to the poor, socially weak, and mentally ill members of society. It was in this way that the eugenic problem of the presumably genetic "inferiors" became for the neo-Malthusians one of the major arguments for the widespread introduction of contraceptives.[119]

This point of view met decided rejection from the American, British, Scandinavian, and German eugenicists. At the very first international congress in London, the eugenicists criticized the neo-Malthusian demand for a general reduction of births. When the engineer Charles Vickery Drysdale, head of the British Malthusian League, announced that the Malthusians would act in a race hygiene manner in that they preferred to urge the "poor" to reduce their families, a vigorous reaction broke out.[120] The German and Scandinavian participants in particular held the opposing view that a broad introduction of contraceptives would contribute to the degeneration of the civilized countries.

Drysdale for his entire career was concerned with a balance between eugenicists and neo-Malthusians in the British Eugenics Society, but the race hygienists around Ploetz and Mjöen questioned his position. In their opinion, a general population policy aimed at a reduction of births would reduce the necessary selection pressure for an improvement of the race, since the number of births would fall below the number of deaths. In the view shared by many American, British, and Scandinavian eugenicists, freely available contraceptives would overwhelmingly be used by intelligent, "racially high value" couples. Since "intellectual defectives" were not prepared to use contraceptives nor were they in a position to do so, the effect of a neo-Malthusian population policy would be destructive from a eugenics perspective.

Ploetz was especially suspicious of the backing of the neo-Malthusians because of their support from a number of women activists in the feminist movement, and he added a race-related component to his arguments in London. He pointed to a serious threat to the "Nordic race" by a neo-Malthusian population policy. In the "great battle of the white race" for ongoing hegemony over the other races, Ploetz felt that Malthusianism was a great hindrance. Drawing on the widespread European anxiety at the beginning of the century about the threat from Asiatics, he warned that the "yellow race" could take over a leadership position through an increase in the number of their children as opposed to the "white race" weakened by uncontrolled birth prevention.[121]

Even though neo-Malthusianism found increasing adherents in the American and British eugenics societies after the First World War, the leadership of the IFEO maintained a strict refusal to entertain freely accessible contraceptives. Leonard Darwin considered the use of contraceptives as destructive for the race, and as president of both the British Eugenics Society and the international eugenics organization, he prohibited any publicity for contraceptives as an instrument of race distinction.[122] Darwin's successor as president of the IFEO, Charles Davenport, joined in this opinion and forbade any form of cooperation with the birth-control movement, which in his view had been undermined by feminism.[123] Henry F. Osborn at the Third International Congress for Eugenics in 1932 in New York summarized this critique by the orthodox eugenicists when he contrasted the ongoing "birth selection" of the eugenicists with the "birth control" of the neo-Malthusians. He claimed that birth control would lead in a "basically unnatural" direction. He did not consider it coincidental that birth control in the Soviet Union was supposedly accompanied by sexual promiscuity.[124]

Even if many French eugenicists supported the campaign by Darwin, Davenport, and Ploetz against birth control, after the World War, they themselves increasingly were criticized by the orthodox eugenicists in Great Britain, the United States, Scandinavia, and Germany. The physician-dominated French Eugenics Society had developed a program that put emphasis on increasing the birthrate of eugenically valuable couples and on general hygienic and medical improvements. The basic Lamarckian conviction of most French eugenicists and the widespread French concern over the declining birthrate resulted in eugenics in France being mixed together with a catalog of social hygiene measures.[125]

The leaders of the international eugenics organization were very upset with what they considered the progressive watering down of eugenics in France. Lapouge, one of the few French eugenicists who represented the position of Davenport and Darwin, supported the fear of the IFEO leadership that France had but very few "real" eugenicists. According to Lapouge, the members of the French Eugenics Society had a completely different understanding of eugenics from the majority of members of the international eugenics organization.[126] Darwin warned that to follow the French eugenicists would mean that the struggle against sex-related diseases and a general increase in the birthrate would remain quite high on the agenda of the international eugenics organization.[127]

In order to reduce the influence of the French Lamarckians in the international eugenics organization, the IFEO leadership even considered accepting as a French member the American A. F. Dupont, who was living in France. As Hodson put it, contact with the "true eugenicists in France" could be maintained through him.[128] Dupont, a close friend of Lapouge, shared the attitude of the majority of eugenicists in the international eugenics organization when it came to questions of race. In a note for the *Eugenical News*, he complained about the "influx" of Poles, Russians Jews, Orientals, Arabs, and Indochinese into France. "Negroes," he said, are already members of the French parliament, and the "black soldiers" stationed in France would marry "French girls."[129]

The growing conflict between the Lamarckian orientation of the French eugenicists and the basic Mendelian outlook of most American, British, German, and Scandinavian eugenicists was the reason that the leadership of the international eugenics organization turned down a tempting offer from the League of Red Cross Societies to make office space available for them in their Parisian offices. The League of Red Cross Societies called for an international policy of social hygiene in which eugenics played an important role. As part of its expansion after the war, it aimed to integrate into its work the international organizations that were active in the field of health.[130]

As part of this "embracing strategy," the League desired to cooperate closely with the international eugenics organization in the areas of health and social policy. It convinced the international eugenics society to name the French eugenicist Lucien March as the delegate for the League of Red Cross Societies and to begin negotiations regarding cooperation between the two organizations. At the 1924 Milan meeting of the eugenicists, René Sand, general secretary of the League of Red Cross Societies, suggested that the international eugenics organization set up a permanent office in the same space as the League of Red Cross Societies.

Although the establishment of a permanent office would have significantly increased the effectiveness of the international eugenics organization, the plan was stopped by a coalition of British and Scandinavian representatives. Darwin warned of the growing influence of French eugenicists within the international eugenics movement. If the headquarters of the international eugenics organization were to be moved to Paris, Darwin thought that this would threaten not only a peculiar combination of the frequently contradictory goals of general hygiene and eugenics, but it would also mean a growing Lamarckian distortion of eugenics with the struggle against tuberculosis and sex-related diseases as the major themes.[131]

The leadership of the IFEO increasingly forced the French eugenicists into the role of outsiders in the international eugenics movement. When the French Eugenics Society in 1926 merged with the French section of the social anthropology–oriented International Institute for Anthropology, once and for all the very close contacts among the British, American, German, and Scandinavian eugenicists and their French colleagues were broken.[132] The French eugenicists from then on were strongly oriented in the direction of cooperation with Italian, Spanish, and Latin American eugenicists, who looked with a critical eye at the Anglo-American–Scandinavian–German variant of eugenics.

The IFEO leadership group was especially concerned by the growing influence of socialist and Social Democratic eugenicists in science and politics. Under the impact of the world economic crisis and the emergence of Fascism in Europe, a series of prominent European and American geneticists grew closer to socialist positions. Their sensitivity to economic injustice and to the danger of Fascism, sharpened by their Marxism, led them to an increasingly critical position against the orthodox eugenicists.[133] Supported by their growing influence in genetics, they increasingly questioned the go-it-alone attitude of the orthodox eugenicists.

Socialist eugenicists like the Americans Hermann J. Muller und Walter Landauer and the British J. B. S. Haldane and Joseph Needham from the very beginning rebuked the eugenicists for simply using eugenics to scientifically legitimize their racial and class prejudices. This critique maintained that the leaders of the American, British, German, and Scandinavian eugenics movements from the beginning had placed the dominant eugenics policy on an unscientific basis. In the eyes of the socialist eugenicists, social inequality had distorted the judgment of geneticists and eugenicists to the detriment of the economically weak and the poorly educated.[134]

The socialist eugenicists demanded that one must first equalize the environmental conditions for all people before one could make judgments about their genetic differences.[135] For Muller, Haldane, and their colleagues, socialism and eugenics were mutually supporting conditions for a more humane society. Biological and social developments were for them two sides of the same coin. By systematically planning for reproduction under the best social conditions, so went the utopia of the socialist eugenicists, a society of genetically and socially valuable humans could be created.[136]

The socialist eugenicists were here not only convinced that eugenics and socialism could easily be brought into harmony with each other, but they were also of the view that one should strive for a synthesis of the two concepts. In this way, in addition to the basic scientific orientation of socialism, the utopias of a socialist society and the derogatory attitude of the workers' movement toward the so-called Lumpenproletariat played an important role. The idea of a socialist society oriented to achievement and the damning of the "asocial," discriminated against, unorganized Lumpenproletariat suggested the use of eugenic measures to genetically and socially prevent reproduction of the low-ranked population groups.[137]

Many socialist eugenicists in Europe and the United States felt drawn to the Soviet Union by their basic Marxist outlook. There, in the 1920s, an active eugenics movement could be created, initially well tolerated by the communist government and financially supported by the State Commissariat for Health and Education. The leaders were the biologists N. N. Koltzoff, S. S. Chetverikov, A. S. Serebrovskii, and Jurius Filipchenko.[138] For some Russian eugenicists, "Bolshevist eugenics" was the logical conclusion of the Marxist credo for a scientific organization of society.[139] The assimilation into Marxist ideology and the differentiation from the racist positions of the orthodox eugenicists resulted in the rapid dissolution of the short relationship arising in the 1920s between the Soviet eugenicists and the international eugenics umbrella organization.[140]

Based on the presumably scientifically supported Marxist policy and the strong promotion of Soviet genetics in the 1920s, a number of European and American eugenicists considered the Soviet Union as a promising starting point for an alternative eugenics policy. Eugenicists like Muller and Haldane were convinced that the true genetic elite could be selected only by the abolition of social injustices as was being practiced in the Soviet Union. Since the Soviet leadership favored general social needs over individual concerns, the hope arose that a consistent eugenics policy would be implemented.[141]

Hermann J. Muller, one of the major defenders of a socialist eugenics, at the Third International Eugenics Congress in New York in 1932, attempted to ridicule what he saw as orthodox eugenics corrupted by class and race prejudices. In a manuscript distributed before his speech to the congress entitled "Domination of the Economy over Eugenics," he claimed that the ruling capitalist elite was attempting to legitimize inequality between classes and races by referring to the genetic differences of these groups, despite all scientific evidence to the contrary. He provocatively claimed that only in a socialist society could eugenic measures truly be carried out. Only the "imminent revolution" would bring eugenicists into a position from which they could judge which human characteristics were the most valuable. Discussions among eugenicists regarding the differential reproduction rights for classes and the differences of the races would be made superfluous by the elimination of exploitation, oppression, and racial discrimination.[142]

When Muller's planned criticism of orthodox genetics became known, the eugenicists on the congress program committee tried with all their might to prevent the reading of the text. Davenport as chairman of the program committee criticized Muller's contribution as too "sociological" and of little worth for eugenics. By reducing Muller's allotted speaking time to ten minutes, Davenport attempted to shorten the text to the point of being unintelligible. Only after the strenuous intervention of the British geneticist L. C. Dunn, who like Muller had "been fed up with the orthodox breed of eugenics," was Muller able to read the complete paper at the congress.[143]

Muller's sharp criticism at the international eugenics congress reinforced the efforts of the eugenicists being attacked to exclude socialist- or social democratic–minded eugenicists from the international eugenics organization. Subsequently, the leadership group of the IFEO denied membership to such critically oriented eugenicists as the Swede Gunnar Dahlberg and the Norwegian Otto Mohr.

That the leadership group of the IFEO pushed aside the socialist, feminist, neo-Malthusian, and Lamarckian eugenicists shows that the international cooperation among scientists does not necessarily lead to a "trans-national objectivity," such as that proposed for molecular biology by the Boston historian of science Pnina Abir-Am.[144] The IFEO claimed to be the sole representative of organized eugenicists; its bylaws allowed for the systematic exclusion of alternative strains of development. These conditions resulted in a situation where efforts at basic reform within the international eugenics movement came much more slowly than in most national eugenics movements. The International Federation was thus a bulwark against any form of alternative development trends in eugenics.

CHAPTER FOUR

THE CRISIS OF ORTHODOX EUGENICS AND
THE RISE OF HUMAN GENETICS
AND POPULATION SCIENCE

In the eyes of the American eugenicists, the Third International Congress for Eugenics, which took place in New York from August 21 to August 23, 1932, was a milestone in international eugenics cooperation.[1] Davenport, crowning his five-year activity as chairman of the International Federation by being named president of the congress, in his opening speech emphasized the advantages of the variety of international contacts for eugenic research and practice. Eugenics societies had been founded around the world; international activities of the eugenicists had stimulated many research projects; national sovereignty in the determination of eugenic immigration quotas had been acknowledged far and wide; and in Denmark, inspired by international contacts, a sterilization law had been passed.[2]

Along with his praise for international scientific and political cooperation, Davenport also attempted to cover up the critical situation in which the eugenics societies in most countries found themselves at the beginning of the 1920s. Even if eugenics societies had been founded in over 40 countries by the beginning of the 1930s, and even if international cooperation within the International Federation of Eugenics Organizations had intensified, Davenport's optimism could not conceal the fact that the Third International Congress for Eugenics was a clear sign that the scientific reputation of eugenics was in danger. Just 73 eugenicists had made their way to New York.[3] While the First International Congress in 1912 had drawn a broad spectrum of scientists and interested laypeople, and at the Second International Congress in 1921 high-powered scientists with varied research interests had come together under the umbrella of eugenics, now in New York only a small group of convinced eugenicists showed up.[4]

Even if the organizers attempted to explain away the small number of participants with the very poor financial situation of eugenics societies amid the world economic crisis, the Third International Congress for Eugenics was a clear indication that the trend of turning eugenics into a science, begun so successfully after the First World War, was now stagnating in many countries. In this chapter, various failed attempts at professionalization by the International Federation of Eugenics Organizations will be presented. They provide clear evidence that eugenicists had great difficulties in establishing their discipline as an independent academic field.

The discrepancy between the scientific aspirations of orthodox eugenics and the reality of slight acknowledgment of eugenics as a science provided an ideal area of attack for animal and plant geneticists, who were increasingly critical of the orthodox eugenicists.

Given this criticism, one group of reform eugenicists in Great Britain and the United States argued that eugenics should be scaled back to a political movement in order to remove the growing pressure from the eugenics societies. "De-scientizing" eugenics, as will be shown, ran parallel to the establishment of the new scientific branches of human genetics and demography in an international context. While the eugenicists involved in the International Federation could still influence these two research areas at the end of the 1920s, during the 1930s, human genetics and demography became increasingly important as independent scientific subjects. The establishment of these two new academic areas more and more was used as a tool by the reform eugenicists to discredit orthodox eugenics.

The Plight of Eugenics Societies and Their Modernization

As already noted, setting up eugenics as an acknowledged science was a central motif for the cooperative work among the international eugenics organizations. The measures taken to establish this scientific status for eugenics—making measurement methods uniform, standardizing the mode of scientific presentation of genealogical trees, increasing the sample size in particular in the case of mental illnesses—were still very strongly tied to an outdated conception of human heredity.

The simple Mendelian hereditary processes had been held out at the First Eugenics Congress as a way to explain an entire array of mental diseases, but it became ever clearer by the end of the 1920s that these processes were useful only for explanation of a rather small number of human characteristics. Since many orthodox eugenicists continued to force the various arbitrary processes into a Mendelian schema, experimentally oriented geneticists kept moving further away. Even the connection of race research with human heredity research, so desired by many eugenicists, was increasingly questionable. It became apparent that the various races defined by anthropologists were very difficult to differentiate from one another genetically.

Basic scientific problems of eugenics were in the end responsible for the fact that many efforts by the international eugenics movement designed to turn eugenics into a science came to naught. For example, confusion about the progress of race research stymied the work of the two committees set up by the International Federation of Eugenic Organizations to standardize human measurements. New measurement techniques were indeed discussed in the working group on psychometry, relating to making uniform the content of intelligence tests. Eugenicists placed great hopes on these, but the goal could not be accomplished.[5] The working group on anthropometry, from 1934 functioning under the auspices of the International Congress for Anthropology, did not succeed in establishing uniform standards for anthropological measurements. The more data that were gathered, the less reliable were the patterns of classification.[6] Even the accelerated race research done for the International Federation of Eugenic Organizations, begun with great hopes, had very limited success. The project initiated by Davenport and Fischer to survey race mixing around the world collapsed because no satisfactory information about the "bastardization areas" was received. Rüdin's Committee for Race Psychiatry succeeded merely in having a few lectures at the meetings of the International Federation and at the Third International Eugenics Congress. These advocates for race psychiatry were not able to set up this field as an independent area of research.

Several other ambitious attempts by the International Federation of Eugenic Organizations at making eugenics a professional academic endeavor were also unsuccessful. The long-desired setup of a permanent secretariat with its own offices collapsed in the 1930s due to insufficient funds of the national eugenics societies.[7] Plans for a scientific journal of the international eugenics society also were not realized. The International Federation of Eugenic Organizations had to content itself with being the copublisher of the *Eugenical News* from 1929 on, along with the American Eugenics Society, the Galton Society, and the Eugenics Research Association.[8] Corrado Gini's plan for an international series of publications, the *Bibliotheque Internationale d'Eugénique*, did not progress beyond the first edition.[9]

The lack of academic acknowledgment of eugenics and the low degree of professionalization of the international eugenics movement became clear to the leadership of the International Federation of Eugenic Organizations when they sought in vain to become a member organization in the International Research Council. This Federation, founded after the First World War, was attempting to organize scientific research in the natural sciences according to internationally uniform guidelines. The International Federation of Eugenic Organizations wished to increase the respectability of eugenics by becoming a member of the Research Council, but the group met with disappointment in that the scientific status of eugenics was far from being comparable to that of chemistry, physics, geography, mathematics, medicine, or astronomy.[10]

The discrepancy between the lofty scientific claims of the orthodox eugenicists and the flimsy basis of eugenic science turned out to be an ideal point of attack for critics. The tight connection between this scientific research program and the list of political demands put out by the International Federation of Eugenic Organizations could be convincing only to the extent that the scientific basis of eugenics was unchallenged. As the scientific basis began to crumble for areas like the division of races, the clear division between heredity and the environment, and the reduction of all hereditary processes to the Mendelian laws, all that was left was the impression that orthodox eugenics was a pseudoscience subverted by race and class prejudices.

The symbiosis of science and politics, which at first had helped eugenics to mobilize resources for its research program and then to scientifically legitimize its political program, proved to be the undoing of eugenics. Eugenics as a scientific discipline threatened to fall apart because of the basically different use of scientific "truth" in its application to political and scientific situations. While in science knowledge is considered to be only provisional because of systematic organized skepticism, in politics the actors demand absolute truth based on generally recognized scientific reasoning.[11] In a situation in which scientific controversy would have been necessary and there was the kind of critical back and forth so typical of sciences, on the political side, important actors within the eugenics movement felt themselves forced by opportunities for political changes to emphasize the supposed stability of the scientific knowledge gained from eugenics research.

Raymond Pearl, who in the second half of the 1920s increasingly distanced himself from the American Eugenics Society, came right to the point in criticizing his colleagues. He noted that eugenics threatened to become more and more a "mixed up muddle of poorly grounded and uncritical sociology, economics, anthropology, and politics."[12] His American colleague L. C. Dunn used the occasion of an assessment from Davenport's and Laughlin's Eugenics Record Office to show that eugenics research was clearly not driven by impartial scientific motives, but by social and political considerations intended to implement scientifically questionable opinions.[13]

Criticism from geneticists like Lancelot Hogben and J. B. S. Haldane in Great Britain and William Castle, Thomas H. Morgan, Herbert S. Jennings, Edward M. East, L. C. Dunn, and Raymond Pearl in the United States pointed out that orthodox eugenicists did not sufficiently consider new findings, such as the multiple-gene theory and the Hardy–Weinberg law. If these critics were far from swearing off the bases of eugenics—that is, the assumption that the inequality of all humans is based on heredity and the possibility and necessity of improving the human hereditary pool—they still called for greater care and restraint in posing eugenic demands.[14]

A group of younger eugenicists in the United States, Great Britain, The Netherlands, and Scandinavia recognized that only a strategic orientation could relieve the pressure of the critics on the eugenics societies, a reorientation in which the positions of the orthodox eugenicists necessarily had to drop out of the running. In the 1930s, 1940s, and 1950s, Frederick Osborn in the United States and C. P. Blacker in Great Britain placed their marks on the American Eugenics Society and the British Eugenics Society. Both of them came from prosperous families, completed their studies at American and British elite universities, and then completely devoted their professional lives to eugenics. After studying at Princeton and Cambridge, Osborn first attempted a career on Wall Street. After a severe illness, he decided at the end of the 1920s to devote himself entirely to eugenics. With the support of his uncle Henry Fairfield Osborn, in 1935, he became one of the directors of the American Eugenics Society. Until his death in 1981, he was a dominant force in the American Eugenics Society as its general secretary, treasurer, and president.

His intellectual alter ego in Great Britain, C. P. Blacker, studied biology at Oxford under Julian S. Huxley. Following work as a physician at Maudsley Hospital in London, in 1931, he was named secretary of the British Eugenics Society on the basis of his services in publicizing eugenics sterilization in Great Britain.[15] Until late in the 1960s, he played a major role in the direction of the Eugenics Society in Great Britain, serving as general secretary, director, chairman, and honorary member.

What distinguished these reformers from the orthodox eugenicists like Davenport, Laughlin, Darwin, und Ploetz? Osborn and Blacker were concerned that the old masters of eugenics were inadequately integrating new knowledge regarding the complexity of hereditary processes and the relationship between environment and heredity into the eugenic body of thought. Relying on the criticism of socialist eugenicists like J. B. S. Haldane and Hermann Muller, they reproached Davenport and others for not acknowledging that simple Mendelian principles were applicable only to a relatively small number of human characteristics. Blacker and Osborn argued for a stronger integration of the environment into eugenic reasoning. In contrast to the nightmare publicized by eugenicists of a disproportionate increase of the genetically infirm, they were of the opinion that social and economic success within a specific group in the population stood in direct correlation with a good basic genetic outfitting. Osborn and Blacker considered it likely that a couple with high genetic qualities would also be the best able to raise children in a healthy and stimulating environment. Starting from this perspective, the promotion of genetically valuable couples could be grounded not only eugenically, but could also be rooted in the supposedly better surroundings.[16]

Beyond the complaint that the orthodox eugenicists used a simplified genetic outlook, Osborn and Blacker also criticized their race and class prejudices. The categories "race" and "class" increasingly disappeared from the perspective of the reform eugenicists.[17] If human hereditary processes were fundamentally more complex than heretofore thought, and if the interaction between environment and hereditary disposition was difficult to determine, then according to the reasoning of the reform eugenicists, the close connection between races and classes with specific genetic qualities was also called into question.[18]

However, the reform eugenicists were interested not only in liberating eugenics from the supposed race and class prejudices and from simplified genetic ways of looking at the world, but also in a complete repositioning of eugenics in the scientific landscape. While orthodox eugenicists at the Third International Congress for Eugenics still tried to show how the tree of eugenics as an advanced academic subject drew strength from its roots—the supporting disciplines of genetics, demography, psychology, and anthropology—the reformers within the eugenics movements now called for giving up the claim that eugenics was an independent academic discipline.[19]

It was Osborn in particular who in the 1930s combated the orthodox eugenicists' notion of eugenics as a science. Influenced by the editor of the *Journal of Heredity*, Robert C. Cook, and the long-time chairman of the American Eugenics Society, Ellsworth Huntington, he declared that eugenics was nothing more than a social movement. In Osborn's eyes, this movement may have been created from a scientific point of view, but it could no longer be considered a science or an academic discipline. Eugenicists could formulate their political demands only on the basis of the scientific knowledge of genetics, psychology, demography, sociology, anthropology, and medicine. If eugenicists were to imply that they were experts in all these fields, they would obviously run into the resistance of specialists in the individual fields.[20] Even Blacker, somewhat more moderate than Osborn in his criticism of orthodox eugenics, declared that eugenics was not a science in the standard sense. Since eugenics was concerned not only with facts but with value judgments, it could hardly be placed on the same level as genetics, demography, and psychiatry.[21]

In conclusion, Osborn, Cook, Huntington, and Blacker called for "de-scientizing" eugenics and simultaneously "repoliticizing" it. They called for a revival of Galton's idea that eugenics should penetrate into human consciousness like a religion. Even though the eugenics societies were to remain as central points for eugenically interested scientists, they would have to give up the notion of uniting knowledge from the various, differentiated disciplines into one overall scientific discipline. The eugenics movement should allow itself to be "broken into academic disciplines" like human genetics, sociology, psychology, demography, and by itself should instead concentrate on realizing its policy demands.

In the 1930s, the eugenicists enjoyed important successes in attaining their political goals. The world economic crisis triggered by the stock market crash of 1929 with its long-lived unemployment and sharply increasing welfare costs resulted in a situation where Protestant countries experienced an unprecedented political acceptance of eugenic sterilization laws.[22] While countries with a majority Catholic population had even increased their rigorous opposition to sterilization based on Pope Pius XI's encyclical *Casti Connubii* of December 1930, in overwhelmingly Protestant countries, the impact of the world depression brought sterilization laws into the realm of the possible.

The parliaments in Denmark, Germany, Norway, Finland, Sweden, Estonia, and Iceland passed eugenic sterilization laws. In Great Britain and The Netherlands, the eugenics societies dominated by the reforming eugenicists came very close to succeeding with their campaigns for the introduction of sterilization laws.[23] In the United States, the number of sterilizations positively exploded after the beginning of the Depression. While up until 1928 "only" about 9,000 people had been sterilized, the number of sterilization operations rose to over 50,000 by 1948.[24] Given the successes of sterilization policy in Protestant countries, one can hardly speak of a general crisis of eugenics in the 1930s.[25] This crisis can be applied only to the idea of establishing eugenics as an independent scientific discipline.

Even the crisis of scientific eugenics must be treated with some caution. Even though the reforming eugenicists met some success in their descientization of the eugenics movement and thus removed a significant defect from the British and American eugenics societies, still the separation desired by the reform eugenicists between political activities and scientific research was an artificial one. The "disciplining" objectively oriented scientists and the scientific "discipline" eugenic activists were generally the same people.

The central role played by eugenicists in the 1920s and 1930s in the founding of international organizations for human genetics and demography shows that the repoliticization driven by the reform eugenicists was accompanied by the diffusion of eugenic research approaches into other branches of scientific endeavor. Eugenic ideology in the 1930s was not really torn apart by new scientific knowledge, as is often still claimed by historians, but instead a reform variant of eugenics stood as godfather at the birth of human genetics and demography in an international context.[26]

Eugenics and International Cooperation in the Study of Human Heredity

The first clear sign that the reputation of eugenics as a freestanding scientific discipline was not the best was seen by the leadership of the International Federation of Eugenic Organizations at the Fifth International Congress of Genetics in Berlin in 1927. At this congress, the discipline of genetics, which in 1921 had presented its results within the framework of the Second International Eugenics Congress, definitively separated itself from eugenics. Over 900 participants in Berlin discussed the new genetic research. For the first time, Hermann Muller announced that the artificial mutations that he had produced through X-rays in drosophila (fruit flies) were of the same nature as natural mutations, and the further inheritance of the artificially created mutations followed a strict Mendelian pattern. The Russian plant geneticist N. I. Vavilov presented his gene centers theory on the development of cultured plants. According to this theory, the area with the greatest genetic variety of a cultured plant is always the area in which it originated and in which its wild initial forms are still extant.[27]

It was only by luck for the International Federation of Eugenic Organizations that Pearl, who wanted to have a general reckoning with the orthodox eugenicists at the International Congress, could not come to Berlin.[28] His lecture, however, later included in the proceedings of the congress, must however have made it clear to every eugenicist that among the general geneticists, a growing criticism of eugenics as a scientific research discipline was obvious. Pearl emphasized that one could not proceed from the hereditary pool directly to a concrete physical image, and he rebuked the orthodox eugenicists for not understanding the complex interactions between heredity and environment.[29] Since these relationships are so complex, in Pearl's view, one cannot determine definitively the genetic quality of the children of specific parents. A quite unlikely but still possible combination of accidents might lead to geniuses developing from a poor heredity pool.[30]

Davenport, taking part in the congress as representative of the International Federation of Eugenic Organizations and in the forefront of the drive to give eugenics an important role at the congress, afterward was quite concerned that further indications had been delivered in Berlin supporting Pearl's criticism of the propagandistic direction of eugenics. In a report to Darwin on the Genetics Congress, Davenport referred to a split between eugenics and research into human heredity. While in the section for research on human heredity innovative methods were discussed for research on twins and blood groups, in the eugenic section, often only personal opinions were expressed that in no way came up to a scientific standard.[31]

The Berlin Congress was a clear indicator to the orthodox eugenicists that their position as scientists among the general geneticists was threatened, and moreover that they were in danger of losing their monopoly on research into human heredity. With their genealogical trees, empirical heredity prognoses, and research on twins, they still dominated the study of human heredity. Nonetheless, the rapid advance in knowledge in plant and animal genetics, the increasing merger of knowledge of cells and general genetics, the development of complex population genetics, and the establishment of research institutes for general genetics all threatened increasingly to call into question the monopoly of the orthodox eugenicists in human heredity research.

Under the impact of increasing criticism from geneticists in the United States, Great Britain, and Scandinavia, and in the face of the threatened loss of monopoly of the eugenicists in human heredity research, the International Federation of Eugenic Organizations resolved to intensify its work on human genetics. A coordinated international initiative of eugenicists in human heredity research would not only help to solidify the central position of the eugenics movement in this area, but it would also reestablish the reputation of the orthodox eugenicists as heredity researchers to be taken seriously. At their meeting in Farnham in 1930, the members of the International Federation decided to set up a "Committee on Human Heredity," which would collate results in human heredity research, initiate and coordinate research in this area, and take care of intensive discussion of this topic at international congresses for eugenics and genetics.[32]

The first fields in which the newly founded committee started to work came in the Third International Congress for Eugenics in New York City and the Sixth International Congress for Genetics in Ithaca, New York. Ahead of time, the organizers of both congresses had agreed that scientists traveling from Europe should be able to attend both congresses.[33] Davenport, the member of the American Working Committee for the Genetics Congress in Ithaca, worked to make sure that the International Federation of Eugenic Organizations would coordinate in Ithaca both the section on human heredity research and the section on genetics.[34]

The decisive connection between the eugenicists and the geneticists was the general secretary of the Genetics Congress Clarence C. Little. After being active as a geneticist and cancer researcher at Harvard University, then as assistant director of the Cold Spring Harbor laboratory, and then as president of the University of Maine, in 1927, he became president of the University of Michigan in Ann Arbor. In 1929, he had to resign because of internal university conflict over his activity as president of the International League for Birth Control. He resumed his cancer research and became director of the newly founded Roscoe B. Jackson Laboratory for Cancer Research. Little served as director of the second International Congress for Eugenics and as chairman of two American congresses for eugenics, and from 1928 to 1929, he was president of the American Eugenics Society.[35]

Even as the Italian geneticist Corrado Gini praised the "twin congresses" in New York and Ithaca as prime examples for the cooperation between genetics and eugenics, the meager participation of geneticists in the eugenics congress made it clear to the organizers how wide was the gap between eugenics on one side and general genetics on the other.[36] Discussions by the eugenicists simply could no longer inspire many animal and plant geneticists. With the increasing scientific differentiation of genetics, the interest of a number of geneticists in drawing direct political conclusions from their knowledge began to decline.

It was just this experience that was the occasion for the eugenicists gathered in New York to institutionalize human heredity research on the international level. The business meeting of the Third International Congress for Eugenics voted to set up a central "Bureau of Human Heredity" under the International Federation of Eugenic Organizations. The model for this central office was the Imperial Bureau of Animal Genetics in Edinburgh. This latter Bureau, directed by Francis A. E. Crew, in just a few years had succeeded in substantially inspiring international research in the area of animal genetics. Following the model of the Bureau in Edinburgh, this central office for human genetic theory was to collect scientific material and to make it available upon request to researchers, health offices, and physicians.[37] It was to sift through, compare, and make material available as quickly as possible for scientific and for practical application.[38]

The central office, established under the direction of IFEO general secretary Cora Hodson, and initially funded by the British Eugenics Society, collected offprints of scientific publications, original family trees of mentally ill families, statistics on heredity, observations on twins, and data on the frequency of hereditary diseases in various countries and among various races. Hodson cooperated closely with the IFEO committees on the standardization of human measurements, on race psychiatry, and on research into race mixing. The office was concerned with supporting the exchange of information among the existing research centers for human genetics. Genetic research under the auspices of the eugenics movement was intended to receive new energy from the exchange of literature among the Berlin Kaiser Wilhelm Institute for Anthropology, Eugenics, and Human Heredity, Rüdin's Kaiser Wilhelm Institute for Psychiatry in Munich, the Dutch Institute for Human Genetics, the Swedish State Institute for Race Biology, the Danish Anthropological Committee, the Czechoslovak State Office for Eugenics, the Hungarian Ministry of Health, the Belgian State Institute for Criminal Anthropology, the French Society for Medical Psychology, the Swiss Julius-Klaus-Foundation for Anthropology and Genetics, and the American Eugenics Record Office.[39]

Although the central office officially prescribed for itself the furthering of "pure" research, and in Hodson's words, understood itself as "a dull machine serving genetics," the ties to the eugenics movement were ubiquitous.[40] The *Eugenics Review* pointed out that the roots of the office were to be found in the British Eugenics Society, and that even with a necessary institutional separation between the "neutral" collection station for heredity information and a publicity campaign directed by the Eugenics Society, the work would be mutually beneficial.[41] Eugenic practice could only make progress if greater clarity were to prevail regarding inheritance of characteristics.[42] In light of the imminent introduction of eugenic sterilization laws in Scandinavia, Germany, and Great Britain, it was hoped that a central office for heredity research would be in a position to supply information about the heritability of mental diseases to health authorities, public welfare institutions, physicians, and concerned parents.[43]

The expectations of the orthodox eugenicists that they would maintain the upper hand in human heredity research through the Bureau of Human Heredity were not fulfilled. Leading geneticists like J. B. S. Haldane and Julia Bell criticized the inefficiency and the mindless collection of data by the office, and turned down a request for a grant through the British Medical Research Council.[44] The most influential British supporters of the Bureau, Sir Arthur Keith, Ronald A. Fisher, Ruggles Gates, and Cora Hodson, were increasingly isolated within the British Eugenics Society because of their orthodox eugenics positions, and in 1936, Blacker was able to achieve a halt in payments from the British eugenicists for the Bureau. In this way, the most important financial sources in Great Britain were shut off. The Bureau degenerated into a one-woman initiative carried on by Hodson.

While the central office for human heredity research, mandated by the orthodox eugenicists, could never fulfill the high expectations for it because of its close connection to orthodox eugenics, the International Committee for Human Heredity Theory developed into an international discussion forum for human genetics. The Committee could take on its important role in human heredity research because it more and more shook off control by the International Federation of Eugenics Organizations, as critics of orthodox genetics escaped the control of the Federation. This was the case with the British geneticists Lancelot Hogben, J. B. S. Haldane, and Lionel Penrose and their American colleagues Hermann J. Muller and Laurence Snyder. The reform eugenicists active in the British and American eugenics societies also were reinforced in their desire to establish human genetics as an independent scientific discipline, since they saw this as a way for the natural science components to be used to take over the repoliticized eugenics movement.

G. P. Frets, chairman of the dutch eugenics society and one of the few reform-oriented eugenicists in the international eugenics organization, utilized his position as chairman of the committee step by step to remove the old guard of the eugenicists and to make way for human heredity research on the international level. He replaced eugenicists like Jon Alfred Mjöen, Charles Davenport, Harry H. Laughlin, Herman Lundborg, and Herman Nilsson-Ehle, who had initially dominated the committee, with younger heredity researchers like the Dutchman Petrus J. Waardenburg, the Dane Tage Kemp, and the American Laurence H. Snyder. In addition, he brought in leading critics of orthodox genetics as members of the committee by appointing the Norwegian geneticists Kristine Bonnevie and Otto Mohr, the Swedish heredity researcher Gunnar Dahlberg, and the leading British human geneticist Lionel Penrose.[45]

As the International Federation of Eugenics Organizations increasingly took on the role of a willing instrument of legitimation of National Socialist race policy, Frets withdrew the International Committee from the International Federation. The International Committee had one meeting in 1938 outside the International Federation, and one year later, the human geneticists at a meeting in Edinburgh formally separated themselves from the IFEO. They called themselves the International Group for Human Heredity and set up their own bylaws.[46]

The struggle for institutional dominance in the area of international human heredity research shows that the orthodox eugenicists were not able to succeed with their idea of establishing human genetics as a subdiscipline of a eugenic science. The Central Office for Human Heredity Research could have no influence on the development of human genetics because of its too strong ties to the orthodox eugenicists and its fixation on family tree research. On the other hand, the International Committee for Human Heredity Theory succeeded in creating an institutional basis for the establishment of human heredity research as an independent science by breaking loose from the International Federation of Eugenic Organizations. For the reform eugenicists and socialist eugenicists who supported the work of the international committee, at the end of the 1930s, the committee was briefly an instrument to deprive the orthodox eugenicists once and for all of their dominant position in the field of human genetics.

The Separation of Demography from Orthodox Eugenics

Research on population as a freestanding branch of science on the international scene came about in the 1920s against the background of a rapprochement between the reform forces within the eugenics movement and the activists in the birth-control movement. Since the efforts of the leading circles of the International Federation of Eugenic Organizations to control the development of demography completely failed, the critics of orthodox eugenics overwhelmingly determined the development of population science. The establishment of demography as a research area, which was to include the "quantitative" as well as the "qualitative" aspects of population on a scientific basis, increasingly eroded the basis of eugenics as an academic endeavor.

The possibility of cooperation between the birth-control movement and the eugenicists had developed because of the growing influence of reform-oriented eugenicists such as Alexander M. Carr-Saunders, Roswell H. Johnson, Julian S. Huxley, C. P. Blacker, Clarence C. Little, Frederick Osborn, and Edward M. East, and because of the increased emphasis on moving eugenics in the direction of family planning by the activists in the birth-control movement.[47] This coming together of the two movements after the mid-1920s led to a close connection between the eugenicists, focused on distinctive quality, and the neo-Malthusians and birth controllers, fighting against "overpopulation." In the opinion of Alexander Carr-Saunders and Henry Pratt Fairchild, key figures in the strategic orientation of the British and American eugenics societies and in the establishment of population science, the quantitative and qualitative aspects of population were two sides of the same problem.[48] The reform eugenicists and leading birth-control activists thought that the topic of demography should not simply be the question of how many people the earth could support, but how many people of which type.[49]

It is an irony of history that the move to "scientize" the study of population was triggered by an initially politically motivated world population congress against a supposedly too strongly politicized orthodox eugenics. Margaret Sanger, who more than any other single person embodied the connection between eugenics and family planning, planned to secure a scientific foundation for worldwide eugenically oriented birth control. She looked to a population congress organized internationally with representatives of birth-control activists and reform-oriented eugenicists.[50] Unlike the earlier international conferences for neo-Malthusianism, which were attended almost exclusively by activists in the birth-control movement, the World Population congress in Geneva in 1927 was to bring together scientists from such varied areas as sociology, biology, demography, geography, and economics for the study of "population problems."

Sanger helped to choose the scientific leaders of the World Population Congress. This group represented the tie-in of the qualitative and quantitative aspects of the recently founded academic discipline of population study and a multidisciplinary approach. Sir Bernard Mallet, previously chief statistician of Great Britain and Wales, served as president of the congress. Mallet, who took over from Darwin as president of the British Eugenics Society, was one of the pioneers in establishing a science of population.[51] The 20 biologists, physicians, sociologists, and economists, who comprised the scientific advisory council of the world population conference were almost exclusively members of eugenics societies.

The four American representatives—the geneticist Edward M. East, the cancer researcher Clarence C. Little, the sociologist Henry Pratt Fairchild, and the population geneticist Raymond Pearl—were numbered among the most prominent eugenicists in the United States. The British representatives, Alexander Carr-Saunders, Julian S. Huxley, John Maynard Keynes, Francis A. E. Crew, and the radiologist Sir Humphrey Rolleston, were all members of the British Eugenics Society. From Germany, two well-known eugenicists were represented on the conference directorate—Richard Goldschmidt, director of the Berlin Kaiser Wilhelm Institute for Biology, and Alfred Grotjahn, professor of Social Hygiene. The Belgian representative, René Sand, was active for the League of Red Cross Societies and was chairman of the Belgian Eugenics Society. The other European members of the scientific council were also internationally known eugenicists. The Frenchmen André Siegfried and Lucien March, the Dutchmen Henri W. Methorst and Marianne van Herwerden, the Italians Corrado Gini and Alfredo Niceforo, and the Norwegian Wilhelm Keilhau were all active in various eugenic societies.[52]

In Sanger's view, the purpose of the conclave was to assemble "a common meeting table" to discuss for several days current "population problems" and to create an organizational framework for international research into population. In this way, Sanger and the other organizers used the study of the population problem to set scientists and politicians thinking about concepts like overpopulation and population explosion. In her welcoming address to the congress, Sanger emphasized that the worldwide growth of population constituted a "menace to the future of civilization" and that it was necessary to stop this development with "concerted international action." Sanger and her co-organizers hoped that an "international point of view" would arise from the conference, which could supply the scientific basis for the solution of the many-sided population problem.[53]

For the participants in the world population conference, there was no question that the scientific consideration of eugenic questions would play an important role in this still developing science of population. Pearl, who had worked closely with Sanger in preparing the congress, proposed in his opening report that birth and death are primarily biological factors. Consequently, the "development of population" of so many different organisms such as yeast and bacteria, fruit flies, and human beings functioned according to similar laws. Pearl went on to say that in the final analysis, every population problem is a biological problem.[54] Huxley, one of the leading British biologists and eugenicists, translated this basic conviction in the *New York Times* into a formula stating that biology is the basis of all "human sciences," and therefore all problems of humanity could be solved only by including biology.[55]

The organizers of the conference in Geneva set themselves the task of placing the emerging population science on a biological basis, but they also had to set themselves off from the approach of orthodox eugenics. Raymond Pearl, J. B. S. Haldane, Julian Huxley, Alexander Carr-Saunders, and C. P. Blacker, leading critics of an allegedly unscientific eugenics based on race and class prejudices, and who were commanding figures at the World Population Congress, desired a eugenically oriented population science without including the direct influence of the orthodox geneticists around Darwin, Davenport, Laughlin, and Ploetz. Hence, Pearl ensured that the eugenic laypeople, whom he so despised, were not admitted to the congress. He succeeded in preventing attendance by Harry Olson, chief judge of the Municipal Court of Chicago and chairman of the committee to prevent criminality of the American

Eugenics Society. For Pearl, the presence of such "a ridiculous, intemperate, unscientific propagandist" would make the American delegation a laughing stock in the eyes of their European colleagues.[56]

It is true that in Geneva, eugenicists like Herman Lundborg and Corrado Gini rehashed the classical thesis of the disproportional reproduction of "inferior portions of the population," and the British eugenicist Ernest Lidbetter repeated his thesis announced at the First International Eugenics Congress in 1912 of the "race of the chronically poor." For the first time, however, research results were also presented that undermined the basis of the orthodox eugenicists that differential reproduction favored the supposedly intellectually and socially "inferior." The director of the Statistical Institute in The Hague, the Dutch eugenicist Henri W. Methorst, reported that in The Netherlands in the meantime, there was a similarly high birthrate among the various social classes. The Swedish statistician Karl Arvid Edin presented his study according to which the spread of contraceptives in Stockholm had already resulted in the higher social layers having more children than the lower layers.[57] The lower birth rate of the upper classes, which had caused sleepless nights for many eugenicists, was from this perspective merely a transitory phenomenon resulting from an earlier acceptance of contraceptives by the middle class.[58] The studies by Edin and Methorst formed the basis of the hope of some reform eugenicists that the massive distribution of contraceptives among the groups considered to be inferior could serve as an appropriate means of negative eugenics.[59]

The World Population Conference in Geneva was the starting point for the founding of an organization designed to set up an international population science. At Pearl's initiative, a committee was set up in Geneva to prepare the founding of an international union for the scientific study of population.[60] Pearl was convinced that the special character of population science had to make it internationally oriented, and for him, political boundaries were merely artificial with regard to the pressing population problems. Pearl went on to write in the *Eugenical News* that the population of a particular country was not an isolated phenomenon, but one that had ongoing interactions with population groups elsewhere.[61]

More than any other scientist in the 1920s, Pearl represented the development from a propagandist of the eugenic movement into one of the chief drivers of scientifically founded population monitoring. As the American historian of science Garland Allen writes, Pearl in the middle of the 1920s had given up the hope that one could convince the "biologically superior classes" to increase the number of their children. He thought that in light of the collapse of his positive eugenics, he should turn to moving "undesired" groups to have fewer children. It was at this point that Pearl's basic eugenics outlook came into contact with Sanger's campaign for birth control and the growing fear of the increase of world population. Pearl's demand for a scientifically grounded population monitoring was, as Allen emphasizes, in the end nothing more than a modernized version of the old eugenic demand for population selection.[62]

Pearl, who was driving to push Sanger out of the planning for the international organization of population scientists, not even 10 months after the Geneva conference called together scientists from 12 countries for the founding meeting of an international union for the scientific study of population problems. The 40 or so scientists in attendance resolved unanimously to found the International Union for the Scientific Investigation of Population Problems (IUSIPP). The major support for the Union were the national committees of population scientists formed in the United States, Canada, Argentina, Brazil, Great Britain, France, Germany, Italy, Spain, The Netherlands, Belgium, Switzerland, Denmark, Sweden.[63] Many national societies and initiatives for the scientific study of population, often with close ties to eugenics and race hygienic societies, took their origin from these national committees.

The American eugenicists and population scientists who turned up in 1930 in the American committee of the IUSSIP started up the Population Association of America (PAA), a still existing national organization for the scientific study of population (in addition to the Population Reference Bureau in Washington).[64] The new association was born at the same time as the American Eugenics Society on May 7, 1931.[65] Until long after the Second World War, very close personal connections existed between the PAA and the American Eugenics Society. Henry Pratt Fairchild, Ellsworth Huntington, Frederick Osborn, Clarence C. Little, Frank Lorimer, and Clyde V. Kiser, all one-time presidents or vice presidents of the American Eugenics Society, for many decades set the policies of the PAA.[66] The American Eugenics Society concentrated on its political propaganda work, while the PAA restricted its activities to the study of the "quantitative and qualitative aspects of human population."[67] As the work of the organizations for eugenics, population science, and population policy in the United States overlapped with one another, it became clear that the American Eugenics Society, Sanger's American Birth Control League, and the PAA were uniting in tight cooperation in the area of "population improvement." Very soon, a merger into one organization was being discussed.[68]

The British Population Society, the committee of the IUSIPP, also consisted almost exclusively of activists from the British Eugenics Society. Of the 21 British anthropologists, biologists, economists, statisticians, and sociologists who formed the British Society for Population Science in 1929, 17 played important roles in British Eugenics Society.[69] Mallet, the first chairman of the British Population Society, immediately after the Paris general meeting of the IUSIPP, strengthened the tight connection of British population science with the Eugenics Society. At his initiative, the British Population Society set up its office in the space of the Eugenics Society, and Eldon Moore, publisher of the *Eugenics Review*, assumed the unpaid office of general secretary.[70] The close cooperation between the Eugenics Society and the British Population Society was the reason that in 1936 an 18-person Population Investigation Committee (PIC) was formed. The Eugenics Society supported the committee financially and organizationally until long after the Second World War.[71]

In the German section of the IUSIPP, from the very first moment, the eugenicists and race hygienists Erwin Baur, Eugen Fischer, Hermann Muckermann, Ernst Rüdin, Fritz Lenz, and Alfred Grotjahn came into contact with demographers like Friedrich Zahn, Hans Harmsen, Friedrich Burgdörfer, and Robert Kuczynski. They were all attempting to merge the qualitative and quantitative aspects of the new science of population. To these were added Paul Mombert and Julius Wolf, two important German macroeconomists. Due to the special developments in Germany, reform-oriented eugenicists did not take control of the German committee, but rather the committee developed in the 1930s much more as a willing scientific instrument of legitimization of National Socialist race and population policy.[72]

The same was true in the national committees of Italy, France, Spain, Denmark, and The Netherlands, where the dominant scientists came from the national eugenic movements. Gini took over the chairmanship of the Comitato Italiano per lo Studio dei Problemi della Popolazione, which continued to exist after the Second World War as the Italian Committee for the Scientific Study of Population.[73] In The Netherlands, the national committee dominated by Frets and Herwerden formed the core of the Netherlands Demographic Society.[74] Cuenot and March played an important role among the French. In Denmark, the psychiatrist August Wimmer and the anthropologist Sören Hansen were leading eugenicists who influenced the work of the national committee.[75]

The establishment of a strong international union of population scientists and the initiation of population science initiatives in various European and American countries stood as a fundamental threat to those eugenicists who attempted to establish eugenics as a freestanding science in the area of research on human heredity and population development. The IUSIPP, which had raised a banner at its meeting in Paris with the words "Research on Population Questions in the World," did not threaten eugenics as a political movement, but it did threaten to cut the ground from under eugenics as a scientific research area.[76] The international population union pushed into scientific research areas that previously had been the domain of eugenics, a development seen in the committees on the topics of the differential birthrate, fertility, and sterility (chaired by Crew), statistics of primitive races (chaired by Gini), and "Ethnogenics," and the interplay of "race-population-culture" (chaired by Pitt-Rivers).

The leadership of the International Federation of Eugenic Organizations was not sure how it should respond to the competition of an international population science union. Should one, as Davenport in particular had in mind, essentially ignore the emerging science and hope that it would not penetrate into the classical research areas of eugenic science, the interconnection between genetics and population research? Or should one, as Gini and Pitt-Rivers argued, attempt to maintain influence on population science in the sense of the International Federation of Eugenic organizations? Or should one, as especially desired by several British eugenicists, cooperate closely with the population union and attempt to have the eugenic movement profit from the boom in population sciences?

The orthodox American eugenicists, who saw the PAA and the IUSSIP grow into strong competitors, attempted to direct population science on national and international levels into the area of classical, quantitative demography. Thus, Harry Laughlin argued that the PAA should be active only in those areas not already occupied by preexisting societies. In his view, it would be a mistake for the newly arising population science societies to attempt to take over the work that had already been performed so successfully by the eugenics societies. Laughlin maintained in a discussion on the founding of the PAA that the field of population science was divided into four areas: on the one side qualitative population research and qualitative population control in the sense of eugenics, and on the other side quantitative population research and quantitative population control in the sense of the efforts to limit population growth. In the United States, qualitative population research was being done by the Eugenics Research Association; qualitative population control was being done by the American Eugenics Society; and quantitative population control was being done by the birth-control movement. There remained for population science only the field of quantitative population research.[77]

The American reform eugenicists understood clearly that the orthodox eugenicists were trying to maintain their monopoly in the area of qualitative population research by forcing population science into the area of quantitative demography. Fairchild and Lorimer were the driving forces in pushing out the orthodox eugenicists around Davenport, Laughlin, and Steggerda from the leadership of the American eugenics movement, and they emphasized the indivisibility of the quantitative and qualitative aspects of population problems.[78] They were convinced that population science had to assume from the eugenics movement the task of the scientific study of population development on national and international levels; they were to leave to the eugenics societies only political activities.

The relationship of eugenic science to population science also played an important role in the clash between the International Federation of Eugenic Organizations and the International Union for Population Science. At the very beginning of the founding meeting of the IUSIPP, Cora Hodson in the name of the IFEO had brought to the attention of Crew and Mallet, the two British delegates, the fact that the international eugenics organization was interested in close cooperation with the International Union for Population Science. In the eyes of the eugenicists, this cooperation was related in particular to the problem of the "quality of the population."[79] What they had in mind was a joint committee made up of members sent by the IFEO and IUSIPP to deal with the question of "population from a qualitative point of view."[80]

At first glance, the leadership of the IUSIPP seemed quite determined to cooperate with the international eugenics movement. On Mallet's initiative, the founding meeting passed a resolution calling on the IUSIPP to cooperate closely with scientific organizations that had similar goals. Directly after the Paris founding meeting, Mallet traveled as representative of the IUSIPP to a conference of the International Federation of Eugenic Organizations in Munich in order to discuss possibilities of cooperation with the eugenicists gathered there.[81] The most comprehensive suggestion was a merger between the International Federation of Eugenic Organizations and the IUSIPP. Using a common office, the scientists hoped to bundle together research efforts and to be able to play a more significant role within the international scientific community.[82]

In light of the weak financial situation and the unclear scientific future of the IFEO, Mallet was convinced that cooperative work with the eugenic population scientists of the Davenport or Laughlin sort would in a very short time be considered meaningless. In a report to Pearl, he explained that population scientists should in no way allow themselves to be forced into a position of simply processing quantitative population issues.[83] Pearl was confident that eugenics of the old type would quickly decline in importance and that public opinion would turn toward the work of the International Union for Population Science.[84] For him, the task of the International Population Union was to introduce scientific "realism into eugenic work."[85]

As the leaders of the International Federation of Eugenic organizations became aware of the marked dominance of the IUSSIP, they increasingly kept their distance. At a meeting of the IFEO in Rome in 1929, they voiced their fear of being swallowed up by the International Population Union. Davenport, Gini, Fischer, and the Austrian delegate Heinrich Reichel in particular worried that the interests of the eugenics organizations would not be sufficiently considered. They also thought that Pearl, as president of the IUSIPP, would extend his crusade against the American eugenics societies to the American-dominated international eugenics movement.[86]

Pearl in particular hoped that in a relatively short time, the IUSSIP could be built up into a strong scientific organization and would be able to deprive the International Federation of Eugenic Organizations of its claim to leadership to research in population issues. This hope was only partially met. For a time a rancorous internal conflict between Pearl and Gini completely blocked the work of the Union set out at its founding meeting in Paris. At this meeting, the IUSSIP had committed itself to abstain from political and religious questions. The work of the International Union was to consist of scientific research in the strictest sense of the word. Every religious, moral, or political discussion would be kept out of the IUSSIP, and no specific population policy would be supported.[87] In this rather general and wordy way, the members committed themselves to scientific objectivity. However, as had occurred in the past with the International Federation of Eugenic Organizations, within a short time, the flexibility of this self-denying ordinance became clear.

In 1929, Gini convinced the board of directors of the IUSSIP that Rome would be the ideal location for the second world population congress.[88] Although Pearl at first greeted this suggestion with enthusiasm, he became increasingly dubious as to whether the political and religious neutrality of the IUSSIP could be maintained in the face of Fascist Italy. His raging arguments with Gini at the board of directors of the IUSSIP made him doubt more and more that Gini was capable of separating his scientific activities within the IUSSIP from his propaganda activity for Mussolini's population policy.[89]

Gini was unable to convince Pearl of the political neutrality of the congress in Rome. Furthermore, Pearl's arch-enemy, the Boston health scientist Edwin B. Wilson, pressured the Milbank Memorial Fund into questioning further support for the IUSSIP because of Gini's close ties to the Fascists. Consequently, Pearl made sure that neither the Second World Population Congress nor the general assembly of the IUSSIP would take place in Rome.[90] As an emergency solution, on the spot, he called a meeting of the IUSSIP for London in June 1931, in which only a few presentations and reports would be given. Gini was so outraged about this decision that he intentionally risked a schism within the IUSSIP.

Supported by the Italian Fascist government and by his colleagues from the International Federation of Eugenic Organizations, Gini was able to set up the International Congress for Population Research without the official blessing of the IUSSIP. Davenport, one of the sacrifices to Pearl's angry attacks on the orthodox eugenicists in the United States, in his role as president of the IFEO was very pleased to give international sponsorship to this International Congress for Population Research in Rome.[91] Other leading members of the IUSSIP, such as Eugen Fischer from Berlin, Leon Bernard from Paris, and Severino Aznar from Madrid, were also not intimidated by Pearl's fears and joined Gini in Rome at the presidential table.

In the end, the character of the international conference in Rome was not basically different from that of the World Population Congress in Geneva.[92] In Rome too, it was clear how closely population research rested on eugenic assumptions. In his opening address, Gini clearly stated that the problems of population quantity and population quality were indivisibly tied to one another.[93] Topics like the biological factors in the decline in the birthrate, the relationship between intelligence and frequency of birth, the results of war for the race, race crossings among humans, the differential birthrates of various classes, and the demography of primitive peoples occupied a meeting place at the conference.[94]

Pearl, worn down by the conflict with the Italian population scientists and by the increased criticism of his research work in America, resigned from his office as president of the IUSSIP and turned over the rudder to two British population researchers and eugenicists. Sir Charles Close, one of the most important British geographers and a member of the scientific council of the Eugenic Society, became president of the IUSSIP. The anthropologist Captain George Pitt-Rivers became the executive director.[95] Pitt-Rivers was the grandson of the founder of the Pitt-Rivers Museum in Oxford, and turned to anthropology after a career in the Army and in the British colonial service. A member of the Scientific Council of the Eugenics Society, Pitt-Rivers prided himself on being the founder of ethnogenics, the study of human history using the concept of race, population, and culture.[96] Close and Pitt-Rivers supervised E. C. Rhodes, a statistician at the London School of Economics, in publishing the IUSSIP journal *Population*, one of the first academic journals devoted to population science.

Close and Pitt-Rivers had their hands full in preventing a split of the IUSSIP and in reviving the research committees that had been crippled by the conflict.[97] The Italian members of the IUSSIP played with the idea of building up their national committee into an international counter-organization to the IUSSIP.[98] Even if personal vanity and personal insults played an important role in the dispute, the basic question still existed as to how far population scientists in their role as scientists should speak out on concrete measures of population policy. Pearl and Close favored the position that the IUSSIP should as much as possible be kept out of political debates and that one should take positions on questions of birth control and immigration policy only as an individual. In contrast, Gini and some non-Italian scientists argued for scientifically based positions on national and international population policy.[99]

Since Gini for quite some time had been an important link between the International Federation of Eugenic Organizations and the IUSIPP, the violent arguments also affected the relationship between the two organizations. At a meeting in June 1933, the board of directors of the IUSSIP agreed to a proposal by Close and the American population scientist and eugenicist Louis Dublin to withdraw the official IUSSIP delegates from the International Federation of Eugenics Organizations. Cooperation with a "propaganda organization" like the IFEO would be possible for members of the Union only as private persons.[100]

It must be pointed out that the grounds for this decision were certainly not an anti-eugenic attitude by the IUSSIP. Both Close and Dublin were influential members of the British and American eugenic societies. What characterized Great

Britain and the United States at the beginning of the 1930s also was carried out on an international level—a division of labor between eugenics organizations and scientific groups in the areas of population policy, medicine, and genetics. Under the influence of the young eugenicists, the eugenic societies scaled back their claims to be scientific organizations and increasingly acknowledged that they were involved purely in political lobbying work, while parallel scientific societies arose in new fields such as population science and human genetics.

The eugenic societies, insofar as they did not hold to their scientific claims in the way that the IFEO did, were not opposed to initiatives in the area of human genetics and population science, but rather they reinforced these claims. For a large portion of the eugenic activists in the 1930s, it was obvious that as scientists they would organize themselves in the IUSSIP or in other national or international professional societies. One need name here only such figures as Osborn in the United States, Blacker in Great Britain, Kemp in Denmark, and Fischer and Rüdin in Germany.

The frequently reform-oriented eugenicists who supported the establishment of population science and human genetics as independent disciplines were confident that the results of these fields would provide the scientific basis for a eugenics freed of race and class prejudices. Osborn emphasized that population science, like human genetics and psychology, would provide "fundamental knowledge to eugenics."[101] The Danish geneticist Tage Kemp reduced the relationship between eugenics and the science of population to a simple headline—population research must supply the scientific basis for race biology, eugenics, and social hygiene.[102]

CHAPTER FIVE
NATIONAL SOCIALIST GERMANY AND THE NATIONAL EUGENICS MOVEMENT

Eugen Fischer wrote in 1936 that at the last moment, "the broad vision and the enormous energy of the *Führer*" have saved the German people from biological collapse. Like most German race hygienists, Fischer made no secret of his admiration for Hitler's race policies and his inspiration by them. To his mind, for the first time, "the entire population policy of the people was being carried out with unprecedented determination" on a biological basis. The German heredity researchers would supply the scientific underpinning of the policy of the "protection of the heredity pool" and the "maintenance of pure race composition... the singular contributions of the National Socialist population policy."[1]

National Socialism raised the "biological and racial perceptions of national life" to the highest level of state doctrine. The fate of Germany lay "solely in the blood and race of the people," according to Walter Groß, director of the Nazi Party Information Office for Population Policy and Race Questions. Since "biological collapse" threatened Germany because of the decline in the number of children, the "reduction of hereditary value," and "race mixing with the bearers of foreign blood," the entire nation was forced "to think and feel biologically." This was the logic of the later director of the Race Policy Office of the party.[2]

The biological character of the National Socialist policy was based on two main themes that supplemented each other—concern with heredity in a sense of eugenics, and concern with race as the basis for anthropological study. The National Socialist program of care for heredity and race did not come into existence out of nothing. In the words of the Jena race hygienist Paul Hilpert, the racial "seed corn" of the National Socialists "fell on well prepared ground." For him, the "National Socialist state revolution brought into reality those things that had been fought for by the "forefathers of race hygiene," Ploetz and Schallmayer.[3]

The basis of National Socialist race policy was the racism that had been given its basic form by scientific and political thought in the first half of the twentieth century, but not only in Germany. Nonetheless, the racist discrimination against minorities in Nazi Germany—be they people with mental or physical handicaps, or people who for some other reason were not to be numbered among the community of the master race—reached a completely new dimension. With the takeover of power in 1933, the Nazis institutionalized their race policy on all levels of the state and made a central dogma of their policies to be discrimination on the basis of race against the so-called racial inferiors.[4]

German race hygienists, psychiatrists, population scientists, and heredity researchers participated in a significant way in the formulation, implementation, and justification of this policy. Here as well the peculiar nature of eugenics in Germany compared to that in other countries was apparent. A German "special way" of race hygiene did *not* exist in the actual content of eugenics, but rather in the possibilities made possible by the National Socialists in implementing a eugenics policy.[5] National Socialism offered the German eugenicists a political structure where it was possible to institute a program that was in its content shared by many eugenicists outside Germany.[6]

It is in a comparison with the eugenicists in the United States and in Scandinavia that one can see that the conflict lines over the content of eugenics were often greater *within* national movements than they were *between* national movements. It was a result of the fact that the National Socialists put into force eugenic measures to an extent hitherto unknown, and unmatched even afterward, that eugenics as practiced by the National Socialists became a reference point for eugenicists outside of Germany.

The scientific legitimation of National Socialist race policies was not at all limited to German scientists. A network of European and North American eugenicists and race hygienists was actively engaged in legitimizing National Socialist race policies both inside and outside of Germany.

The Special Relationship of National Socialism to Eugenics

In 1931, two years before the Nazi takeover of power, Fritz Lenz praised Hitler's National Socialist party in the *Archiv für Rassen- und Gesellschaftsbiologie* as the "first political party" in the world that promoted racial hygiene as the central part of its program. In Lenz's eyes, Hitler had adopted the "fundamental thinking of race hygiene with great spiritual receptiveness and energy." One might expect that the "National Socialist movement would do great things for an effective race hygiene." Lenz was convinced that Hitler's 1924 *Mein Kampf* was shaped by a careful reading of the standard German work on race hygiene, the so-called Baur, Fischer, and Lenz, and the works of the Nordic race theorists Houston Stewart Chamberlain and Hans F. K. Günther. He went on to claim that Hitler's anti-Semitic and race hygiene thoughts were not at all new for the eugenicists.[7]

Hitler's demand that the "*völkisch*" state be placed at the center of daily life was not far from the ideas that Ploetz and his colleagues had put forth at the beginning of the twentieth century. The entire palette of race hygiene demands is found in *Mein Kampf*—from the sterilization of "defective humans" through the "promotion of fertility of the healthiest bearers of *Volkstum*" all the way to preventing "counter-selection" due to war.[8]

Given this "core affinity" (Lenz) between National Socialism and orthodox race hygiene, it is not surprising that leading German race hygienists even before 1933 made no secret of their sympathy for Hitler. Eugenicists like Fritz Lenz, Eugen

Fischer, and Hermann Muckermann spoke out in private circles as early as 1932 with positive comments about Hitler and his race hygiene program.[9]

After the Nazi rise to power in January 1933, one could observe regular attempts by a large number of the German race hygienists to curry favor with the political leadership. People like Alfred Ploetz, Fritz Lenz, Friedrich Burgdörfer, Otmar Freiherr von Verschuer, Ernst Rüdin, and Eugen Fischer were seeing their dream come true of implementing eugenics in practical policy and in establishing race hygiene as a major academic discipline. Suddenly, their expertise stood in the center of state activity.

Ploetz in April 1933 wrote a humble letter to Hitler in which "with heartfelt honor he shook the hand of the man whose steadfastness has led German race hygiene out of the jungle of its earlier path into the wide field of free activity."[10] A few months after the takeover of power, Fischer, whose influential position had initially been viewed skeptically by some representatives of the Nazi regime, showed himself to be quite taken with the National Socialist measures for "healthy heredity" and "race purity."[11] In a declaration of subservience to the Reich Ministry of the Interior, he noted he had placed "from the first to the last" his strength in the "service of the most important portion of the National Socialist theory and policy, the theory of race, the theory of heredity, race hygiene, and population policy."[12] His Kaiser Wilhelm Institute for Anthropology, Human Heredity, and Eugenics stood "completely and fully ready to serve the tasks assigned by the state."

In front of 300 race hygienists gathered in Berlin in 1934, Rüdin celebrated that the "significance of race hygiene in Germany was first made public to all reawakened Germans through the political work of Hitler." He went on: Finally the "dream of 30 years" of the race hygienists had come true in the work of Hitler—race hygiene was no longer nothing more than a research program; now it had become implemented in fact.[13] Like his mentor Ploetz, Rüdin joined the National Socialist German Workers' Party (NSDAP) in 1937 and a short time later was awarded the Civil Service Faithful Service Medal in Gold.[14] Rüdin's enthusiasm for Hitler was so extensive that his close colleague Agnes Bluhm remarked that the "free Swiss" Rüdin conducted himself like a "120% National Socialist."[15]

By June 1933, the German Society for Race Hygiene had integrated itself into the Nazi regime and had submitted to the "*Führerprinzip*" (Leader Principle) demanded by the National Socialists. The old board of directors, consisting of Fischer, Muckermann, and Verschuer, resigned. Reich Interior Minister Wilhelm Frick named Rüdin as "Reich Commissioner for Race Hygiene" and as the first chairman of the Society for Race Hygiene.[16] In his memo of nomination, Frick explained that the theories of heredity and race hygiene were of great importance for differentiating the race of the German people, and therefore a close collaboration between the Reich Interior Ministry and the German Society for Race Hygiene was necessary.[17]

The first meeting of the German race hygienists after subordinating themselves to the new regime consisted of an address of greeting to Hitler, the singing of the Nazi anthem (the Horst Wessel Lied) and of the German anthem, and Rüdin's call for "Sieg-heil" salutes, followed by a reorientation of the society's program. In his speech, Rüdin emphasized that the German race hygienists would completely support both the eugenic and race anthropological aspects of National Socialist race policy. "Enlightenment and recruitment for all the principles coming from race research in favor of the German people on its soil" would be the central task of the association. "The German Society for Race Hygiene would represent the view that the Nordic race stands in first place in world history" and therefore it was imperative "to maintain it and protect it."[18] The journal of the National Socialist Confederation of Physicians, *Ziel und Weg,* called the relationship between the National Socialist state and the core of race hygienists and demographers exemplary. In contrast to the many scientists who had not yet come to understand the "sense of the times," the German scientists had shown themselves to be true supporters of the political task involved in the "area of race and population policy."[19]

The National Socialist administration rewarded this behavior with a plethora of appointments to university positions and of grants for research. For example, the Reich Ministry of the Interior assumed a portion of financing of the *Archiv für Rassen- und Gesellscbaftsbiologie,* and named it the official organ of the Reich Committee for the People's Health Service.[20] While in other countries the eugenics societies saw themselves forced "to descientize themselves" in order to develop into a scientifically oriented political movement, the special circumstances in National Socialist Germany led to the completely opposite development. In Germany, the integration of human genetics, psychiatry, and population science into an academic endeavor that understood itself politically to include eugenics and race hygiene met with success and renown. As a "lead science" for the National Socialist race policy encompassing various areas of science, eugenics and race hygiene received the full support of the state.

It is true that tensions among the German race hygienists and within the National Socialist bureaucracy existed that continued to burden the relations between race hygienists and National Socialist race politicians. Nonetheless, after 1933, both blocks worked ever more closely in a kind of symbiotic relationship.[21] In any case, the leading German race hygienists acted as though there was no difference between the claims of science and the claims of national policy. The tenor of their pronouncements was that Nazi race policy was a direct result of the scientific knowledge gained by eugenics, and therefore the promotion of human genetic, demographic, and psychiatric research was of immediate use for the national race policy. In the eyes of Hitler's assistant Rudolf Heß, National Socialism was essentially nothing more than applied biology.[22]

In this context of a broad-ranging decline in the distinction between science and politics in National Socialist Germany, the leading scientists among the German eugenicists assumed important functions for Nazi race policy. They became active in the formulation of various laws and gave advice on the wording of the official legal commentaries. They participated in training the state personnel who were to implement the government's race policies and to publicize these policies in schools and universities.[23] Finally, the German race hygienists were able to use the international context that they had built up over decades to form close connections and to be official sources of information between the National Socialist race politicians and the often critical outside world.[24]

The International Eugenics Movement in the Service of National Socialism

In 1939, the magazine *Ziel und Weg* celebrated the fact that in recent years, population and race policy changes around the world had occurred that as recently as 1930 had been considered completely impossible. In the opinion of the author of the article, Erich Berger, the National Socialist policy of "race differentiation" and the prevention of "race mixing" had won an increasing number of supporters abroad. This jubilee edition is only one example of the confidence of the National Socialist race politicians in 1938–1939 that Hitler's race policy was meeting increased understanding around the world and that the "German model" was finding imitators everywhere.[25]

In a widely publicized speech in January 1939, Groß at the Hamburg College of Politics praised the fact that "German race thought and the German race policy had been successful in the world." In his opinion, this was one of the greatest accomplishments of National Socialist Germany. Gone were the hard times in which the National Socialist sterilization law was conceived as a "battle tactic against political foes" and when National Socialist race policy was understood as defaming all "non-Aryan" peoples. He went on to say that after a five-year battle about race theory, now scientists and politicians from other countries were coming to Germany in order to study National Socialist race policy. They returned home convinced that the "the story was a sober and orderly" one. Only negative public opinion in their home countries kept them from implementing similar measures there as well.[26]

The myth of the "triumph of German race thought throughout the world" was undoubtedly an intentionally planned action of the Race Policy Office of the NSDAP. As the director of this office, Groß continued to push for acceptance abroad for Nazi race policy and to use the foreign support as domestic legitimization. In 1939, German race policy among scientists was not nearly as uncontroversial as the Nazi race politicians wanted to make their German public believe. Still, at first glance, it is no wonder that the National Socialists during the 1930s saw an international swing in opinion in favor of their race policy. Groß is an example of someone who could proudly point out that it was with the help of foreign scientists that the public image of Germany held by the outside world was improving.[27]

A closer look shows that the foreign supporters of German race policy consisted of a small but not insignificant group of scientists from various countries in Europe and North America.[28] So, for example, the two leading Swedish race biologists Herman Nilsson-Ehle and Herman Lundborg openly and unreservedly came out in favor of the National Socialist race policies.[29] The Norwegian Jon Alfred Mjöen extolled the National Socialist takeover of power as the "last gigantic effort to save Western culture from collapse" and praised German race policy as an example for all Nordic countries.[30] The British eugenicist George H. L. F. Pitt-Rivers made himself a lackey of German foreign policy and joined the British National Socialist party.[31] The general secretary of the American Eugenics Society, Leon F. Whitney, announced that Hitler had shown "great statesmanship...and the courage of the knowledge to put sterilization to work." Harry H. Laughlin praised the German sterilization law as the "most significant legislative act of this type that a nation has ever brought about."[32]

Draper's Pioneer Fund played an important role in propagating National Socialist race policy abroad. Harry Laughlin organized American screenings of the film *Inherited Suffering [Erbkrank]*, a eugenic and anti-Semitic propaganda film from the Race Policy Office of the NSDAP that was shown on Laughlin's initiative at various American schools, colleges, and church congregations.[33]

What made German National Socialism so attractive to this group of eugenicists and race hygienists? The National Socialists presented their ideology as a logical political implementation of biological principles. The restructuring of the German state along supposedly biological natural laws astounded quite a few eugenicists and race hygienists, though it was really only a radical implementation of their demands for making politics biological.[34]

While receptive politicians like the Italian dictator Mussolini, the Belgian minister of state Emile Vandervelde, the British Prime Minister Winston Churchill, and the US presidents Theodore Roosevelt and Herbert Hoover would not or could not go beyond mere lip service to eugenics, it was the National Socialists who carried out an extensive eugenics policy with great determination.[35] For members of the International Federation of Eugenic Organizations, which in 1930 had unsuccessfully attempted to draw government representatives into their international work, the receptiveness of the Nazis to eugenics resounded wonderfully.[36]

The National Socialists within a few years were able to implement eugenic measures only by setting aside fundamental rights, having the state penetrate into the private sphere of its citizens, and by centralizing power over the race, social, and health environment in the hands of a scientific and political elite. In many other countries, such matters had been shuffled off to committees or stopped by courts. The dictatorial implementation of eugenic measures in National Socialist Germany fascinated quite a few eugenicists, who in their own countries were involved in complex democratic negotiation processes. The antidemocratic move, increasingly visible in the eugenic literature of the 1920s, was the basis for the fascination of many eugenicists for the ruthless dictatorial implementation of eugenic laws.[37]

These were the reasons that a strikingly large number of non-German eugenicists supported National Socialist race policy. The collaboration of 20–30 influential foreign eugenicists with the National Socialist race politicians and race hygienists was based as a rule on previous cooperative work within the international eugenics movement. The channels that the National Socialists used for propagating their race policies abroad were in particular the International Federation of Eugenic Organizations and the International Union for the Scientific Investigation of Population Problems.

When on July 14, 1933, the Nazis passed as one of their first political measures a law to prevent the birth of children with hereditary diseases, the reaction of foreign public opinion was far from positive. The general criticisms were as a rule not about the eugenic core of the law—the sterilization of those with schizophrenia, Huntington's chorea, epilepsy, blindness, deafness, physical disfigurement, alcoholism, the "hereditarily ill" with "feeble-mindedness," or "manic-depressive madness"—but rather its potential "misuse" against political, religious, or ethnic minorities.[38]

The German race hygienists and race politicians had to struggle with these reservations at the first conference of the International Federation of Eugenic Organizations after the National Socialist assumption of power. At the conference in Geneva in 1934, the Dutch delegate G. P. Frets and the French representative Georges Schreiber raised doubts about National Socialist race policy. Schreiber had written in 1933 in the French *Revue Anthropologique* that Rüdin's support for the anti-Semitic policy of the National Socialists disqualified him from being president of the International Federation. The writer did not doubt that Rüdin undoubtedly was a "highly respected and renowned scholar," but he would find it difficult to be scientifically objective under the totalitarian system of the National Socialists.[39]

Rüdin was in the vanguard of those who, as he expressed it, called attention to Schreiber's "Jewish machinations" and disapproved of Frets.[40] In light of the criticism and due to the public interest at this first international meeting of eugenicists and race hygienists after the Nazis took office, Rüdin took care to arrange a large, carefully selected German delegation. He prevented a long-planned speech about the inheritance of criminality by Rainer Fetscher, a Dresden eugenicist considered by the Nazis to be politically unreliable.[41] Only the race hygienists favored by the Nazis, like Otmar Freiherr von Verschuer, Alfred Ploetz, Ernst Rodenwaldt, Lothar Tirala, Heinz Kürten, and Lothar Loeffler, were allowed to travel to Zurich with the leading lights of National Socialist race policy, Walter Groß, Falk Ruttke, and Karl Astel.

This coordinated action was possible because shortly after the National Socialist takeover of power, German academics could no longer attend scientific conferences without the permission of the German Central Congress Office (*Deutsche Kongress-Zentrale*). This office, founded in 1934 and set up as part of Goebbel's Propaganda Ministry, grouped the German scientists into delegations led by a group leader authorized to exercise control and to make reports. This "conference policy," later codified in the "Guidelines for the Supervisors of German Delegations to Conferences Abroad," forbid German delegates from making any public criticisms. As a result, the Germans often appeared as a homogeneous block, while scientists from France, Great Britain, the United States, and Sweden participated in the normal academic conference manner and often assumed controversial positions toward each other.[42]

Led by Rüdin, the ten-person German delegation to Zurich succeeded in getting the International Federation to take a positive attitude toward the eugenic aspects of National Socialist race policy. In his opening speech, Rüdin had already pointed out that only "energetic measures to improve mankind" could prevent the "decline of the peoples of culture." "The care of our valuable heredity pool" and "liberation from incapable persons" could not be accomplished with "wise anthropological speeches," but with "well-aimed actions and rock-hard consequences."[43]

Ruttke was a member of the expert committee for population and race policy in the Reich Ministry of the Interior and National Socialist chief legal adviser for all questions of race policy. In his report, he noted how the "well-aimed actions and rock-hard consequences" could be expected to look. He explained to the assembled eugenicists from Denmark, Austria, Hungary, Norway, Poland, Switzerland, France, Great Britain, Czechoslovakia, The Netherlands, and the United States how "total legislative action with race hygienic viewpoints" was being instituted in Germany. The institution of loans for marriage, which could be given only to "hereditarily sound, non-Jewish true Germans *[Volksgenossen]*," and the national law on inheritance, according to which "hereditarily sound" non-Jewish peasants could count on special support from the state, were both seen to lead to the "selection" of high-value members of the race. Together with the sterilization law, the law on castration of "habitual criminals," and the institution of counseling offices for biological heredity, Ruttke claimed that they were "creating the preconditions for a hereditarily sound German people."[44]

The only criticism directly at the National Socialist race policy presented by Ruttke came from the French and Dutch delegates. Frets explained that public opinion in The Netherlands would never accept a sterilization law like the one in Germany.[45] Schreiber pointed out to the Nazis present that in France, sterilization was considered as "a severe attack on individual freedom." Even before the conference, Schreiber and Frets had criticized the compulsory character of the sterilization law. To these two, sterilization as a eugenic measure was possible only on an individual and voluntary basis.[46]

Despite these doubts, even Frets and Schreiber voted for a resolution submitted by Alfred Mjöen, according to which the International Federation of Eugenic Organizations, "with its great variety of political and philosophical viewpoints," was united in the idea that the "practice of race hygiene is extremely important for the life of culture peoples and is unavoidable." The scholars meeting in Zurich advised the "governments of the world" that "in an objective fashion similar to what had already occurred in several countries in Europe and America, they would study the problems of hereditary biology, population policy and race hygiene, and apply their results for the common weal of their people."[47]

Although German policy was not expressly mentioned, with this resolution, the participants in this IFEO conference issued a scientific letter of approval to National Socialist race policy.[48] Immediately after the conference, the NSDAP press service proudly reported on the "international renown of the German race hygiene legislation." *Neues Volk*, the propaganda magazine of the race policy office of the party, announced the "breakthrough of the new spirit" outside Germany as well. "International experts," as shown by the Zurich declaration, here increasingly acknowledging the National Socialist point of view, would "inevitably" bring about a "spiritual adjustment" beyond German borders.[49] *NSK*, the National Socialist press service, called on German newspapers to rely on the Zurich declaration to report that "leading scientists in the entire world" had acknowledged the correctness of the German measures.[50] Even at the national party conference, the leader of the Reich physicians and Hitler confidante Gerhard Wagner referred to the Zurich decision of the IFEO, and boasted that in foreign countries, the "understanding for the population and race political tasks" was growing unstoppably.[51]

The magazine *Eugenical News* presented National Socialist race policy quite in the sense of the Zurich declaration. Davenport and Laughlin, the editors of the official organ of the International Federation of Eugenic Organizations, the Galton Society, and the Eugenics Research Association, took care to report only positively about German population policy. So *Eugenical News* reported that in no other country was eugenics coming to fruition more strongly as "applied science" as in

Germany. The National Socialist state had undertaken to strengthen the German population in size and in quality. In that way, Germany had become the first of the "great nations of the world" to directly implement practical race hygiene knowledge in its practice.[52] *Eugenical News* printed without comments contributions from Kurt Thomalla, an expert from Josef Goebbels's propaganda ministry, and from the National Socialist Minister of the Interior Frick.[53]

The position that was repeated again and again in *Eugenical News* was that even though one could have various opinions about the total program of the National Socialist policy, in any case, one must recognize that Germany was in the vanguard of other countries in biological funding of the "national character." The *Eugenical News* wrote in a featured article in 1933 that the law on the prevention of hereditarily sick children was, together with the American sterilization laws, a milestone in the eugenic "monitoring of human reproduction."[54] Up until 1939, when *Eugenical News* was taken over by the American Eugenics Society, the journal showered praise on National Socialist race policy. Without commentary, it printed an article from the *Rassenpolitische Auslands-Korrespondenz* about Jewish physicians in Berlin, and praised Germany as the country in which the results of the study of human heredity for the improvement of the race were being implemented most rigorously. It wished Otmar Freiherr von Verschuer great success on the occasion of the dedication of his Institute for Biological Heredity at the University of Frankfurt.[55]

While the report in *Eugenical News* and the favorable vote on the resolution of the Zurich conference constituted a first step out of international isolation for the National Socialist race politicians, they achieved an even greater propaganda success with the Berlin World Population Congress in 1935. The International Union for the Scientific Investigation of Population Problems had decided in 1931 that the next World Population Congress would take place in Berlin.[56] Although the American population scientists in particular had doubts after Hitler's takeover of power as to whether an international scientific congress should be held in National Socialist Germany, the board of directors of the IUSSIP decided unanimously to confirm Berlin as the site of the conference.[57]

For the directors of the IUSSIP, the influence of the German National Socialists on science was in no way comparable with the suppression of science in the Soviet Union, on the basis of which the board had refused in 1933 to accept a Russian demographer into the IUSSIP.[58] The president of the IUSSIP Charles Close and his predecessor Raymond Pearl were confident that the IUSSIP board could exert control over the shape of the congress and that such a "distinguished and broad-minded scientific man" as Eugen Fischer in the role of congress president would maintain the scientific level and the political neutrality of the congress.[59]

Close and Pearl misjudged how much Fischer, whom they expected to be the next president of the International Union, had already become a lackey of the Nazis. As the chairman of the organizing committee, consisting of members of the Nazi German Society for Race Hygiene, the German section of the IUSSIP, and the German Statistical Society, Fischer was not prepared to bring the board of the IUSSIP into the planning of the congress.[60] Since the congress was being financed basically by the German government, Close could only accept the inevitable, that "he who pays the piper has a right to call the tune."[61]

In line with the claims and accepted forms of the international organizations for population science and eugenics, Fischer as congress president officially called for very strict adherence to science, but as in 1927 in Geneva and in 1931 in London and Rome, in Berlin too, the close connection between politics and science in questions of population was clear. Finally, the National Socialists also understood, as emphasized by Groß, that their race policy consisted simply of "practical conclusions from the knowledge of objectively researched science."[62]

Even before the conference, Fischer pointed out that the foreign participants should not only be convinced that "they could work free and unrestrained like any individual in the Third Reich," but also that the National Socialists were instituting the "findings of population science" in political measures.[63] In his opening address, he explained that population science is not "a worldwide and foreign work at one's desk or in the statistical office," but work on the quantity and quality of one's own people. Population science with its statistical methods that was open to the ups and downs of birth and death rates, marriages, and migratory movements must first capture the major questions of population in all their breadth in the tight connection between human heredity and race hygiene.

The "population statisticians" and the "heredity researchers" must in their work contribute to a situation where a people can fulfill its "holy duty" of keeping its "völkisch existence in its quality of race in the way it existed in the forefathers' time," and to care for and pay lavish attention on "healthy hereditary lines." Fischer proclaimed to the 500 population scientists assembled in Berlin that Adolf Hitler clearly recognized the deep and consequential sense of population science and was ready to draw the necessary conclusions. He went on to say that one must be thankful to this man for success in "turning the German people away from the fate of a population that had led to the death of past cultures and past peoples."[64] Fischer called on the congress to send a telegram of greeting to Hitler expressing its confidence that the "visionary heredity and race hygiene population policy" of the National Socialists would assure the future of the German people.[65]

These shows of devotion to Hitler came from others beside the German race hygienists and population scientists. Leading members of the international eugenics movement did not conceal their enthusiasm for Hitler's race policy—these included the British Pitt-Rivers and Cora Hodson and the Norwegian Jon Alfred Mjöen.[66] The 180 academics present in Berlin numbered among their ranks important leaders of the International Federation of Eugenics Organizations and the International Union for the Scientific Investigation of Population Problems. Those who came from Great Britain were Close, Pitt-Rivers, and Hodson as well as Alexander Carr-Saunders and the birth-control activist Marie Stopes. The French delegation included two influential population scientists and population policy activists, Fernand Boverat and Adolphe Landry, while the Italian delegation included Corrado Gini, Livio Livi, and Franco Sovorgnan, the president of the Italian Statistical Institute. Others who came to Berlin were the host of the Zurich IFEO conference, the Swiss Otto Schlaginhaufen, and the Dutch human geneticist G. P. Frets. From further afield, there came Harry Federley from Finland, Jon Alfred Mjöen from Norway, and Herman Lundborg and Herman Nilsson-Ehle from Sweden, as well as almost all the Scandinavian members of the IFEO.

Only the reform-oriented eugenicists stayed away from the conference, though they were becoming increasingly important in Great Britain and the United States. Among the no-shows were the British eugenicists C. P. Blacker and Julian Huxley, who even before the conference had criticized the forced character of the German sterilization law and had spoken out against the anti-Semitic racist policy of the Nazis. Nor did the American reformers come.[67] The American section of the IUSIPP, dominated by the reform-oriented members of the American Eugenics Society, decided decisively against participating in the Berlin conference. They sent to Berlin only Frank H. Hankins, a professor from New York open-minded about National Socialist policies, as an official observer.[68]

The American demographers thus left the field open to a small group of American eugenicists who were very well disposed toward Nazi race policy: Clarence C. Campbell and Harry H. Laughlin as vice presidents of the congress, and one of the major financiers of the American eugenics movement, Wickliffe Draper, representing the Eugenics Research Association.[69] Campbell wound up presenting himself as the official American representative at the congress, and flattered the German race policy of the National Socialists with a very laudatory speech.[70] He noted that the heredity quality of a race is primarily responsible for human development. This had already been understood in the previous century by Joseph Arthur Gobineau and Francis Galton. One should be thankful, he said, to Sir Arthur Keith, Leonard Darwin, and R. A. Fisher in Great Britain, Georges Vacher de Lapouge in France, Harry Laughlin, Charles B. Davenport, Madison Grant, and Lothrop Stoddard in America, and Alfred Ploetz, Eugen Fischer, and Ernst Rüdin in Germany for moving the theory forward. It was certainly among the German National Socialists that the approaches of race anthropologists and eugenicists were brought together in a unified race policy. For Campbell, other nations and race groups needed to follow the German pattern if they did not want to hopelessly fall backward in their "race quality" and in their "chances for survival."[71]

Even if Campbell's clear and blunt speech went well beyond the standard praise of Hitler's race policy, it was typical of the eugenic and race anthropological orientation of the entire congress. In contrast to the Geneva World Population Congress in 1927, at which the fear of possible overpopulation was the main subject, in Berlin anxiety reigned over the "growing danger that the white peoples would die out because of the decline in the birthrate."[72] Charles Close warned of a further decline of population in Great Britain. Arthur Linder of the statistical office of the city of Berne spoke of a "gap" in Swiss births. The Vienna population scientist Wilhelm Winkler confirmed that in Austria, the "one-child marriage" had become the most common type of marriage, and in Vienna, it was even the "no-child marriage." For the Scandinavian countries, according to Knud Asbjörn Wieth-Knudsen, professor at the Technical University of Norway, "intellectual disgraces" like liberalism, radicalism, and feminism were responsible for the decline in births.[73]

In contrast to these developments, expressed by Hans Harmsen as a "biological crisis for the people of all European countries," Friedrich Burgdörfer announced an increase of the birthrate since Hitler's assumption of power. Burgdörfer, along with his colleagues Friedrich Keiter, race anthropologist at the University of Hamburg, and Siegfried Koller, the leading German statistician of heredity, expressed the opinion that the increase of the German birthrate could be traced to the population policy of the National Socialists and to the growing trust of the Germans in their economic and political future.[74]

In Berlin, there was broad agreement that only the birthrate of the "hereditarily sound and racially valuable families" should be increased. That "the hereditarily sick and asocial children" would have to be prevented with all means possible was the representative opinion of the Nazi race politician Arthur Gütt.[75] It was clear from the contribution of the German anthropologist and race hygienist Friedrich Keiter that this basic eugenic principle was becoming increasingly mixed with an open anti-Semitism. Keiter, who joined the NSDAP in 1940, flagellated "Jewry" as the "pacemaker of biological collapse." He gave voice to the hope that the "exclusion" of "the Jews" would have a "significant healing effect on the German people."[76]

Even though the majority of the foreign participants did not share Keiter's open anti-Semitism, certainly not in its radical form, it was still possible for the German population scientists to approve German race policy, given the reigning eugenics consensus in Berlin. Hans Harmsen and Franz Lohse, two very significant race hygienists who participated in the organization of the congress, expressed the opinion that the congress gave "German population scientists the possibility of presenting population laws that had been worked out by them and put into effect by the German national government and presenting in part the results of this scientifically funded population policy to the experts from all countries."[77]

Only one participant of the congress, the French eugenicist Jean Dalsace, who stood close to the Communists, attacked the German sterilization legislation and, as reported by David Glass, the official observer of the British Eugenics Society, raised a great deal of consternation.[78] Under the motto "Man is what he eats [*Der Mensch ist, was er isst*]," he opposed the predisposition of the eugenicists toward behavior being determined by heredity. He claimed that the human has essentially been formed by his environment and his education. He excoriated sterilization and castration as a "return to barbarism" and drew attention to the severe psychological consequences arising for the sterilized person as a result of this unnecessary operation.

Leadership circles of the International Federation of Eugenic Organizations and the National Socialist race politicians felt themselves particularly provoked by Dalsace's presentation. Rüdin, Mjöen, and Hodson rebuked Dalsace for misunderstanding the influence of heredity on human beings and the value of an active hereditary hygienic program. Mjöen justified sterilization by referring to the enormous costs that arose due to "hereditarily sick" families. Blithely ignoring the thousands of "accidental" deaths as a result of sterilization, Rüdin pointed to the absolutely benign nature of the operation. Cora Hodson referred to the positive experiences with eugenic sterilizations reported by Laughlin in Berlin, and reported on the attempt of British eugenicists to introduce in Great Britain at least the possibility of "voluntary" sterilization. Falk Ruttke chided Dalsace for not being informed about the "most recent development of sterilization legislation," and noted that other countries in Europe besides Germany had introduced eugenically motivated sterilization laws—Denmark, Sweden, Norway, Finland, and the Swiss canton of Waadt.[79]

Despite this single criticism of the German sterilization law and a boycott by some American scientists, the Berlin World Population Congress was a complete success for the National Socialists. They could proudly note that at the reception hosted by the Reich government for congress participants, Campbell gave a toast to the Reich Chancellor Adolf Hitler, while Mjöen flattered Germany with the words that, eugenically speaking, "world history" had been made in Germany.[80] In his closing address, Fischer called attention to the fact that the World Population Congress had shown that "in the fourth decade of our century the thought of man received a concept that was the focus of great interest: Race. The fact that race has so penetrated into the consciousness of today's spiritual life can be definitely attributed to the National Socialism of the new Germany."[81]

Immediately after the congress, the National Socialist propaganda machine began to exploit the positive position of the population scientists with regard to National Socialist race policy. Together with the Congress on Criminal Law that had taken place in Berlin shortly before, one that had passed a resolution greeting eugenic sterilization, the World Population Congress served National Socialist propaganda for domestic justification of German race policy.[82] The *Deutsche Allgemeine Zeitung* noted that the "major portion of the Congress" had acknowledged the perception that the laws on heredity were recognized in principle and that it would be justified to use the sterilization laws to save "the people's patrimony squandered in the service of the hereditarily ill and generations that deserve to be wiped out."[83] Heinrich Schade, Verschuer's scientific colleague and a German race hygienist, a National Socialist, and a Hauptsturmführer in the Schutzstaffel (SS), reported in *Erbarzt* that the congress had shown that the "most renowned researchers from all countries acknowledged and greeted the path beaten by the German National Socialist government as a useful and promising one."[84]

While the National Socialist propaganda celebrated the World Population Congress as a great success, among the non-German scientists, a dispute was raging as to how far the Berlin Population Congress had been made into an instrument of Nazi policy. The Dutch eugenicist Franz Schrijver in the journal *Erfelijkheid bij de Mens* lauded the congress for "having observed the rules of complete objectivity." It had been easy to present dissenting opinions. Schrijver, who in 1933 had been a scientific guest at the Kaiser Wilhelm Institute in Berlin, remarked that he was impressed that a mere four years after the last IUSIPP Congress in London, the German population scientists could report concrete measures against the dangers of a declining birthrate and of counterselection.[85]

Campbell, who after his visit to Berlin started a regular campaign for race policy in German and American journals, praised the fact that through the World Population Congress, the "representatives of almost the entire world" had been made aware of the "significance of race hygiene measures."[86] In an article in *Eugenical News*, which was eagerly taken up by the NS press, he pointed out that National Socialist race policy was not the product of political opportunism, but involved the "well considered demands" of German anthropologists, biologists, and sociologists. During his visit in Germany, Campbell had become convinced that almost the entire German nation enthusiastically supported this policy.[87] Hermann Lundborg was just as euphoric with regard to the Berlin Congress. He praised the race hygiene "heroic battle" of the German nation, and stated that in "a time of need and confusion on every side of life," every people must follow the path laid out by "the Germans under the leadership of their Führer."[88]

In contrast to this view, the British, American, and Dutch population scientists in particular reacted critically to the Berlin Congress. Unlike Close, Pearl made clear that Campbell and his ultra-racist theses in no way represented American population science. For Pearl, the Berlin Congress had not been a forum for scientific discussions about population problems, but instead had served as political propaganda for the National Socialists under the umbrella of the International Union for the Scientific Investigation of Population Problems.[89] A similar criticism was voiced by the Englishman David Glass. In *Eugenics Review*, he complained that besides the "race prejudice" presented by the German delegates, foreigners had delivered papers laced with Nordic smugness.[90] Because of the strong politicization of the population congress in Berlin, both the American and the Dutch committees of the IUSSIP demanded henceforth a stronger influence of the International Union on the shape of congresses. Both committees used the Berlin Congress as the occasion to demand a selection of the lectures by the board of directors of the IUSSIP with an eye to their scientific character.[91]

The international controversy over what happened at the Berlin Congress made it quite clear to the Nazis how important it was to have influence on discussions of science and science policy in the areas of eugenics, population science, and genetics. In the second half of the 1930s, the rising influence of the eugenic reform forces and the socialist eugenicists at the international congresses for population science and genetics made the situation for the National Socialist race hygienists and their foreign sympathizers more difficult. On the other hand, the race hygienists brought an important international organization under their control. The German race politicians profited from a situation where Rüdin, true to the National Socialists, was the chair of the IFEO; crucial roles were played by old eugenicists of the stripe of Laughlin, Davenport, Lundborg, and Mjöen and by enthusiastic fellow travelers of National Socialist race policy like the British Pitt-Rivers and the Swede Torsten Sjögren. The coalition among German race hygienists, orthodox eugenicists of the first generation, and younger sympathizers of National Socialist Germany was successful in preventing any critical voices coming through within the IFEO.

Consequently, the leadership of the IFEO refused membership to eugenicists critical of the National Socialists, like the Swede Gunnar Dahlberg and the Norwegian Otto Mohr. Dahlberg, a social democratic–oriented eugenicist and critic of National Socialist anti-Semitism, had taken over the Swedish Race Biological Institute in Uppsala over the opposition of Lundborg and Nilsson-Ehle, and had restructured it into a research center for human and medical genetics. Although he was entitled to membership in the IFEO as director of the Swedish Institute, the processing of his membership application was drawn out for practically five years.[92] Mohr ran up against similar opposition. The Norwegian heredity researcher, who combined his basic eugenic convictions with progressive social political positions, stood close to the Norwegian Social Democrats. Although he was one of the most important human geneticists in Scandinavia, there could be no question of his becoming a member of the International Federation because of his criticism of the National Socialists and their mouthpiece in Norway, Jon Alfred Mjöen.[93]

Mjöen called his arch enemy Mohr a "fanatical opponent" of National Socialist race hygiene. In opposition to IFEO policy under its presidents Darwin, Davenport, and Rüdin, Mohr was seen to speak out for eugenic "preventive propaganda," "legal abortus provocatus," and "the killing right of mothers." Mjöen explained to Rüdin that Mohr in the IFEO would do "great damage." He warned that the "bitterest enemies" of the current federation policy could continue their "destabilizing work" within the organization. He felt that the members of the IFEO had enough to do in combating the "destructive forces" outside the organization; if one were to admit the "destabilizing forces" into the organization itself, the organization "would be finished."[94]

Because of this policy of the IFEO leadership group and the lack of interest on the part of more moderate American and British eugenicists in the work of the International Federation, the National Socialists and their foreign supporters were able to constantly build up their position in the IFEO.[95] Rüdin used his position as president of the International Federation to bring further representatives of the German Race Hygiene Institute into the IFEO alongside Ploetz and Fischer. The first of these was Verschuer entering the IFEO as representative of the Frankfurt Hereditary Biology Institute, followed immediately by Kurt Pohlisch as representative of the Hereditary Biology Institute of the University of Bonn. These were joined in 1936 by the Hygienic Institute of the University of Heidelberg, represented by the Nazi party member Ernst Rodenwaldt.[96]

This strong presence of German race hygienists paid off for the first time at the IFEO conference in Scheveningen in The Netherlands. The major objection lay in the fears circulating among the Germans that the Dutch eugenicists Frets and Sanders, who were so critical of National Socialist race policy, would be able to raise their voices against German anti-Semitic policies. For that reason, the NS bureaucracy considered forcing the transfer of the conference from The Netherlands to Germany with very little notice. They hoped to keep better control there. In the end, the Reich Ministry of Education opposed the shift to Germany, but it saw as absolutely necessary to have a "quantitative and qualitative outstanding representation of German academics." The National Socialist government should not defend itself personally at the conference, since the "German standpoint" was said to be purely scientific. According to the bureaucrat in charge at the Reich Ministry of Education, this was much more promising than having a conspicuous representation of National Socialist politicians.[97]

Rüdin also saw the "very great significance" of a very strong German delegation at the conference "for the prestige and reputation" of German race hygienic legislation.[98] As the leader of the delegation, he coordinated the speeches of "German participants who were reliable philosophically and politically" and obtained references from the NS Federation of Lecturers regarding the political "suitability" of those making presentations. The Dutch eugenicists desired to have the Catholic Hermann Muckermann, but for Rüdin, Muckermann as well as any other religiously affiliated scientist was out of the question as a presenter.[99] The carefully chosen 15-member German delegation arrived in Schevenigen, and formed the numerically dominant group of the total of 40 scientists from America and Europe. Despite the bitter opposition of Frets, the National Socialist race hygienists forced through their position on two important questions, the relationship of the IFEO to race hygienic measures in Germany and the choice of a new president.

In light of the sterilization laws presented in Scheveningen from the United States, Denmark, Norway, Sweden, Switzerland, and Germany, the National Socialist law on the prevention of hereditarily sick children played a particularly important role. Hodson praised the conference presentations of Rüdin, Ruttke, and Astel as a "revelation" for all those who were still not familiar with National Socialist policy.[100] National Socialist Germany was said to have set new standards internationally with its sterilization policy. By 1936, Germany had sterilized some 200,000 people, far more than all other countries combined, under the rationale of inferiority. In addition to sterilization by operation, in the mid-1930s, the Nazis also instituted a procedure using X-rays. They also legalized abortion on eugenic grounds.[101]

Once again, it was Frets who came forth with criticism of German race hygiene. He found fault with the German system of "sterilization by diagnosis," that is, the classification of so-called hereditarily sick people into such vague categories as feeblemindedness or circular insanity. He demanded that every case of "inherited degeneration" be examined individually in detail. In his view, sterilization should be performed only in exceptional cases and with the full understanding of the patient or the patient's guardian.[102] However, since there were so many representatives of countries that had successfully instituted sterilization laws at home—the American Charles M. Goethe, the Swede Torsten Sjögren, the Norwegian Alfred Mjöen, the Danes Gunnar Wad and Tage Kemp, and the Swiss Hans W. Maier—Frets's criticism fell on deaf ears.[103] Since the Scandinavian laws did not basically differ from the German law, criticism of National Socialist sterilization measures alone would have had little credibility.[104]

The *Nieuwe Rotterdamische Courant* reported that despite the differences in details, all the congress participants were completely agreed upon the "necessity of sterilizing the mentally ill." "Hereditary illness was enemy number 1 of humanity."[105] The NS press announced that "despite the differences in philosophical points of view," the "leading position of the German heredity research and the practical measures in Germany in the area of caring for the race were acknowledged." Articles with a similar tone appeared in various German newspapers, noting that "Germany was the only country in the world where comprehensive measures to care for the race had been instituted." Other countries would certainly have "made decisions quite similar to those of the German race laws," but their adoption often foundered "on the lack of unanimity of the people's spirit."[106]

Concerned that eventually a critic of National Socialist race policy might become president of the IFEO, German race hygienists had before the Scheveningen meeting come to an agreement that only a convinced supporter of National Socialist Germany was acceptable as a successor to Rüdin. Since the previous presidents had come from Great Britain, the United States, and Germany, representatives of these three countries were not viable as candidates. Since Lundborg and Nilsson-Ehle, the most prominent Swedish race hygienists and enthusiastic supporters of National Socialism, could not be considered for the office because of ill health, Fischer hit on the idea of nominating Mjöen, the longest-serving member of the IFEO. Fischer recognized that Mjöen's scientific accomplishments were questionable, but it was also true that the erstwhile IFEO President Leonard Darwin had not exactly been one of the leading scientists of his country. As Fischer wrote in a confidential letter to Rüdin, the things going for him were "officially his age" and for the Nazis "his friendliness to Germany."[107]

Mjöen declined to be a candidate precisely because of his age. In order to prevent the presidency of Frets, Rüdin felt himself forced to nominate the Frenchman Georges Heuyer, the Dutchman Petrus Johannes Waardenburg, the Austrian Heinrich Reichel, and the Swede Torsten Sjögren. Although these men did not have the same blind enthusiasm for Hitler's race policy as Mjöen, at least they were well disposed to the eugenic measures of the National Socialists. It turned out to be quite lucky for the National Socialists that Heuyer, Waardenburg, and Reichel declined to be nominated, and that finally Sjögren was elected.[108]

Sjögren was a student of Lundborg and a psychiatrically oriented human geneticist. He was the director of a psychiatric institution in Lillhagen near Göteborg, and was a convinced defender of National Socialist race theory. Like Rüdin, he labeled the hereditary biology research and practice in Germany as his "pathbreaking model."[109] In 1935, Rüdin, Davenport, and Federley had attempted to push him through as director of the Swedish Race Biological Institute in Uppsala. Despite their support, along with that of Lundborg and Nilsson-Ehle, they were unable to overcome the candidacy of the medical statistician Gunnar Dahlberg, who was sponsored by the Swedish minister in charge.[110]

During Sjögren's presidency, the IFEO turned into a submissive propaganda tool of the National Socialist government. Immediately after being elected, Sjögren turned to Rüdin to receive "reliable suggestions and information" for his activity as president of the IFEO.[111] During a several-week stay at the Kaiser Wilhelm Institute for Psychiatry in the summer of 1937, he worked out with Rüdin a strategy for the federation. Sjögren came out very strongly for making the *Archiv für Rassen- und Gesellschaftsbiologie* the official organ of the IFEO. Along with *Eugenical News*, the German journal was now to print on a regular basis short news items and reports from the IFEO.[112] He also did everything in his power for a large race hygiene congress to take place under the sponsorship of the IFEO in Berlin.

At the meeting in Scheveningen, Rüdin had arranged for an invitation from the National Socialist government to the IFEO to have its congress in Berlin in 1937.[113] The Dutch eugenicists, who as members of the International Union for the Scientific Investigation of Population Problems had protested against the political orientation of the Berlin World Population Congress, were alarmed. To Rüdin's great disappointment, they were able to prevent the IFEO from declaring itself ready to serve as a platform for an international eugenics congress in Germany. After a lively discussion, the majority of those eugenicists present supported the idea of an international eugenics congress in Germany, but they expressly rejected IFEO sponsorship for this event. The next official conference of the IFEO was to be held independent of the German congress in 1938 in Poland, Hungary, or Estonia.[114]

In consultation with the German race hygienists, Sjögren took steps so that, despite this resolution, the next meeting of the IFEO would take place in Germany as part of a world congress of racial hygiene.¹¹⁵ He cleverly prevented Hodson's planned 1938 IFEO conference in Estonia,¹¹⁶ making the only alternatives be no congress at all or a congress in Germany. The conflict between Hodson and Sjögren did not take place over the role of the IFEO at a race hygiene congress in Germany or over any substantial difference in their attitudes toward National Socialist race policy. Hodson too considered fortunate the fact that under the National Socialists, "a theoretical science had become something real," and she lauded many eugenic measures of the National Socialists.¹¹⁷ While Sjögren still promoted an exclusive position of the IFEO in favor of the National Socialists, Hodson did her utmost to hold both the National Socialist race hygienists and their critics in the IFEO. Clinging to a middle way that avoided very clear positions, from the time of the Scheveningen meeting, she attempted to prevent a threatened schism in the IFEO.¹¹⁸

When Sjögren at the beginning of 1939 succeeded in depriving Hodson of her office and installing Mjöen as the temporary new business director, no obstacle any longer existed to an international congress for race hygiene in Germany under the sponsorship of the IFEO.¹¹⁹ Without consulting with other members of the IFEO, on January 9, 1939, Sjögren "humbly" informed the Reich Minister of the Interior that the International Federation was ready to create a "platform for an international congress on eugenics" in Germany.¹²⁰

A mere four weeks later, agreement was reached at a meeting in Berlin regarding the scope of the Fourth International Congress for Eugenics and Race Hygiene. The meeting included representatives of the Reich Ministry of the Interior, the Ministry of Propaganda, the Ministry of Finance, the Reich Health Office, the staff of Rudolf Heß, the Reich Committee for National Health Service, the Statistical Office of the Reich, and the German Society for Race Hygiene. This meeting was also attended by Ernst Rüdin, Friedrich Burgdörfer, Falk Ruttke, Kurt Thomalla, and Herbert Linden, and it thus comprised the central figures of the National Socialist race bureaucracy and the race hygiene movement. The group decided to hold the congress not as originally planned in Berlin, but in Vienna. The participants hoped that through a congress in an area recently annexed by Germany, they would be able to conduct intensive "recruitment for race thinking in Austria."¹²¹

Reich Interior Minister Frick took over official sponsorship, and a working committee of Ernst Rüdin, Karl Astel, Friedrich Burgdörfer, Eugen Fischer, Fritz Lenz, Herbert Linden, Lothar Loeffler, Hans Reiter, Falk Ruttke, and Fritz Wettstein assumed responsibility for concrete preparation for the congress planned for August 26, 1940. In the eyes of the organizers, "research and progress in the area of the scientific bases of race hygiene, including the resultant possibilities for application" were to be discussed. The meeting was also to deal with the relationship of race hygiene and eugenics to race research, heredity research, race hygiene population policy, medicine, and legal theory. On the model of the world population congress, the organizers intended to offer travel to the meeting to "important institutes and organizations serving race hygiene."¹²² Rüdin, who was planning to be the officiating president of the congress, arranged to send out invitations to almost all leading eugenicists in the world. On his address list, he marked off the eugenicists critical of the National Socialists like Dahlberg from Sweden, Mohr and Bonnevie from Norway, and Schreiber from France with question marks or with notes like "Jew" or

"Jew Bolshevik."[123] Rüdin and his co-organizers did not want to have the possible success of the Vienna Congress endangered by the presence of critics.

In the end, the war and the connection of National Socialist race and health administrations in the mass murder of the mentally handicapped and the psychiatrically sick prevented the congress from taking place. Herbert Linden, one of the chief organizers of the mass murder of the mentally handicapped during the Second World War, reported on October 11, 1940, to the Reich Minister for Science, Education, and National Training that the Fourth International Congress for Race Hygiene (Eugenics) in Vienna had not taken place "because of political considerations" and was being postponed until after the war.[124]

The Further Splintering of the International Eugenics Movement

The attempts of the German race hygienists and their foreign supporters to make the IFEO a puppet of National Socialist race policy led to a split in the international eugenics movement in the second half of the 1930s. Although between 1925 and 1935, the IFEO had been able to bring together most of the leading representatives of the national eugenics movement, it was no longer able to do so because of the increasing radicalization of the race policy of the Nazis. Because of the inactivity of the IFEO under Sjögren's presidency, the British Eugenics Society even considered withdrawing completely from the organization.[125] In parallel to the Nazi-oriented leadership of the International Federation of Eugenic Organizations, other international groups of eugenicists organized themselves, such as the International Group of Human Heredity and the Fédération Internationale Latine des Sociétés d'Eugénique.

Under the chairmanship of Frets, the International Human Heredity Committee had detached itself from the International Federation of Eugenic Organizations and had taken along Dahlberg and Mohr, leading critics of National Socialist race policy. However, since Verschuer was one of the vice presidents of the group and since German race hygienists were still strongly represented, the group had to refrain from discussions of science policy and could not act as a corrective to the pro-Nazi IFEO.

In 1935, Corrado Gini, who had fallen out in the 1930s with the leadership of the International Federation of Eugenic Organizations, founded in Mexico the Fédération Internationale Latine des Sociétés d'Eugénique (FILDSE).[126] This organization represented the leading eugenicists from the primarily Catholic Southern European and South American countries. The Europeans who belonged to it included the Belgian Albert Govaerts; the president of the eugenics section of the Parisian Institut International d'Anthropologie, Eugene Apert; the leadership group of the Societa Italiana di Genetica ed Eugenica around Corrado Gini and Agostino Gemelli; and the presidents of the Rumanian and Catalonian eugenics societies, G. Marinesco and Puig Sais. The South American members knew each other mainly from working together in the pan-American eugenics organizations. These included Mariano R. Castex, the president of the Associacion Argentina de Biotipologia, Eugenesia y Medicina Social; Renato Kehl, the president of the Comissao Central Brasileira de Eugenia; Adrián Correa, president of the Sociedad mexicana de Eugenesia; and Carlos A. Bambaren, the president of the Peruvian Liga Nacional de Higiene y Profilaxia Social.

The FILDSE did not oppose National Socialist race policy, as the American historian William H. Schneider claims, but rather came out against the negative eugenics that had originated with the Anglo-Americans.[127] The roots of the FILDSE organizationally lay in the row between Gini and his American and British IFEO colleagues and ideologically in the rejection of a eugenics sterilization policy by Catholic eugenicists.[128] The tight tie of South American and Southern European eugenicists to the Catholic Church led to their hesitant position regarding contraception among the so-called hereditarily sick.

Instead, the FILDSE propagandized for the forced increase of "valuable" couples, a policy that was quite similar to the pronatalist measures of the National Socialists. For example, Schreiber, one of the sharpest critics of the German sterilization law, at the first and only congress of the FILDSE in 1937 expressly praised the National Socialist policy of giving marriage loans to genetically "viable" couples. With the encouragement of marriage between intellectually and physically sound young people, the National Socialists had implemented demands that echoed the ones that he had made in 1932 at the Third International Congress for Eugenics.[129]

Radicalization of National Socialist race policy and the submissiveness of the IFEO chairman Sjögren to the wishes of the German race hygienists contributed to the disintegration of the international eugenics movement in the second half of the 1930s. The controversy between the National Socialist race hygienists and their critics interestingly enough took place not in the international eugenics organization, but at the congresses for genetics and population science.

International Criticism of National Socialist Race Policy

As shown by the conferences of the International Federation of Eugenic Organizations in 1934 and 1936 and the 1935 Berlin World Population Congress, the National Socialist race hygienists and race anthropologists were able to achieve significant international successes. In the international organizations for eugenics and population science, they were cleverly able to use two important institutions for their goals.

As opposed to them, the attempts to scientifically discredit the Nazi-based race policy up until 1937 consisted primarily of individual initiatives by concerned American, British, Czechoslovak, and Scandinavian scientists. It is true that many European and American academics were ready for united action against the limitation of academic freedom and against the expulsion of Jewish scientists, but very few attempted to undermine National Socialist race policy from a scientific point of view.[130] These few included the New York anthropologist Franz Boas and his student Melville Herskovits, the British biologists Julian S. Huxley, Alfred C. Haddon, and J. B. S. Haldane, the Prague anthropologist Ignaz Zollschan, the Swede Gunnar Dahlberg, and the Norwegian geneticist Otto Mohr.[131]

The most active of all was Franz Boas. After studying at the universities of Heidelberg, Bonn, Kiel, and Berlin, Boas had emigrated at the age of 29 in 1887 to the United States because as a Jew he saw no chance in Germany for an academic career.[132] As a professor at Columbia University in New York, Boas became the most important American anthropologist and the godfather of cultural anthropology. In contrast to physical anthropology, which limited itself to the study of race anatomy and pathology, the cultural anthropologists around Boas emphasized the independence of race, language, and culture. In Boas's view, a person could be of any race, could in principle speak any language, and develop any form of culture. Race was therefore for him a cultural category. From his anthropological approach, Boas very early on drew the scientific conclusion that there was no basis for objecting to race mixing nor for the superiority of the white race.[133]

The assumption of power by the Nazis in Germany was the occasion for Boas to take up the fight against National Socialist race policy. In journal articles, books, public speeches, and private correspondence, he indefatigably disputed any scientific basis for National Socialist race politics and described Hitler as an "intellectually limited fanatic," giving evidence with the first edition of *Mein Kampf* of his "complete ignorance" about race.[134] Despite mainly vain attempts between 1933 and 1936 to set up a broad front of American scientists against the National Socialists and their American supporters, he never grew tired of flagellating the theory of the superiority of specific races.[135]

Boas's counterpart in Europe was the Czech anthropologist Ignaz Zollschan. Zollschan, a Jew born in Austria, at the beginning of the twentieth century migrated to the Czech territories. At the beginning of his scientific career, he had attacked anti-Semitism as racist and scientifically ungrounded, but still he had shown a certain sympathy for the thesis of the genetic superiority of the Jewish race.[136] In the 1920s, however, he distanced himself from Jewish race theories.[137] Directly after the takeover of power by the Nazis, Zollschan renewed his struggle against National Socialist race policy. On the basis of his initiative, the Czech Academy of Sciences and Arts in 1935 published a book about the "equality of the European races," in which leading Czech scientists sharply criticized German race policy.[138]

In Great Britain, it was particularly Julian S. Huxley and Alfred C. Haddon, who with their book *We Europeans* contributed to undermining scientific racism in general and National Socialist race policy in particular. Even though the British anthropologists could not agree on a single definition of race, let alone complete a statement against racism, this book, written for the broad public, showed the scientific simplifications of National Socialist race ideology. Although Haddon and Huxley in the 1920s had themselves expressed racist hypotheses, they now argued that nowhere in the world do "pure races" exist. They admitted that in the past, there may have been pure races of men, but through the migrations of peoples and mixed marriages everywhere, mixed races had arisen.[139]

The initiatives by Boas, Zollschan, Huxley, Haddon, and a few other scientists did cause a sensation in Europe and the United States, but all attempts at making the scientists protest internationally came to naught before 1937. So race questions played only a subordinate role, when in July 1934, for the first time since 1912, scientists came together for an International Congress for Anthropology and Ethnology in London. Attempts by Boas and Zollschan to put through an antiracist position paper collapsed due to the lack of interest of their colleagues and the opposition of the organizers of the congress.[140]

Zollschan experienced the same lack of success in his attempt to bring together scientific experts for a conference on race ideologies. Due to the increasing threat to Czechoslovakia by Germany and the National Socialist dominance of the Sudeten German minority, the Czechoslovak state leaders around President Tomás Masaryk and the private Academy of Sciences initially supported Zollschan's initiatives. On a trip around Europe, he also received moral and financial support from the president of the International Institute of Intellectual Cooperation, Edouard Herriot. Nonetheless, his plan to use the Institute, an organization of the League of Nations, to stop the conference collapsed in the end due to the international appeasement policy toward Nazi Germany and the lack of support from his academic colleagues.[141]

It was only with the radicalization of the National Socialist race policy through the Nuremberg laws in the fall of 1935 that the scientific critique came together internationally. These laws forbade marriage and sexual intercourse between Jews and non-Jews as well as marriage between the "hereditarily ill" and the "hereditarily healthy." Reading these laws made it clear to many foreign academics that the National Socialists were prepared to go beyond any scientific caution in implementing their eugenically and anthropologically oriented race policies.

The open conflict over National Socialist race policies erupted in 1937 at the Paris world exposition, which included the First International Congress for Child Psychiatry, the Second International Congress for Mental Health, and the Third World Population Congress. The congresses for child psychiatry and for mental health proceeded satisfactorily for the German delegation. Because of the proximity of psychiatry and mental health to care for hereditary health and thus to the racially motivated sterilization policy of the Nazis, this congress was very important for them.[142] In his presentation on the "Conditions and Role of Eugenics in the Prophylaxis of the Mentally Disturbed," Rüdin defended the sterilization policy of the National Socialists. A few non-German participants raised the persistent objections to the logic of eugenics—the danger of sterilizing crazy geniuses, the questionability of forced measures, and the danger to the population caused by excessive negative eugenics—but on the whole, German sterilization policy met with a good deal of sympathy, as Rüdin remarked in his report to various ministries.[143]

In contrast to the First International Congress for Child Psychiatry and the Second International Congress for Mental Health, which had moved along quite calmly, at the Population Science Congress, there was a rancorous confrontation between the German delegation and prominent critics of the National Socialist race policies. The French group Races et Racisme, an initiative of French scientists against National Socialist race and foreign policies, had held a small conference on racism in the framework of the Paris world exposition and then had joined the World Population Congress as a participant. Together with Boas and Zollschan, the French members of the board of directors of Races et Racisme—Henri Laugier, Célestin Bougle, and Paul Rivet—formed a strong counterweight to the German scientists and race politicians present in Paris.[144]

The Paris planning committee around the French Minister of Labor and population scientist Adolphe Landry was well aware of the imminent clash regarding National Socialist race policy. Given the experiences at the Berlin Population

Congress, Landry very much wanted to prevent a politicization of the Paris Congress. Paradoxically, the preselection of contributions based on criteria of scientific value, practiced for the first time here, did not work against the German scientists, but against their critics. In the same manner in which Muller's presentation to the Third International Eugenics Congress in 1932 was to have been prevented, Landry refused a lecture originally planned by Boas regarding the contribution of Jewish and non-Jewish Germans to general culture as too sociological and not sufficiently scientific.[145]

In their lectures and contributions to discussions, the critics of National Socialist race policy concentrated their attacks on two aspects—the relationship of race to the environment and the relationship of race to culture. Boas maintained that the division of humans into various racial types did not rest on biological principles, but on purely subjective outlooks. His anthropological researches were intended to show that there are no innate characteristics of human types. The human types determined by the race anthropologists were much too strongly influenced by their environment for them to be separated from each other by ability, character, disposition, and intelligence. Boas pointed to his research from 1909 on Jewish immigrants into the United States, in which he had shown that within a very short time, the immigrants had adapted to the population average in intellectual ability and in their physical constitution. Drawing on further research of his own and of his students on the intelligence of Indians, Africans, and whites, he showed that the abilities of a race, a concept in itself very difficult to define, were completely dependent on the cultural framework.[146]

Zollschan's presentation on the "Significance of the Race Factor for the Genesis of Culture" followed in the same vein. He noted that Arabs who had migrated to cities behaved quite differently from their forebears living in the desert, and he went on to point out that in Denmark, the children of the warlike Vikings had become a people of conventional dairy farmers. In contrast to the "stability and eternal unchangeability" put forth by the race anthropologists, Zollschan showed that one might explain the dynamics of human societies by drawing in the milieu and the environment.[147]

His countryman Maximilian Beck focused on the error made when politicians attempted to transfer the race development possibilities of plants and animals to human beings. With plants and animals, one could determine the importance of all varieties. The effect of nature on a bloodline and arbitrary intervention in the laws of nature occur here on several levels. However, in the attempts by humans to breed the race, the paradox arises that while the race hygienists subject themselves to "fate determined by blood," they also maintain that human breeders are "masters of the blood." If everything is determined by race, the paradox arises of how can one intervene in determining the race? Like Zollschan and Boas, he underlined the fact that mental works of culture do not arise "from themselves" as an automatic growth product of some drive existing in the blood, but are the product of arduous cultural creativity.[148]

These lectures and the systematic criticism of all the race hygiene and race anthropology contributions of the National Socialist scientists unleashed aggressive reactions from the German delegation. Ernst Rodenwaldt roundly condemned the criticism from Boas, Zollschan, and Beck as "rabbinic" arguments that had nothing to do with "the standard mode of European academic debate."[149] Elisabeth Pfeil, a Berlin population scientist and supporter of National Socialist race theory, castigated the contributions from Boas, Zollschan, and Beck as politically motivated "diversions."[150] Karl Thums, an associate of Rüdin at the Kaiser Wilhelm Institute in Munich, proclaimed that the presentations by the "Jews Beck, Prague, Zollschan, Prague, Boas, New York" and contributions to discussions by "their race colleagues" were unworthy of a scientific congress. According to Thums, the arguments did not match the "confirmed facts and results" of serious scientific research, but sprang only from "hatred of National Socialist Germany."[151]

The criticism from Boas, Zollschan, and Beck and some other scientists did not only question National Socialist race policy, but their sharp criticism of the way that National Socialists had made politics into a biological issue also was undermining the eugenic premises of population science, even though this latter point was largely unnoticed at the time. To be sure, congress president Adolphe Landry had noted in his opening address the dangers for European and American peoples from a quantitative and qualitative point of view, but for the first time at a population science congress, Boas, Zollschan, and Beck and the other scientists fundamentally questioned the significance of inheritance.[152]

The German participants were astonished that the criticism was directed not only against the National Socialist discrimination against other races, but also against the eugenic aspects of National Socialist policy. In their report, Pfeil emphasized that there were connections between the race idea and race hygiene, but the "attacks" from Boas and Zollschan work clearly against all "programs of a eugenic sort." These programs however were considered by Pfeil as clearly not "a specialty of the Germans," but were practiced in several other European and American countries.

Rüdin also observed that not only was the race anthropological contribution by Robert Ritter on Central European Gypsies and that of Ernst Rodenwaldt on "non-shared race elements of the Baltic castes" aroused opposition, his own contribution on the eugenics of mental illness was sharply attacked by Beck and Schneerson.[153]

The polarization at the World Population Congress in 1937 was also apparent within the executive committee of the International Union for the Scientific Investigation of Population Problems at their meetings. Bitter arguments broke out when Frets, speaking in the name of the Dutch delegation, rebuked the German race hygienists for having organized the Berlin World Population Congress as a political event. Only the mediation of Frederick Osborn brought the scholars together in agreeing that in future population congresses, an international committee would monitor the scientific quality of the presentations.[154] At Rüdin's suggestion, the directors were able to agree on the Frenchman Adolphe Landry as the new Union president and on Eugen Fischer, Ernst Mahaim, director of the Scripps Memorial Foundation for Population Research, Warren S. Thompson, Charles Close, H. W. Methorst, Livio Livi, and Karl Edin as his vice presidents. General Secretary Pitt-Rivers was removed from office because of his scandalous behavior at the conference and his close connections to the National Socialists.[155]

Pitt-Rivers during the 1930s increasingly allowed himself to be a puppet of the National Socialists, and attempted to downgrade the IUSSIP into an instrument of German foreign policy. In September 1936, he traveled to Czechoslovakia on his own authority in order to study the situation of the local German minority. While he avoided any contact with Czech population scientists, he sought out the National Socialist head of the Sudeten German party, Konrad Heinlein. In a side discussion at a propaganda exposition for the merger of the Sudeten German areas with the German Reich, Henlein demanded that an international commission of population scientists should study the living conditions of the German minority in Czechoslovakia.[156]

For Pitt-Rivers, Czechoslovakia was part of a British–French–Russian conspiracy against Germany. The Czechs were said to have become a "tool" of the Communist International. As evidence for his thesis, he pointed to the strength of the Jews in Czechoslovakia.[157] Completely in line with Henlein, Pitt-Rivers opposed this stand of the Board of Directors of the IUSSIP taken at the Paris Congress regarding the situation in Czechoslovakia.[158] In his role as general secretary, he cast doubt on the possibility of purely scientific research in that country, questioned the legitimacy of the Czechoslovak IUSSIP committee because there were no German population scientists represented on it, and demanded, as he had agreed with Henlein, that an international delegation be sent to "Sudeten Germany."[159] As Close expressed it, Pitt-Rivers had run "amuck" and had made his role as general secretary an unacceptable one. He was replaced by the Frenchman Georges Mauco.[160]

If the critics of the National Socialist population policies were disappointed that the majority of the population scientists in Paris could not come to a clear position against National Socialist race doctrine, the congress did show a clear change in the climate in comparison to the one that had prevailed at the Berlin World Population Congress and at the conferences of the International Federation of Eugenic Organizations. While the National Socialist press after the conferences in Zurich, Berlin, and Scheveningen had been able to report on the international renown of German policies, now they could report only about very strong disagreements. Instead of an international victory parade of German race policy, now the discussion was that the critics had been unable to prevail over "German scientific basics" and that the critics had not succeeded in "unsettling the scientific basis of the German point of view."[161]

The critics of National Socialism who had appeared at Paris came exclusively from outside the International Union for the Scientific Investigation of Population Problems. Within the IUSSIP, except for Frets and a few American population scientists, all discussions regarding the scientific basis of National Socialist racism were suppressed. The choice of Landry as chairman and Mauco as general secretary strengthened this position. The attempt to mediate between the various forces within the IUSSIP led to a complete stoppage of the work of the Federation after the Paris Congress. From 1937 until after the Second World War, the IUSSIP essentially went into hibernation.

As opposed to the international conferences for eugenics, population science, psychiatry, and anthropology, only a small group of scientists expressed criticism of National Socialist race policy.[162] Among the geneticists, however, the front against the race anthropological aspects of National Socialist policies was much broader. This was connected to the fact that progressive American, British, and Scandinavian geneticists were gaining influence. In contrast to the scholars like Boas, Zollschan, and Beck, who increasingly included a criticism of the very principles of eugenics within their criticism of National Socialist race policies, the Socialist geneticists, such as Hermann Muller, Walter Landauer, Gunnar Dahlberg, and J. B. S. Haldane, were concerned with another form of eugenics. They condemned the Nordic race theory spread by the National Socialists and some other eugenicists of the first generation, but at the same time, they pleaded for a eugenics free of class and race prejudice.

Outside the Permanent International Eugenic Committee and its successor organization, the International Federation of Eugenic Organizations, in various European countries and the United States, a string of prominent Socialist and Social Democratic eugenicists had set forth their own theses. After the Nazi takeover of power, some of them formed a loose circle in the center of scientific critics of National Socialist race polemics. In Sweden, Gunnar Dahlberg unceasingly questioned the scientific basis of Nazi race policy. In Great Britain, in addition to Julian Huxley and the Marxist biologist J. B. S. Haldane, the Socialist biochemist Joseph Needham and the social biologist Lancelot Hogben unremittingly criticized not only the policies of the National Socialists but also that of the orthodox eugenicists in their own country as unscientific. In the United States, it was the hitherto unknown geneticist Walter Landauer of the University of Connecticut who tried to mobilize his colleagues against National Socialist race policy. In the Soviet Union, those involved against the National Socialists included Hermann Muller, who as an enthusiastic supporter of Marxism took up residence in the Soviet Union in 1933, and Julius Schaxel, a German geneticist driven out by the Nazis to Moscow.[163]

The critics had initially worked primarily on the national level, but during the 1930s, under the pressure of the broader radicalization of German race policy and the propaganda successes of the German race hygienists, they began to work together internationally. The Seventh International Congress for Genetics provided the concrete occasion for the first international initiative against the National Socialists. This congress initially had been scheduled for 1937 in Moscow. One year before the planned congress, a group of scientists living in the Soviet Union and a group of American geneticists agreed to oppose the scientific bases of National Socialist race policy at the congress. The geneticist Schaxel made a proposal to make the Moscow Congress a scientific forum against National Socialist race policies, a proposal that aroused great interest among his American, British, and Russian colleagues.

Schaxel had been a professor at the University of Jena and representative of a Marxist biology before the Nazis took power, but in 1934, he had been deprived of his citizenship. On the initiative of his Jena colleague, the geneticist Otto Renner, his doctoral degree had been voided in 1935, and he was excluded from the German Society for Genetic Science because of his political propaganda against Germany and his "gross insult to German science."[164]

Schaxel used the occasion of his exclusion to write to various scholars in Europe and the United States. His supposed "political propaganda" consisted of "the theoretical and practical opposition to National Socialist race theory," which demanded "hundreds of thousands of innocent sacrifices." Because "geneticists of the stripe of Mr. Jenner place themselves in the service of barbarism," there should be a "forum of serious scholars conscious of their responsibility [to deliver] an objective assessment" of National Socialist race theory at the international genetics congress.[165]

Schaxel's suggestion was taken up by Walter Landauer, a development geneticist and Socialist eugenicist who had emigrated from Germany to the United States in 1924.[166] He convinced 30 members of the Genetics Society of America to actively work through the program committee of the Seventh International Congress of Geneticists to set up a discussion forum. This group included Clarence C. Little, the editor of the *Journal of Heredity*, Robert C. Cook, Harrison R. Hunt of Cornell University, Leon J. Cole of the University of Wisconsin, and four members of the American Eugenics Society. Julian Huxley, a member of the British Eugenic Society, also supported this suggestion. What they had in mind for the discussion forum was the topic of the genetically provable differences between the various races, whether there is scientific proof for the superiority of a specific race, and to what extent eugenic measures might contribute to the progress of society.[167] Solomon G. Levit, a close colleague of Muller and founder of the Moscow Biological-Medical Institute and chairman of the program committee, was quite favorable to the suggestion; he set in motion on the preparation committee for the congress a special section on the topic of "Human Genetics and Race Theories."[168]

This initiative by American, British, and Russian geneticists aroused the fear among the responsible German authorities and race hygienists that the congress would be used for "propaganda against German race policies."[169] When Rüdin learned of the plans, he immediately contacted his colleagues Mjöen, Lundborg, Nilsson-Ehle, and Sjögren in Scandinavia about developing a joint strategy against Schaxel and his colleagues.[170] The Gestapo in an internal memo warned that the German researchers in Moscow might be seduced into ongoing political discussions. According to an official in the Gestapo, the Moscow Congress threatened to be "a demonstration against the National Socialist world view." In light of the propaganda effect in the Soviet Union, the Gestapo therefore urgently suggested that the Germans should not take part in the congress.[171] This strategy was completely agreed to by representatives of the Race Policy office of the NSDAP, the Reich Ministry of Education, the Foreign Office, the congress central office, and by the representatives of the Führer at an emergency meeting in August 1936. The German geneticists urged their foreign "friends of Germany and the German race theory" to join a boycott since the congress simply was going to serve the "political aims of Bolshevist propaganda." Only if a broad international boycott proved to be unsuccessful should a small German delegation of "specially picked and trained scholars" be sent to Moscow.[172]

National Socialist authorities felt palpable relief when the Soviet government at the beginning of 1937 postponed the congress indefinitely and then completely cancelled it. The basis for this decision lay in the ever stronger influence of the Communist state leadership over genetic research and the imprisonment and later execution of leading Russian geneticists and eugenicists as part of the Stalinist purges.[173] Although the Soviet eugenicists had been quite clever strategically in raising the unity of eugenic measures and socialist state doctrine, in the 1930s, they fell ever more strongly under political pressure from the Stalinist regime. Following Stalin's declaration of neo-Lamarckian theory as the political and scientifically correct doctrine, the Soviet authorities used the supposedly strong connection between eugenics and Mendelian genetics to discriminate against the leading Russian geneticists and eugenicists as eugenic mass murderers. When the Genetics Society of America protested against the state suppression of genetic research in the Soviet Union, the official answer of the Soviet Union was quite concise: In the Soviet Union, genetics did not have the same "freedom" as in certain countries where "freedom" was understood to mean the killing of people or the elimination of supposedly inferior peoples.[174]

Because of the severe suppression of Mendelian genetics in the Soviet Union, the permanent committee of the International Genetics Congress decided to hold its international congress in Edinburgh from August 22 to August 31, 1939. Over 600 participants registered. The scheduled reports promised to give genetic research the same kind of impetus as had the contributions at the Fifth Annual Genetics Congress in 1927. However, because of the threat of the outbreak of war, the scientific discussion moved into the background. In the meantime, the international situation was as intensely discussed as were the scientific reports. In the foyer of the university, radio broadcast loudspeakers were set up, and large groups of scientists gathered around them at each special announcement.

Scientists from the USSR, including the geneticist Vavilov, who had been designated as the congress president, were forced by the Soviet government to cancel their participation at the last minute as a result of international tensions.[175] The 32 members of the carefully selected German delegation did indeed travel to Edinburgh, but they kept their bags packed in preparation for an early departure. After the signing of the German Soviet Non-Aggression Pact on August 23, the German Foreign Office ordered the immediate return of the German scientists. At the same time, most other European scientists left the congress, so that on the fifth day of the meeting, the only remaining scientists were the British and American geneticists and a small group of continental European ones.[176]

It was an enormous achievement of Hermann Muller, the person responsible for the congress program, that in this situation a number of leading geneticists were convinced to undertake a new direction for eugenics. The occasion was a request by Watson Davis, the head of the American news agency *Science Service*, who wanted to know the opinion of the scientists in Edinburgh on how genetics could most effectively improve world population. For the geneticists assembled in Edinburgh, this question presented the possibility of defining a eugenics policy beyond the racism-oriented German hereditary notions and the traditional genetics à la Davenport and Darwin.

Muller noted from an earlier memo (written in 1935) in favor of a socialist eugenics against the National Socialist oriented manifesto; he obtained 32 high-ranking geneticists as co-signers.[177] Under the pressure of the international situation, the signers of the so-called *Geneticists' Manifesto* distanced themselves from the aggressive racism of the National Socialists. The group included Muller, Landauer, Haldane, Huxley, Needham, Dahlberg, and Crew as well as such renowned geneticists as Theodosius Dobzhansky, Cyril Dean Darlington, and Rollins Adams Emerson. According to the signers, the differences between races and between classes were culturally determined. "Good or bad genes" were not the monopoly of specific peoples or classes. An assessment of the genetic quality of specific individuals would first become possible when all the social framework conditions were equal for all people. The end of the class society, the geneticists left no doubt, would be the precondition for selection according to "objective" criteria such as health, intelligence, and social behavior. With a better understanding of biological principles, the average of the population could be raised to the intellectual, psychological, and physical level of currently living outstanding contemporaries.[178]

As the historian Karl Heinz Roth has convincingly demonstrated, the signatories of the *Geneticists' Manifesto* wished to save the eugenic perspective on progress with a new balancing of the relationship between heredity and the environment. Systematic selection according to criteria free of race or class prejudice was presented as an alternative to the racist policies of the National Socialists. This program, aiming toward the differentiation of the entire human race, reached back to such classic instruments of eugenics as voluntary sterilization and contraception, but also opened the way to new methods, such as artificial insemination.[179] At the beginning of the Second World War, at the moment when the National Socialist race policies were showing themselves in their entire radical nature, scientific critics of this policy presented a eugenic counterversion. In a society free of oppression, war, and exploitation, state "monitoring of human reproduction" would create a group of humans with a "better" genetic hereditary pool.

Chapter Six
The Second World War and the Mass Murder of the Sick and Handicapped

"All eugenicists know that war means a frightful extermination of the most virile and most viable elements of a nation, both for the victor and the conquered." Consequently, Ernst Rüdin, the author of this 1934 comment, went on to say that all eugenicists are bound together by the "fervent desire for peace among all nations."[1] At the International Congress for Mental Health in 1937, he was even more moving: "Not loud enough, not grandly enough, not often enough can eugenics raise its voice of warning against mass destruction, which is a hostage of war." From the standpoint of race hygiene in general and from that of "eugenic prophylaxis of the mentally disturbed" in particular, modern mass war is to be condemned as a phenomenon that "carries off the most hereditarily virile persons before they have children" and "leaves behind the intellectually hereditarily inferior with all their progeny."[2]

Despite the pathos, the question at that time did not seem to be so "basic" for Rüdin. His unconditional propaganda for the German war of aggression after 1939 made his earlier comments seem mere mockery. Shortly after the march of German troops into Poland, in a lecture with the title "The War That Has Been Forced Upon Us and Race Hygiene," he complained that "it was Great Britain's jingoistic classes and Jews and their French toadies" who had drawn Germany into a war, despite the earlier efforts at peace by German race hygienists and their foreign colleagues. Despite all race hygienic scruples about war, Germany could respond to this British "war for hegemony" only with armed force. In the last analysis, this would make eugenic sense, since the race hygienists had to defend their goal of the "eternal Germany."[3] On the occasion of the death of Alfred Ploetz in 1940, Rüdin regretted the fact that his mentor, who had believed so earnestly "in the leadership of Adolf Hitler and in the holy national and international mission of race hygiene," had not lived long enough "to experience the solution of the problem of mutual understanding and cooperation of the Nordic peoples." It would however be a comfort for his fellow warriors that Ploetz "up to his dying breath maintained with unshakeable hope his belief in the victory of arms of the German people and in the continuing march of victory of race hygiene that would come in the peace that followed."[4]

The Second World War showed the limits of the eugenics peace policies so tirelessly propagated by the German race hygienists. The warning voices were silent and were replaced by the demand that because of the losses in the war one should tighten the race hygiene reins even further. In what follows it shall be shown that the swing in eugenics peace policies over to a sharpening of the race policies came as a reaction to a suddenly unavoidable dysgenic war policy. This new attitude was a central element in the framework for the murders by the tens of thousands of the mentally handicapped and psychologically ill in the areas controlled by the National Socialists. During the war years, the German race hygienists participated not only in legitimating this mass murder but were also actively engaged in spreading race hygiene and race policy ideology in the occupied territories.

The involvement of eugenics in this stage of mass murder and the racist motivated war of aggression in foreign eyes discredited the orthodox variant of eugenics that was interested in race research. Particularly in the United States, Scandinavia, and Great Britain, National Socialist policies accelerated moves by the reformers to displace the orthodox eugenicists of the type of Darwin, Davenport, and Ploetz.

National Socialist "Peace Policy" and the Murder of the Mentally Handicapped and the Psychologically Ill

The official National Socialist position on the topic of war was two-faced. On the one hand, the leadership of the state threatened to use armed force to release Germany from the "chains of the Versailles treaty," but on the other, they emphasized that National Socialism, due to its biological and eugenic underpinning, was deeply peace-loving. Minister of the Interior Wilhelm Frick grappled with the eugenic thesis of the dysgenic effect of modern war at the World Population Congress in Berlin in 1935. "The healthiest and best of the nation in the prime of their youth, not having reproduced," would lose their lives on the battlefield and thus would give the "sick and the weak an increased measure of reproduction possibilities." Every war is therefore paid for with great sacrifices of "the most valuable blood." Recognition of this fact turned "National Socialists into opponents of war."[5]

Walter Groß fiddled this same tune in 1935 at a reception by the Foreign Office of the NSDAP for foreign diplomats accredited in Berlin. Under the motto "race policy is a peace policy," he referred to the "destruction of the most valuable carriers of the breed" in war. Since National Socialism placed its central focus on the "maintenance of the race substance of the people," it was the "conception of the state most ready for peace" that one could imagine. Only "rootless ignorance" could assign bellicose aims to National Socialists. Freedom was a basic precondition for the "development of the population and the desires for distinction so necessary for life" of Germany. For Groß, nothing could endanger National Socialist race policy more than a thoughtlessly provoked war. The "race standpoint of National Socialism" was therefore the "best guarantee of respect for the independence of peoples and races."[6] Adolf Hitler too did not tire of presenting the same arguments to portray National Socialism with a freedom-loving view of the world. Instead of having a war of conquest destroy the "selection of the best," the nation should be strengthened through an "increase in the joy of giving birth." That is why, said Hitler in 1937, "National Socialist Germany...desired peace as part of its deepest convictions."[7]

The National Socialist leaders in their speeches referred directly to the discussions of peace within the International Federation of Eugenic Organizations. At the meeting in Zurich in 1934, a resolution addressed the "highest levels of government of the civilized countries" with concern about an imminent war. In the eyes of the eugenicists assembled in Zurich, a second world war would "carry off the more virile men in masses." Because of the "difficult and slow regeneration," there would be a disastrous "further loss in virile material for Western culture." Completely in the sense of the programs of most eugenics societies in the 1930s, they called on government leaders to do all they could to prevent another war between the "civilized states."[8]

This Zurich resolution was used by the National Socialists to prove that their calculation of "race politics as peace politics" was scientifically well-founded.[9] The poster boy of the National Socialists, Alfred Ploetz, was the founding father of German race hygiene. As Ploetz got older, he increasingly made eugenics peace policy the main area of his activity. After the positive experiences with the peace resolution that he had brought forward in Zurich, various National Socialists pressed him to be involved even more strongly as a missionary for a race hygiene idea of peace.

In May 1935 the Munich National Socialist professor Wirz revealed to him that the NSDAP considered him an appropriate candidate for the Nobel Peace Prize and would support his candidature. Even though Ploetz himself considered his chances to be very poor because of his anti-Semitism and his engagement for National Socialism, he declared himself ready to be a part of the campaign. After discussing the matter with Frick, he scheduled for the Berlin World Population Congress a speech about race hygiene peace policy and an antiwar resolution on the model of the one in Zurich.[10] During the congress, Ploetz gave up on the idea of the antiwar resolution, but he did give a speech titled "Race Hygiene as the Basis of Peace Policy."

He maintained that not only specific individuals but also specific anthropological races would be especially affected by counterselection in wartime. The "Nordic race elements," which according to Ploetz were the "unifying tie between all European nations," in particular suffered during a war. Men with a great deal of Nordic blood were more willing to volunteer for a war with an idea, and because of their larger body size were more frequently called to the Army. For Ploetz, the ones who would most profit from the counterselection would be the Jews. They were reputed to have a weaker physical constitution, an inborn aversion to being soldiers, and frequently a "deficient capacity of enthusiasm for their host people and the state." The "reduction of the Nordic race elements" brought on by war would lead to "an impoverishment of the culture peoples in the qualities that were necessary for the blood and for subsequent standing firm against colored races and against Bolshevism." The work of the race hygienists could prosper only in peacetime. Ploetz felt that German involvement in a new war would destroy the race hygiene achievements of National Socialist Germany from one moment to the next. Because a second world war would move like a "crushing steamroller over the young seed of new life," Ploetz concluded to the Berlin Congress, the German race hygienists were seeking "with all their might [to create] a sincere peace."[11]

Because of his contacts with influential Norwegian circles, Jon Alfred Mjöen undertook the task of promoting Ploetz's candidature to the Oslo Nobel Prize committee.¹² Working with Wirz as a facilitator, Swedish and Norwegian scientists worked out with Ploetz the best strategy for supporting his candidacy.¹³ The National Socialist press did its part in the campaign. For example, under the title "Peace Prize for a German?" the *Berliner Illustrierte Nachtausgabe* gave Ploetz space to spread his ideas.¹⁴ When it turned out that a German other than Ploetz won the Nobel peace prize in November 1936, the outrage was great among the German race hygienists and their foreign colleagues. The awarding of the Nobel prize to Carl von Ossietzky, a critic of the National Socialist regime who had been interned and tortured over a number of years in concentration camps by the National Socialists, was a slap in the face to the Ploetz supporters. Agnes Bluhm, who at first believed rumors that Ploetz had won the prize, was furious after the clarification of the misunderstanding and learned that the peace prize had been given to the "traitor" Ossietzky. This was an "impudent provocation against Germany," which could not even be atoned for by awarding the prize to Ploetz in the following year.¹⁵ When Ossietzky's prize was announced, Mjöen emphasized that in no way did all Norwegians agree with this. From the point of view of the race hygienists, Alfred Ploetz was the best candidate for the Nobel Peace Prize.¹⁶

It was clear that Ploetz had no serious chance of winning the Nobel Peace Prize,¹⁷ but the intensity with which the German, Norwegian, and Swedish race hygienists worked for his candidacy shows that the propaganda of race hygiene as a guarantee for the National Socialist desire for peace was more than an intentional covert plan for a German war of aggression. The tolerant and often supportive position of German race hygienists for the bureaucratically organized death of the mentally and physically handicapped must be seen against the background of this eugenic discussion of peace. The slaughter of well over 200,000 handicapped and psychologically ill persons during Second World War was a segment of a radical final solution of the race question pursued by the National Socialists, a pursuit that claimed as victims around six million Jews plus many Roma and Sinti and Slavic groups.¹⁸

Before the war, the vast majority of German race hygienists had spoken out against the killing of the intellectually handicapped and the psychologically ill under the name of "euthanasia"; they were much more interested in preventing reproduction of the handicapped and not in their death. The killing of the handicapped and the sick would have been a eugenic measure only if it had been intended to exclude these groups from reproducing by killing them. However, the most severely handicapped, who were to be eliminated by what was called euthanasia, were not considered to be especially sexually active, and reproduction by the slightly handicapped could have been handled just as effectively by sterilization as by euthanasia. For the eugenicists, the killings were not the first choice.¹⁹

For this reason the euthanasia debate opened up in Germany by Karl Binding and Alfred Hoche in 1920 echoed well beyond the race hygiene movement.²⁰ In general, members of the German Society for Race Hygiene were critical of euthanasia. Lenz declared in 1923 that in no case could euthanasia be a race hygiene "effective means" that the race hygienists should support. Rüdin spoke out similarly that same year in a memo to the Prussian Ministry for Welfare against the "destruction of the lives of people who were inferior, but who had been put on the world, born into it and therefore were feeling and sensitive beings." Hans Luxenburger, his colleague at the Kaiser Wilhelm Institute for Psychiatry, also rejected euthanasia with a reference to "respect for the life of the human individual."[21]

What motivated the German eugenicists after 1939 to tolerate the slaughter of the intellectually handicapped and mentally ill, and even to participate in this drastic form of negative eugenics? What were the reasons that race hygienists integrated the murder of sick people into the politics of race and health, even though sterilization and the ban on marriages had accomplished enough to satisfy the eugenic and race hygiene demands?[22]

Even before the outbreak of the war there had been in Germany eugenic and racially motivated "unintended" deaths during sterilization, individual killing of handicapped newborns, and the murder of Jews as a part of anti-Semitic pogroms. Well before 1939 the National Socialist leadership had discussed the systematic killing of humans who were considered inferior in the eugenic or anthropological sense. Hitler had revealed the calculation in 1929 that by "doing away with" 700,000 to 800,000 of the weakest people, a biological "strengthening" could spread to the entire German nation.[23]

With the Second World War came the "qualitative switch" from a policy of "care for hereditary patrimony and the race" that consciously took account of deaths to one of bureaucratically organized extermination of life. It is my hypothesis that the German war of aggression was in theory and practice directly related to the murder of the handicapped, and beyond that to the slaughter of religious and ethnic minorities.[24] The close connection of the war with the murder of the handicapped and the sick had developed even before 1939. The handicapped were stigmatized with military expressions—they were the "army of the inferiors" and the "enemy #1 of humanity"; therefore a "war against the defectives" was absolutely necessary.[25] The scholar Ernst Bergmann spoke of a "world war against idiots, cretins, the feebleminded, habitual criminals, and other degenerates and contaminated persons. [Of the] human trash in the big cities, certainly one million [could be] shoveled aside."[26]

The victims of forced sterilizations were justified by reference to much greater losses of frontline soldiers. Falk Ruttke declared to the IFEO in 1934 that the state could demand of the handicapped the sacrifice of fertility; after all, in war situations the state demanded "the sacrifice of one's life by the best citizens."[27] In the same way Groß defended sterilization: "As long as the state in the eyes of all people has the right to demand the life of its soldiers for the good of the nation, then in our opinion it also has the right to demand renunciation of reproduction, if this reproduction would bring sorrow and misery to the individual and to the nation."[28] Eugen Stähle, medical expert at the Württemberg Ministry of the Interior, in 1935, justified the loss of life during forced sterilization of handicapped victims with a reference to the frontline soldiers in the World War, who also lost their lives for the common good.[29]

The association of the war with the struggle against the handicapped and the psychologically ill took on ever more extreme forms in National Socialism. In 1936 Ploetz made clear that in an emergency the "counter-selective effect of the war [would have to be made good] by increasing the extermination quota and especially by raising the selection quota."[30] Gütt, Ruttke, and Rüdin went so far as to call for the wiping out of the handicapped as a true condition for a true peace. For this trio of commentators on the sterilization law, "The struggle of Germany for a hereditarily healthy new generation [should unite] the hereditarily healthy and race conscious around the world." Only a true "community of the strong and healthy" would be in a position to give "the world a new and better shape [and to] bring about a true peace through the mutual high regard of the virile people."[31]

It was in this atmosphere of a close connection between war and race-oriented eugenic politics that Hitler in the fall of 1939 spoke of the parallel nature of the "declaration of war against the outside" with the euthanasia decree and a "declaration of war internally."[32] Hitler himself made the connection quite clear with the "decree on euthanasia," which he assigned in 1939 to the manager of his chancellery, Philip Bouhler, and his personal physician, Karl Brandt. The decree was backdated to September 1, 1939, the day German troops marched into Poland. As shown as early as 1948 by the psychiatrist and psychoanalyst Alice Platen-Hallermund, Hitler with this decree on September 1 had shown that this was the "beginning of a new order for Germany" and "the beginning of the internal cleansing of inferior elements."[33]

In the context of the racist and eugenically founded assignment of the mentally handicapped to "lives not worthy of living," the rationale of saving money on care and on the creation of hospital beds gave a "practical" reason for the murder action. Supported by Herbert Linden, within a few weeks Bouhler and Brandt organized a central office at Tiergartenstrasse 4 in Berlin for the registration, selection, transfer, and gassing of the handicapped. The psychiatrists Werner Heyde and Paul Nitsche coordinated and supervised the physicians responsible for the "selection" of the victims. In August 1941, that is, at the beginning of what was called Action T4, the scheduled figure of 70,000 handicapped to be killed was reached. According to the official report of the centrally managed Action T4, physicians and caretakers in a decentralized "euthanasia action" killed a planned 100,000 handicapped in Germany alone by lethal injection or by starvation.

Pohlisch and Rüdin, the two psychiatrically oriented members of the International Federation of Eugenic Organizations, both participated in the euthanasia action. Pohlisch was one of 40 medical experts who looked at paper medical records and within a few seconds checked off with a plus or minus the life or death of a handicapped person. Like the Munich-based hygienist Fritz Lenz, he participated in discussions on the euthanasia law, which was to make legal the Hitler-ordered murder action. Pohlisch together with his colleagues considered this to be "a mercy death for a person who suffers from an incurable or strongly handicapping...illness."[34] Rüdin was informed of the killing of handicapped persons shortly after the beginning of action T4 but was never directly involved in the deaths. Still, in October 1942, he expressed his agreement with the "extermination of beings not worthy to live." In response to a questionnaire from the minister of health regarding especially urgent research questions during the war, Rüdin suggested the question as to which children at a very young age "could unquestionably be characterized as clinically and hereditarily biologically (based on descent) candidates for elimination," and as a result they would be "recommended for euthanasia in their own interest as well as in the interests of the German people."[35]

Even though only the psychiatrists among the German race hygienists were directly involved in the killings, the race hygienists as a group as well as their foreign colleagues contributed significantly to reducing the level of inhibition against killing the handicapped. Without the eugenic ideology as justification, the participating psychiatrists, physicians, and medical personnel would have had no scientific justification for the killings. Without the reduction of human beings to "excess baggage" and to "empty human hulks," pressure for justification for the medical murders would have been much stronger.

Even before 1939 the German race hygienists tolerated exceptions to their rejection of euthanasia, but now they became outright supporters. This change of heart was directly related to the outbreak of the Second World War. The counterselection related to the war, against which the race hygienists had warned again and again, threatened to destroy the "race hygiene work of construction" of the National Socialists; it legitimated the radicalization of negative eugenics. While in peacetime the scientifically grounded genetic distinction of a particular race stood in the foreground, with the outbreak of the war another aspect came to the front—the subdivision of the population into "inferior" and "superior" beings. Because of the counterselective effect on "higher-value" individuals, the war could no longer guarantee the best possible results, and suddenly radical, scientifically questionable countermeasures came into the discussion.

Without taking a direct position on the secret killing of the mentally handicapped, race politicians and race hygienists during the Second World War again and again demanded a heightening of population and race political measures. Walter Groß declared that the "eradication of hereditary diseases" and the "incarceration of the asocial types" were urgently necessary in order to assure the "productive capability" and "inner strengrh" of the German people. Groß expanded his remarks by saying that if "tens of thousands of young men and thereby very worthy hereditary stores are lost, [then the] eradication of non-virile and burdensome groups is doubly important."[36] Friedrich Keitel, in 1941 promoted to associate professor for race biology at the University of Würzburg, called for meeting the counterselection of the war with increased efforts in the population policy and race political areas.[37] In the same sense, Rüdin's coworker at the Kaiser Wilhelm Institute, Hermann Ernst Grobig, declared that race hygiene and race political measures in no way should be subordinated to what were considered as more important war policies. Just the reverse! According to Grobig, the war called for greater accomplishments in the race policy area.[38] In 1944 Otmar Freiherr von Verschuer expressed satisfaction that the "purification" of the German "physical body" from Jews and Gypsies in recent years had been completed recently through the race policy measures of the National Socialists. It was in this context that Verschuer understood Second World War to be a "race war" in which the final conflict was taking place with "world Jewry."[39] Friedrich Burgdörfer felt that the war had placed an obligation on every German to improve the race value of the German people. Every hereditarily healthy married couple that did not take advantage of its reproduction duties was committing "völkish running from the field," an action that was no less humiliating and embarrassing than was military flight.[40]

The attitude of German race hygienists to killing the sick must be seen in this context of "race hygiene mobilization." The slaughter of the handicapped and psychologically ill as the end toward which eugenics was driving became conceivable at the moment when the supposed "counter-selection by the war" seemed to make radical measures necessary in the eyes of the German race hygienists. The destruction of the peace utopia of eugenicists allowed these actions, which previously had been considered too extreme, now seem to be necessary.

German Race Hygiene in the Second World War

At the beginning of 1941, Eugen Fischer wrote a basic paper regarding the future of his Institute for Anthropology, Eugenics, and Human Heredity Theory where he laid out that the war and "the mighty expansion of the Greater German Empire" also posed "great new missions" for human geneticists and race researchers. In the light of the new challenges, research in the area of "hereditary health, the race, human selection, and effects on the environment" had two tasks. From one angle, he maintained that research "had to be placed directly in the service of the people, the war, and of politics," but in addition to applied research, basic research should not be neglected. This basic research could in the long run also serve race policy. No one could have known that Gregor Mendel's studies on peas would at some time in the future form the basis for legislation on hereditary health and that Fischer's own "bastard studies" would underpin National Socialist race legislation. Since any race policy was based "on the reference to the heritability of racial characteristics," on race policy grounds, human genetic basic research had to be supported alongside race anthropology research.[41]

Fischer, who was involved in the renaming of his institute to the Kaiser Wilhelm Institute for Heredity and Race Science, set up two new departments in his institute after the beginning of the war. Wolfgang Abel, party member and member of the SS, took over the section for race science. Hans Nachtsheim, one of the few geneticists who did not constantly parrot the Nazi party line, became the director of the section on hereditary pathology.[42]

The members of the Kaiser Wilhelm Institute did not hesitate to make use of new "research material" that resulted from the war. In 1943 the head of the SS approved a project of the anthropologist Wolfgang Abel to carry out anthropological studies on Russian prisoners of war in the Sachsenhausen concentration camp.[43] Otmar Freiherr von Verschuer, who in 1942 took over the directorship of the Berlin Institute from Fischer, starting in 1943 received samples of blood and organs from his one-time assistant Josef Mengele in Auschwitz. Mengele entered the SS as a physician after completing his doctoral work with Verschuer on the heritability of cleft palate and was encouraged by Verschuer to exploit his position as the camp physician in Auschwitz to follow up on the possibilities of human genetic research. Many of the twins that he used for his gruesome medical experiments in Auschwitz died.[44]

During the war most international scientific relationships were broken off. The close working relationships of German race hygienists, geneticists, and population scientists that had existed up to 1939 with foreign colleagues were lost in wartime. To be sure, individual American eugenicists, such as Lothrop Stoddard and T. U. H. Ellinger, visited Germany, but as a whole, Germany was scientifically isolated by the war.[45]

Contacts between the German race hygienists and race researchers were limited to areas occupied by Germany. Since research on human heredity and eugenics was one of the cornerstones of National Socialist science policy, cooperation of the race hygiene scholars in Germany was extremely important for the new order of the European scientific landscape. The initial German successes in the war and the proposed greater German Reich demanded planning for science policy in the conquered territories and a new definition of relationships to international scientific organizations.[46]

German race hygienists and heredity researchers played a central role in advertising scientific and political accomplishments in the conquered territories and in the countries that cooperated with National Socialist Germany. Directly after the resumption of scientific contacts with the Soviet Union in October 1940, Fritz Lenz and Eugen Fischer were requested to report in Moscow on the status of race hygiene and human hereditary theory in Germany.[47] In Hungary, in February 1942, Fischer gave papers on "the concept of race" and "heredity experiments in the service of medicine."[48] In Croatia, in April 1942, he gave a lecture and at the same time advised the Croatian minister of education on the organization of an academic chair for anthropology and race hygiene. Since at that moment there were no appropriate German candidates for such a position because of the founding of similar academic chairs in Strassburg, Prague, Graz, and Posen, Fischer felt compelled to suggest a Croat for the position.[49]

Fischer and Verschuer both gave lectures in occupied Paris.[50] Starting in 1940, Fischer was involved in the dissolution of the Parisian Institut International d'Anthropologie, founded in 1917 and supervised since 1926 by the former French Eugenics Society as an independent section. In a report to the Ministry of Education, he called for the "final liquidation" of the institute, which had always been a thorn in his side. That institute was oriented to cultural anthropology, and its eugenics members took a neo-Lamarckian position; its longtime director Louis Marin had been a constant critic of the Nazis. Fischer assigned his one-time assistant Horst Geyer, a neurologist and member of the SS and the NSDAP, to work with the occupation authorities to provide proof of the anti-German past of this "notorious" institute so that it would "forever disappear from the face of the earth."[51]

The dissolution of the Institut International d'Anthropologie formed part of a program of the Ministry of Education to "reorganize international scientific federations" after the beginning of the war. As part of National Socialist foreign and cultural policy, the dominant influence of Germany in these federations was to be assured.[52] While scientific federations "inimical to Germany" were to be dissolved or transferred to Germany, organizations where Germany had the dominant influence were to be maintained in their existing form. The National Socialist science bureaucracy and the German race hygienists involved in the discussion considered the German influence in both the International Union for the Scientific Investigation of Population Problems and in the International Federation of Eugenic Organizations to be strong enough so as not to require a transfer to Germany.

After 1945 members of the International Union were pleased to point to the successful wartime resistance of the then Parisian general secretary Georges Mauco to the shift of the headquarters of the Union to Germany.[53] That the central office of the IUSSIP remained in Paris was however not to be traced to Mauco's opposition, but rather to Nazi confidence that they retained control of the organization. In an internal report of the Reich Ministry of the Interior, it was stated that "with the successful outcome of the war for Germany, the leading role of Germany in the Union was assured," and the originally planned transfer to Germany was thus not necessary.[54]

The National Socialist authorities were even more certain of German influence on the International Federation of Eugenic Organizations. On the basis of a report by Fischer on the significance of German race hygienists in the IFEO, the organization was formally left in Sjögren's hands. The self-satisfied report of the Ministry of the Interior noted, "At the present time the president of this Association is Dr. Sjögren in Göteborg, who is extremely friendly to Germany." Since there was no "French influence" in the federation, a transfer to Germany was not necessary.[55]

Reorientation of Eugenics Outside Germany

The radicalization of National Socialist race policy accelerated the modernization of the various European and American eugenics organizations. In Scandinavia, the orthodox eugenicists such as Mjöen, Sjögren, and Lundborg kept losing influence, and social democratic welfare state oriented eugenicists such as Dahlberg, Mohr, and Kemp steadily gained political power. The eugenic sterilization laws in Sweden (1934), Norway (1934), Finland (1935), Estonia (1937), and Iceland (1938), as well as the heightened Danish sterilization law of 1935, already bore the imprint of social democratic eugenicists.[56] In The Netherlands G. P. Frets had significant influence on the Dutch eugenics society with his liberal ideas. In Great Britain, the reform forces around C. P. Blacker and Julian Huxley increasingly displaced the orthodox eugenicists such as Darwin and Hodson. In the United States the reform eugenicists around Osborn took control by the end of the 1930s.

In contrast to the orthodox eugenicists, who represented an international viewpoint within the IFEO, the reform forces within the eugenics societies in the main developed in national isolation. The reform eugenicists lost the motivation for the transnational organization of the "white race" and for internationalization as an instrument to make eugenics into a science. It was only at the end of the 1930s that closer contacts emerged between reform eugenicists of the various national eugenics movements. It came as a great surprise to Blacker when he became aware at the Paris World Population Conference in 1937 of how extensively his views coincided with those of the American reform eugenicists who were present—Frederick Osborn, Frank Lorimer, Warren Thompson, and Frank Notestein. In comparison to Leonard Darwin, Blacker observed that it was astounding how similar the eugenics societies had developed in Great Britain and the United States. With reference to the political agreements between British and American reform eugenicists, Osborn noted that he and Blacker could exchange positions, and no one would notice the difference.[57]

Since contacts among the British, Dutch, American, and Scandinavian reform eugenicists developed only slowly, they did not develop a unified position either regarding the National Socialist race policies or regarding the International Federation of Eugenic Organizations. As a rule they followed the "great eugenics experiment" in Germany with great interest, but very early on in various ways they expressed their skepticism about the racist orientation and the compulsory character of the German policy. The Dutch eugenicists G. P. Frets and Jacob Sanders expressed their criticism of German racial policies mainly within the IFEO. The social democratic Swedish eugenic group around Dahlberg coordinated their criticism with the American socialist eugenicists such as Landauer and Muller. Blacker and other reform-oriented British eugenicists feared in particular that the sterilization law introduced by the Nazis would harm their campaign for a sterilization law on a "voluntary" basis. Huxley and Blacker indefatigably communicated to the British public that one should not paint British eugenics with the same brush as the racist oriented eugenics in Germany.[58]

Unlike their colleagues in Europe, the reform eugenicists in the United States followed the eugenic aspects of National Socialist race policy not only with interest but also with a certain sympathy.[59]

Frederick Osborn also combined sympathy for the eugenic measures in Germany with a decided rejection of the Nordic race theories. While he may have agreed with Franz Boas's evaluation of the "absurdity of the pseudo-scientific theories of the Nazis," he also praised the eugenic measures in Germany as "perhaps the most important social experiment which has ever been tried."[60] In that same year, 1938, while working with Franz Boas to found a population committee directed against the Nordic race theory, he also warned the annual meeting of the American Eugenics Society that one should not damn the "excellent" sterilization program in Germany simply because of its National Socialist origin.[61]

The sympathy expressed by Osborn for the eugenic measures in Germany receded as the National Socialist race policy became more radical. It also slowly became clear to Osborn that what they had considered to be exemplary eugenic policies in Germany were actually a central building block of the National Socialist racist "policy of a distinctive race." As in The Netherlands, Sweden, Norway, and Great Britain, developments in the United States were influenced by what was happening in Germany, leading to a polarization within the eugenics movement. In the end it led to an accelerated modernization of the eugenics societies. Eugenicists such as Mjöen, Lundborg, Nilsson-Ehle, Pitt-Rivers, Hodson, Campbell, and Laughlin had tied the fate of their own eugenic concepts to the future of National Socialism by unconditionally supporting Nazi race policies. With German race policy becoming more radical and with the outbreak of the war, they were discredited not only within the international scientific community but also within the reform eugenics societies.

By around 1940 the reformers had taken over the eugenics movements in Great Britain, the United States, The Netherlands, and the Scandinavian countries. When Frederick Osborn took over the editorship of *Eugenical News* from Davenport, Laughlin, and Steggerda in 1939, the editorial point of view changed overnight. Osborn thought that the earlier editions had included "unscientific" material and had seriously harmed the acceptance of eugenics in the United States.[62] In the first edition under his control, Osborn, in an article titled "The American Concept of Eugenics," outlined a eugenics program that included the complexity of genetic processes in political considerations. Osborn made clear that there were no scientific proofs for the view, still represented by his own uncle Henry F. Osborn, that races differed genetically. Nor were there any basic genetic differences between the various classes. Consequently, he went on, a modern eugenics must attempt to influence the individual birth rate without any regard of a particular race or class.

Instead of reaching for forced measures, eugenicists should attempt to improve the conditions for reproduction for "human beings with above average genetic tendencies." At the same time, people who did not have these tendencies should be given easier access to contraceptive means so that they could reduce the number of their children in the framework of a "new eugenic social consciousness." Such a eugenically oriented system of marriage and health consultation, together with proper social and medical supports, appeared to Osborn and his colleagues as more proper than the forcible measures of the National Socialist dictatorship.[63]

Scandinavian countries served Osborn and his colleagues as a model where in their opinion a welfare state social policy had been successfully joined with eugenic measures. The reform eugenicists showed special interest in Sweden, where social democratically oriented eugenicists such as the geneticist Gunnar Dahlberg, the economist Gunnar Myrdal, and the sociologist Alva Myrdal helped develop a qualitative and quantitative population policy on a "voluntary" basis. Forced sterilization, as Alva Myrdal wrote in an article in 1939 for the *Eugenical News*, should be limited to a very small, clearly "biologically inferior" group. The number of children to people whose "genetic value" was dubious should be reduced essentially by propaganda and indirect pressure.[64]

The change of strategy in Scandinavia, The Netherlands, Great Britain, and the United States in no way meant a departure from the founding principles of eugenics. Outdated, politically and scientifically discredited race and class prejudices were thrown overboard, while the methods to achieve eugenic ideals were refined. Eugenics was embedded in a democratic and welfare state design. With the exception of Norway and Sweden, where the split between the orthodox and social democratic eugenicists had come very early, in the other eugenics societies there was no break between the reform forces and the eugenicists of the first generation. The eugenic founders, who had made policy of the societies over many decades, kept their place of honor. Modifying the policies of Laughlin, Davenport, and Darwin did not mean to Blacker and Osborn that they should keep their distance from the old masters. Quite often, only after a "biological solution"—that is, the death of a prominent eugenic personality—were that person's earlier hypotheses questioned.[65]

Chapter Seven
On "Good" and "Bad" Eugenics: Refocusing on Human Genetic Counseling and the Struggle against "Overpopulation"

In July 1948, the management board of the French Institut National d'Études Démographiques (INED) inquired of the British Eugenics Society as to whether the British eugenicists could undertake the reestablishment of the International Federation of Eugenic Organizations.[1] The interest of the demography institute in the revival of the international eugenics organization did not come out of the blue. The INED was an offshoot of the Fondation française pour l'étude des problèmes humaines, started by Alexis Carrel, a French Nobel Prize winner and champion of National Socialist race policies.[2] The founding document of this Institut National of October 24, 1945, had expressly called for studying the means to bring about "quantitative growth and qualitative improvement of the population." The course of the institute was set primarily by two French eugenicists, Alfred Sauvey, the director of the INED, and Jean Sutter, the head of the division of qualitative demography.[3]

Blacker, the undisputed leading postwar figure of the British Eugenics Society, showed little interest in a revival of the International Federation of Eugenic Organizations. For him, this organization was the expression of an outdated version of eugenics that had no justifiable place in the postwar period. Still, Blacker informed the erstwhile general secretary of the international eugenics organization, Cora Hodson, about the interest of the French eugenicists. He asked her if she were in a position to follow up on the initiative of the French population scientists. Hodson was however so severely ill and so worn down by the criticism for her support of the Nazis that she absolutely refused to take on responsibility for reestablishing the international eugenics organization.

After the attempt of the French eugenicists had collapsed, one year later the eugenicists Corrado Gini and Ruggles Gates tried to revive the International Federation of Eugenic Organizations. They hoped to bundle the various eugenic initiatives of the postwar period under their leadership through the resurrection of the international eugenics organization.[4] Gini brought up this idea in 1950 in Rome at the Congress of the International Institute of Sociology, but it foundered on the refusal of many eugenicists. In spite of their sympathy for eugenics, human geneticists such as Tage Kemp and Laurence Snyder felt that the time was not yet ripe. Kemp explained, "Eugenics before and during the World War had been so misused" that at the present time it was not tactically a good idea to rebuild an international eugenics organization.[5]

The two unsuccessful attempts to revive the IFEO showed how far orthodox eugenics had been discredited by the war. When the Nazi horrors committed in the name of eugenics and racial purity became known, the entire classical variant of eugenics fell into disrepute among scientists and in the general population. Eugenicists in the tradition of Ploetz, Davenport, and Darwin could hardly hold on to influential positions in postwar Germany. In National Socialist Germany there was no independent development in race hygiene, psychiatry, human heredity research, and demography, and the orthodox German eugenicists ran into internal opposition when they attempted to take on leading positions after the war. For this attempted resumption of leadership positions, they used the international relationships that they had built up under the international federation.

Even if the eugenicists in the tradition of the International Federation of Eugenic Organizations could hold onto their positions only in Germany, this by no means signifies that one can conclude that there was an immediate and general decline of eugenic ideology after the Second World War. Hitler's policy was not, as the American scholar Thomas M. Shapiro claims, the death knell for eugenics as a whole.[6] The eugenic policies of uprooting and differentiating by the Nazis were responsible only for an accelerated decline of the racist variant of eugenics. A resolution of leading geneticists and anthropologists within UNESCO against the scientific legitimization of racism shows that the close relationship between race research and genetics increasingly fell apart after the Second World War. Even if this transition was not problem free, and even if some geneticists continued to think it was possible that the races differed genetically according to their intellectual qualities, the declaration of the scientists was a clear sign of the change in general opinion after the war.

Interestingly, the major proponents of this resolution were the driving forces behind various eugenic initiatives and movements that continued to exist after 1945, particularly in the United States, Great Britain, and Scandinavia.[7] The eugenics societies dominated by the reform eugenicists were especially active in Great Britain and the United States. These eugenicists tried with all their might to set off the "good" eugenics that they represented against a "bad" eugenics that had been corrupted by the Nazis and by racist "pseudoscientists." By a firm rejection of traditional race research, the American Eugenics Society gained influence among human geneticists and demographers. Since eugenics of the new type no longer was claiming to be a separate academic discipline, after the war the leading eugenicists supported the establishment of independent scientific societies in the area of human genetics and demography.

A New Beginning That Really Wasn't: The Race Hygienists in Postwar Germany

Toward the end of October, 1952, Kurt Pohlisch, one-time member of the International Federation of Eugenic Organizations and one of the expert advisers on the T4 murder campaign, stood at the grave of the recently deceased Ernst

Rüdin. To the colleagues assembled at the funeral, he eulogized Rüdin as one of the "outstanding founders of eugenic research in psychiatry." He went on: People with the "enormous energy" of Rüdin accumulated not only friends and colleagues, but enemies as well. Whenever one tries to help other human beings, this is always regarded variously by some people. Among the National Socialists, there were those who had taken the ideas of race hygiene in a direction quite different from the one desired by Rüdin. For Pohlisch, that is where the "deep tragedy of eugenics" and the deep tragedy "in the life of Ernst Rüdin" lay. On an optimistic note, the speaker closed with the remark that just as Rüdin in the years after the war had overcome the "tragedy of his life," so the "surviving world would balance out the history and the substantive criticism from that which was tragic in eugenics."[8]

Even as Pohlisch was making his speech at Rüdin's graveside, many of the race hygienists active under the Nazis were assuming or were about to assume professorships in human genetics, anthropology, and psychiatry. Since 1947 Fritz Lenz had held a professorship for human genetics in Göttingen with the permission of the British military government.[9] Despite massive criticism for his ties to the concentration camp doctor Josef Mengele, Otmar Freiherr von Verschuer had become full professor of human genetics in Münster in 1951. Hans Nachtsheim, someone regarded by the American authorities as politically unburdened and supported by Jewish immigrants, such as Richard Goldschmidt and Hans Grüneberg, from 1949 on occupied the chair for human genetics at the newly founded Free University of Berlin. He also continued to direct the Berlin based Kaiser Wilhelm Institute, which kept functioning as a much smaller institution.[10]

The second ranks of German race hygienists as well could continue their careers unhindered. The National Socialist race hygienist Heinrich Schade, Verschuer's assistant in Frankfurt and Berlin, was helped by his mentor to become associate professor for human genetics at Münster in 1954, and starting in 1964 directed the Institute for Human Genetics and Anthropology in Düsseldorf. Hans Weinert, a race researcher protected by Walter Groß (NSDAP official), since 1937 had been a Nazi party member and at one time a scientific member of Kaiser Wilhelm Institute in Berlin, held on to his professorship in Kiel from his appointment in 1935 without interruption until retirement in 1955. Lothar Loeffler, party member since 1932 and director of the Race Biology Institute in Königsberg and then professor for race biology and race hygiene in Vienna, in 1953 was able to take the post of professor for social biology at the Technical University of Hanover. Siegfried Koller during the war had cooperated in a draft of a "law on the loss of völkisch civil rights for the protection of the national polity" and then rose to the post of professor in Berlin in 1944. Following his transitional employment as division director of the Federal Statistical Office in the 1950s, he became a professor in Mainz. The race hygienist Karl Valentin Müller received a teaching position at the University of Dresden in 1938 and in 1941 was promoted to professor of sociology at the German university in Prague. Directly after the war he became director of the Institute for Research on the Gifted for the Ministry of Education of Lower Saxony. In 1955 he was named to a professorship in the Nuremberg College for Economic and Social Sciences.[11]

Even scientifically isolated race researchers such as Friedrich Keiter and Wolfgang Abel and humanities-oriented race theoreticians such as Hans F. K. Günther and

Ludwig F. Clauß plus many other leading German race hygienists continued to be influential after 1945. They were successful in bringing along their assistants from the Nazi period with them into positions that opened up. It is therefore not surprising that in the reconstruction of human genetics, anthropology, and psychiatry after the war many of their previous doctoral students and assistant professors became full professors. Horst Geyer, formerly a scientific worker at the Kaiser Wilhelm Institute in Berlin and later assistant to Loeffler in Vienna, after the war became professor of psychiatry at the University of Oldenburg. Hans Grebe, member of the NSDAP since 1933 and assistant to Verschuer in Frankfurt and Berlin, after the war was named to the professorship of human genetics at the University of Marburg. Gerhard Koch, member of the Nazi party and the SS since 1932 and an external scientific worker at the Kaiser Wilhelm Institute in Berlin from 1942, after the war became director of the Institute for Human Genetics in Erlangen. Peter Emil Becker, received his doctorate under Fritz Lenz and joined the NSDAP in July 1940; he followed his doctoral supervisor to Göttingen and in 1957 became director of the Institute for Human Heredity Research. Wolfgang Lehmann, party and SS member and professor for race studies in Strassburg (*now Strasbourg*) from 1943, directed the Institute for Human Genetics at the University of Kiel from 1948 to 1975.[12]

How is it possible that the German race hygienists, who so frequently occupied central positions in the hereditary health and race policy of the National Socialist regime, were able so quickly to reconstruct their scientific network? In postwar Germany there were in fact no alternatives to the academics active under National Socialism. Any development independent of the Nazis did not exist, particularly in the politically sensitive disciplines of race hygiene, psychiatry, human heredity, and population research. Because of the likelihood of clashing with their earlier colleagues, the academics who had been exiled in the 1930s had no motivation to return to their previous positions. And so the military authorities had the choice of either dissolving entire areas of academic studies such as human genetics, demography, psychiatry, and anthropology or allowing the academics active under the Nazis to return to their old positions.

The impact of the early days of the Cold War, the conflict between the United States and Great Britain with their one-time ally Soviet Union, allowed many denazification processes in Germany to fall by the wayside. The Lysenko affair—the imposition of a Lamarckian position in the Soviet Union by political force—also deflected attention from the central role that the German psychiatrists, population scientists, and human geneticists had played in National Socialism.

In this climate the German race hygienists used the earlier international recognition of National Socialist hereditary health policies to justify their actions between 1933 and 1945. The American military authorities charged that Rüdin, as an acknowledged authority in the area of race hygiene, had covered up National Socialist race policies and had contributed significantly to the strengthening of the rule by terror. He responded by pointing to international agreement to the National Socialist eugenics policies; according to him, the internationally recognized race hygiene legislation in Germany had nothing to do with the excesses of the Nazis. Race hygiene legislation had been in line with strictly scientific criteria, and the "extreme excesses" had been instituted without his knowledge or against his will.

In Rüdin's interpretation, the National Socialists had "burdened" a classical race hygiene that he represented with the "odium of race hatred." The Nuremberg racial laws, the killing of the sick and handicapped, and the extermination of religious and ethnic minorities were according to him not the result of race hygiene or race anthropology theories, but were perversions that in the end only harmed his research. He continued his self-justification by saying that race hygiene had been led by several representatives of National Socialism "completely in the opposite direction." Instead of becoming an honorable area of study, "it had been led down the road to "criminal modes of thought."[13]

At Rüdin's denazification trial, his attorney rejected the accusation that "pseudoscientific activity" was a reference to the international eugenics movement. He asked the rhetorical question of what would be said by the "researchers of all culture nations" who had supported the Kaiser Wilhelm Institute in Munich, such as the Rockefeller Foundation, or by other members of the International Federation of Eugenic Organizations.[14] The entire array of events of international recognition for National Socialist race policies was trotted out to serve as a defense—from the Zurich IFEO resolution of 1934, which had promoted the worldwide political use of knowledge gained from hereditary biology, population policy, and race hygiene, to the enthusiastic support from Sjögren and Hodson for the German hereditary biology and hereditary eugenics research, down to the hoped-for Nobel Peace Prize for Ploetz.

Faced with a possible guilty sentence from the American authorities, Rüdin molted from a convinced National Socialist into a hard-line opponent of the regime. On the grounds of his "resistance to the SS and Gestapo and against measures of the party," Rüdin claimed to the military authorities that he had had to suffer many personal blows. He had been harassed and threatened; his assistants were able to complete their habilitations only with great difficulty; and the Kaiser Wilhelm Institute in Munich had lost much of its research money because of the criticism he had leveled at National Socialism. In the end, Rüdin went on, the "many years of anxiety because of the SS and Gestapo" had led to severe damage to his health, including a case of angina pectoris.[15]

Rüdin was certainly a very striking example of the belated self-mutation from a convinced National Socialist to a resistance fighter, but his justification strategy was used by many of his colleagues. Sterilization laws in Scandinavia and the United States, the recognition of the eugenic aspects of the National Socialist race policies by the IFEO, and the great success of the Berlin Population Congress helped the German race hygienists to separate an internationally accepted eugenic core of German race policies from the supposed "excesses" of National Socialism.

Shortly after the end of the war German race hygienists attempted to use foreign colleagues to set up the myth of the supposedly misused science and in this way to rehabilitate themselves from a scientific point of view. Obviously in agreement with each other, Verschuer, Lenz, and Rüdin turned to meetings of foreign heredity researchers, whom they knew from international scientific congresses or from cooperative work in the International Federation of Eugenic Organizations and the International Group for Human Heredity. They requested their foreign colleagues for testimonials regarding their scientific qualifications. They hoped in this way to be able to convince the American and British military parties to allow research in the established framework.

Fritz Lenz lamented to Ruggles Gates the dysgenic effects of the World War, and expressed his hope that scientific relations would quickly be normalized. Unfortunately, he went on, the National Socialists had "misused" the concept of eugenics, but he had great hopes that eugenics could be revived independent of political prejudice.[16] Verschuer shamelessly presented himself in 1946 in a letter to various prewar foreign colleagues as "a public opponent of the National Socialist race fanaticism."[17] In a letter to Ruggles Gates he asked to be quiet "about all the horrible things that we had to experience in the meantime." He expressed the hope that "the many things separating us could be overcome by taking account of our common interests that existed earlier through the pursuit of similar scientific goals and through personal encounters."[18] To Muller he emphasized that it was his "most urgent wish" to cleanse science "of the cloud that had descended over it" due to experiences under National Socialism. In rebuilding German genetics and human genetics, individual improvements would not suffice; it would be necessary to lay a new foundation. According to Verschuer, it was unfortunate that the American and British military parties did not give sufficient flexibility—not a single German human geneticist was able to continue his work unhindered; no research centers had been completely rebuilt; not a single university chair had been set up again.[19]

The foreign scientists reacted variously to the pleas of their German colleagues. Anti-fascists such as Otto Mohr, Gunnar Dahlberg, J. B. S. Haldane, and Hermann Muller did not make a move to help their colleagues; the memories of the conflicts with the German race hygienists before the war were still foremost in their minds. Otto Mohr commented that the "damn German race hygiene" had formed the core of Hitler's race ideology. Verschuer and his German colleagues could shrug this off. After all, rather than acknowledging their guilt, they were merely asking their foreign colleagues for help in rebuilding their careers.[20] Other human geneticist and eugenicists had fewer scruples about protecting their German colleagues. For example, Ruggles Gates and Tage Kemp willingly testified to the scientific integrity of National Socialist race hygienists such as Lenz and Verschuer. Gates expressed his hope in a letter to Lenz that the German would "soon be able to resume his eugenic work." The improvement of the hereditary patrimony of all peoples was for Gates "fundamental for the future of the human race."[21] Kemp too wrote a positive recommendation for Lenz to the effect that Lenz's research seemed to be important for the future of medical genetics and should be supported to the greatest extent possible.[22] Verschuer received from the Danish eugenicist a testimonial to his "noble character." Kemp called Verschuer's work on twins extremely valuable for medicine and genetics.[23]

The UNESCO Position Paper on the Race Question: The Temporary End of Orthodox Eugenics

The German war of aggression and the racially motivated crimes of the National Socialists had shown what a consistently applied, race oriented eugenics could lead to. The worldwide horror at the mass murder led to a fundamental change in opinion regarding orthodox eugenics, which had been shaped by racist prejudices. Eugenics and race hygiene became in public opinion a synonym for the murderers in Hadamar (home of the euthanasia program) and Auschwitz. The social climate of the 1920s and 1930s had been quite well disposed to the race outlook of the orthodox eugenicists, but by now it had undergone a fundamental change.

Before the war, serious opposition had been unleashed at the criticism against race theories from academics such as Julian Huxley, Joseph Needham, J. B. S. Haldane, Hermann J. Muller, Otto Klineberg, and Ashley Montagu. Even after 1945 there was no unanimous support for this critique, but at least to the broad public, it remained the dominant scientific opinion. Research on races in the 1950s completely lost any significance not because there was new scientific knowledge in genetics but overwhelmingly because scientific race orientation had become a synonym for National Socialist race policies.[24]

The legitimation of political racism by scientists in the first half of the twentieth century now prompted the initiation of an international campaign against race prejudice by the United Nations Educational, Scientific, and Cultural Organization (UNESCO). UNESCO was founded in 1946 with the goal of furthering cooperation of countries in the areas of education, science, and culture for peace and security in the world. The UNESCO charter was anchored by human rights, regardless of race, sex, language, or religion. Under its first general director Julian Huxley, an international program of quantitative and qualitative population control took over. Huxley made it clear that after the experiences of Germany no radical eugenic policy had any prospects of being implemented in the foreseeable future. However, he furthered the idea that UNESCO should survey the possibilities for eugenics measures within the framework of a "scientific humanism," if only to now make "thinkable" what had previously been unthinkable.[25]

Huxley was unable to implement his vision of a progressive eugenics, free of race prejudices, in the framework of a scientific humanism, against the conservative and Catholic forces within UNESCO. Still, along with Joseph Needham, another signer of the Geneticists' Manifesto and division director of UNESCO for scientific questions, he set the science policy of UNESCO in its early years. The scientific humanism of Huxley and Needham formed the basis for the politically motivated attempt to scientifically discredit racism through UNESCO. Even when the two of them no longer held official positions at the time of the beginning of the UNESCO campaign against racism, the UNESCO initiative against race prejudice clearly showed their fingerprints.[26]

The goal of UNESCO was to start a scientific campaign against race prejudice in the sense of the 1948 Resolution 116 VI B of the United Nations.[27] The director of the division for social sciences, the Brazilian anthropologist Arthur Ramos, and his successor, the American sociologist Robert C. Angell, gathered together eight well-known cultural anthropologists and social scientists from December 12–14, 1949, to draw up a scientific position on the race question. Those attending included the Frenchman Claude Lévi-Strauss, the Mexican Juan Comas, the Englishman Morris Ginsberg, the Indian Humayun Kabir, and the American Ashley Montagu.

Geneticists were not represented at the meeting, but afterwards Montagu as facilitator integrated the remarks of the geneticists Dahlberg, Dobzhansky, Dunn, Conklin, Needham, Huxley, and Stern into the final version of the declaration.[28]

UNESCO took up its task as a continuation of the anti-racist and anti-Fascist efforts of Ignaz Zollschan in the 1930s. Faced with the National Socialist race policies, as pointed out by a background paper from UNESCO, Zollschan had had the idea of scientifically destroying the foundation of racism through an international position paper by anthropologists, geneticists, and social scientists. Despite the support from the later Pope Cardinal Pacelli and the International Institute for Intellectual Cooperation, the precursor of UNESCO, the project unfortunately fell victim to "the spirit of Munich," the policy of appeasement of National Socialism that set in during the mid-1930s. Fifteen years later the time was right to carry out the original idea.[29]

In the Statement on Race issued by UNESCO in July 1950, the eight scientists unambiguously declared that there is no scientific justification for discrimination based on race. In their view, race is not a biological fact, but a social myth, in whose name millions of people have lost their lives and as a result of which people had been prevented from exercising their fundamental rights. Since races often incorrectly have been made equivalent to national, religious, geographic, linguistic, and cultural groups, it was thought to make sense to drop the concept "race" and to substitute "ethnic group." Testing had shown clearly that the intellectual capacities of all existing ethnic groups were the same. Innate differences in intelligence, temperament, and social behavior between the various ethnic groups could not be shown scientifically. Under similar cultural and social conditions, people from various ethnic groups would develop similar abilities. Consequently, the ban on marriages between people from various ethnic groups could not be justified biologically. According to the signatories, comprehensive studies had shown absolutely that race mixing has no negative biological consequences. Neither physical disharmonies nor intellectual degeneration had ever been proved convincingly. In conclusion, the eight scientists remarked that biological studies should not support the idea of race hatred but should support the ethics of general brotherhood. The human being is a social animal, who can achieve his greatest potential only by cooperative interaction with fellow humans.[30]

On the occasion of the presentation of the declaration in July 1950, UNESCO tried to give the impression that this position paper reflected a broad scientific consensus. Although there were no new basic pieces of information on race research since the prewar period, UNESCO in a press release declared the rejection of any biological basis for race prejudices to be the conclusion of current knowledge garnered by biologists, geneticists, psychologists, sociologists, and anthropologists in this area.[31] The international press picked up this primarily politically motivated statement. The *London Times* published the headline "'Race' a Social Myth." The *New York Times* ran the headline "No Scientific Basis for Race Bias Found by World Panel of Experts."[32]

Despite the broad positive reaction among the general public, the declaration met massive criticism from a large number of scholars.[33] After only a short time following the publication, it was clear that no broad consensus existed among scientists on the race question in any form. British anthropologists and geneticists in particular attempted to undermine the basis of the statement. C. R. Blacker, who previously had tirelessly worked for eugenics free of race prejudice, criticized the premature conclusions of the resolution. He voiced the opinion that "subtle but nonetheless important inborn differences" between the races could exist with relation to their intellectual capacity.[34] A group of British scientists around the anthropologists Herbert J. Fleure, Sir Arthur Keith, and M. L. Tildesley criticized the resolution as a product of Ashley Montagu, one that even after subsequent reworking of criticism from American and British geneticists was still not convincing.[35]

Montagu, a student of Franz Boas, was a red flag for most British anthropologists. In 1942 he had published a condemnation of National Socialist race policies with a fundamental criticism of anthropological race ideas. Despite clear physical differences between ethnic groups, he maintained, human beings were impossible to distinguish from one another by race. The transitions between the groups were much too fluid. For him, the genetic differences within the groups were much more significant than the differences between the groups. Since race was merely a "dangerous myth," according to Montagu no science of race existed.[36] Since the UNESCO position paper essentially expressed Montagu's opinion as the position of most scientists, it threatened to remove scientific and political legitimation from the work of the race anthropologists.

The British anthropologists were successful in discrediting the UNESCO declaration with the reproach of inadequate genetic foundation. In particular they criticized the reduction of race concepts to a social myth, the replacement of the concept of "race" by "ethnic group," the negation of any genetically conditioned intellectual differences between "races," and the supposed tendency of human beings to cooperative behavior. These anthropologists protested against the social science and cultural anthropology basis of the resolution and demanded that the biological concept of race should be retained at least in its outlines.[37] Using the journal *Man,* the organ of the Royal Anthropological Institute, they forced UNESCO to rework the statement with significant participation by geneticists.

Both Montagu and Alfred Métraux, the anthropologists in charge of race questions at UNESCO, saw the danger of a new version of the declaration being a "victory for racism and defeat for naïve humanism." Nonetheless, they agreed to a reworking.[38] Both of the anthropologists were aware that a scientifically well-founded antiracist declaration by UNESCO required the broad agreement of the scientific community. In order to retain an antiracist core in any reworking, in June 1951 Métraux invited a select group of geneticists and physical anthropologists to draw up a second UNESCO declaration.[39] With Huxley, Dobzhansky, Haldane, and Dahlberg, all four signers of the Geneticists' Manifesto were involved. These were complemented by the two scientists Montagu and Dunn, who belonged to the group of the most active opponents of National Socialist race policy. The only German geneticist to participate in the reworking of the resolution was the politically unsullied Hans Nachtsheim.

Due to the antiracist basic position of a large proportion of the participants, the tenor of the first declaration could in large part be retained. The authors of the second declaration agreed that there was no scientific basis for talking about race unity or about the superiority of specific races. No convincing proof existed that the races differ in intellectual qualities such as intelligence, social behavior, or mental condition. In contrast to the first statement, the authors however gave up on replacing the concept of race with the concept of "ethnic group." In their opinion, the concept of race could be retained to describe groups with well-formed, heritable physical characteristics. In that way Montagu's statement that the concept that "race is a social myth" was rejected. In contrast to the social science authors of the first declaration, the geneticists and anthropologists were not prepared to get rid of the concept of race once and for all.[40]

The limitation of the second version was made clear by the criticism of the human geneticist Lionel Penrose. In 1945, Penrose took over the chair once held by Francis Galton and Karl Pearson at University College in London. In 1954, he changed the name of the journal of the Galton laboratory from *Annals of Eugenics* to *Annals of Human Genetics* and assumed a consistent anti-eugenic research position. Penrose declared his complete support for UNESCO's intentions, but he criticized the second declaration for holding on to a race distinction of human beings. In his eyes, the idea of human races was inexact and archaic. It belonged to an "unscientific epoch" and could no longer be used without causing confusion and opposition. If scientists like the signers of the declaration continued to retain the "mythical term 'race'," they would support superstition and prejudice in public discussion. In the interest of scientific and social progress, therefore, the use of the term "race" should be completely stopped.[41]

Confident that the second declaration with both its cautious balancing between genetic and environmentally related factors and the retention of physically based race factors would gain the support of a large majority of geneticists and anthropologists, Métraux asked over 100 scientists for their comments. In that way, he hoped that a declaration would not simply be a manifesto of a select group of geneticists and anthropologists, but would stand for the opinion of science as a whole.

With this provocative statement, UNESCO opened itself up to criticism from all sides. It was clear that the second declaration simply reflected the position of only a small portion of human geneticists and anthropologists.[42] This small group of scientists had succeeded in using the general antiracist feeling after the Second World War to force through a resolution on their own terms. However, the more the revelations about National Socialist crimes receded into the past and the more intense the discussion of the race question among scientists, the more did the originally clear rejection of the concept of race by UNESCO unravel. The moral outrage of scientists who were the driving force behind the first and second declaration increasingly became the subject of a detailed discussion of the supposedly scientific basis of differences among the races.

Even two signers of the Geneticists' Manifesto—Hermann Muller und Cyril Darlington—considered that differences between races in psychological and intellectual ability probably do exist. After the war Muller was named to a professorship in genetics at Indiana University in Bloomington; in 1946 he received the Nobel Prize for biology for his studies of mutations. He noted that the psyche as well as the physical body is influenced by hereditary material. If now the outer appearance of races could be determined by a frequently occurring gene, then the same conclusion could be drawn for the intellectual abilities of specific races.[43]

Darlington, who after the war proclaimed a close relationship between eugenics and race research, made similar comments. While Muller however attempted to move in concert with the sense of the UNESCO declaration toward a broad genetic agreement of all human beings, Darlington considered the antiracist efforts by UNESCO to be fruitless. According to the Oxford geneticist, there has never been any success in bringing together various races in the same environment. Any attempts to allow "Indians, Gypsies, or Negroes" to live in the same environment as Europeans had never been successful and would never have any success. Darlington adamantly denied that various population groups had similar genetic concentrations for the storing of intelligence. Just as one did not come upon "idiots" and "angels" in the same frequency in Milan as in Naples, so one could not attribute the same distribution of gifts among the various races.[44]

Several British heredity researchers issued criticisms similar to those of Darlington. After 1945, Sir Ronald A. Fisher distanced himself ever more from what he considered to be the too moderate view of race questions of the British Eugenics Society, and he emphasized that there was a "secure scientific basis" for the assumption that various human groups had various inborn capacities for intellectual and emotional development. The differences among the races should not be minimized, but people should learn to share the resources of the planet with humans of another type.[45]

The sharpest criticism came from the German anthropologists and geneticists, who in large measure completely rejected the declaration. Lenz complained to Nachtsheim that the UNESCO declaration was "shot through with ideological and political beliefs." He complained that scientific issues received short shrift. Since the declaration ignored the "enormous genetic differences between humans [as well as] the significance of poor selection for the decline of civilizations," it absolutely contradicted "eugenic science."

Lenz disputed that science had given evidence that human beings belong to the same species. "An unprejudiced scientist [comparing the] West African Negroes, Eskimos and Europeans living in the Northwest" would find it impossible to speak of the unity of humankind. On the basis of their racial instincts and their spatial isolation neither "African pygmies nor Bushmen would cross-marry Negroes or Europeans." They formed a distinctive race. According to Lenz, racial differences did not prevent him for having sympathy for other races: "I also have sympathy for chimpanzees and gorillas, and I am sorry that they face extinction, like other species of animals and so-called peoples of nature. I am also affected by the fate of millions of Jews, something that is very painful." All of this however did not prevent anthropologists and geneticists from "treating biological issues in a purely objective manner." Just a few years after the racially motivated mass murders in the National Socialist extermination camps, Lenz asked rhetorically why "it looked bad [for German anthropologists to] deal with the race question of the Jews in an objective manner." It was true that the Jews did not form a uniform race, but "on the average [they differ] racially, that is in hereditary characteristics, from the particular population among whom they are living."[46]

Karl Saller, one of the German anthropologists who had kept his distance from the National Socialists, criticized the rejection of mental differences between specific population groups. To him it was of secondary importance whether one called such population groups races or not. In the final analysis, however, all eugenics was based on such mental differences between population groups. He argued that even if one had to be very careful with the concept of race, in the light of "eugenic motivations" one would have to at the very least leave open the possibility of the existence of hereditary differences between various population groups.[47]

The Hamburg anthropologist Walter Scheidt, a convinced anti-Semite who went through a phase in which he supported the National Socialists, but then, like Saller, criticized the static race concept of many Nazis, did not hesitate to criticize the UNESCO declaration as adopting "the disastrous errors of the National Socialist period" with the same items turned on their head. Just as in National Socialist Germany and in Soviet Russia, so also UNESCO had resolved scientific questions by a political manifesto, and he as an anthropologist wanted nothing to do with that. UNESCO apparently was not ready "to give Germans credit for objective discussion." "Any type of objection from the German side against this manifesto" would certainly be interpreted there as a "Nazi residue."[48]

Eugen Fischer came out against any political definition of scientific positions. Just as National Socialism had attempted to set up certain hypotheses as the only correct "results of race research" and had attempted to suppress the opposite opinions, and in the same way that the Soviet Union was allowing Lysenko's theory of inheritance to be the only correct one, so now too UNESCO was trying to force through its position on the race question as the only correct one.[49] Fischer's one-time assistant Hans Weinert did not see in the declaration any justification for the persecution of other people because of "their race, religion, or politics," but he nevertheless was a harsh critic. Why was it that up to that time only "members of the white race" had produced scientific knowledge, if in fact "all races had quote similar abilities for intellectual developments"? He also asked, "Which of the gentlemen who had signed the declaration would be happy about having their daughter marry a Bushman or an Australian aboriginal?"[50]

Although Nachtsheim was upset regarding the comments made by his German colleagues against the UNESCO declaration, during this conflict over the declaration he adopted part of their line of argument as his own. In opposition to the American geneticist Alfred H. Sturtevant, he retreated from his original support for the second UNESCO declaration: "It appears to me to be absolutely necessary to oppose the Nazi standpoint, but it is no less necessary to oppose the opposite extreme, whereby all mental differences between human races are to be ascribed to the environment."[51] Nachtsheim supported Hermann Muller, who wanted to remove from the declaration the denial of all intellectual differences between human races. Muller wanted to prevent the notion of the negation of racial differences in an otherwise "praiseworthy declaration" from reaching the public as the authoritative consensus of the universe of the best and most representative thinkers in genetics, anthropology, and related fields.[52]

The coalition of German anthropologists, racist hardliners from Great Britain, and one-time signers of the Geneticists' Manifesto caused the UNESCO declaration to collapse. In a letter to Nachtsheim, Métraux appeared surprised about the change in the discussion. The position of the German anthropologists had put his hopes for "harmonious cooperation" with German scientists to a difficult test. He felt insulted by Fischer's equation of the UNESCO initiative with Communist and National Socialist methods and Weinert's racist assumptions about the behavior of colleagues with regard to marriage with Bushmen or Australian aborigines. The "insinuations and the doubts by people like Walter Scheidt cast upon the intellectual honesty of their colleagues in the democratic world" were deeply wounding to him. Métraux could not understand why suddenly a common front had been formed of German anthropologists and convinced racists like Darlington with such "honest" scientists as Nachtsheim and Muller. In opposition to Nachtsheim, he declared that it would be better to "throw the entire declaration into the trashcan" and to declare to the world that the scientists on the committee believed in the "inequality of races" than to have the passage about the equal intellectual capacity of ethnic groups changed or deleted.[53]

The American geneticist and eugenicist Theodosius Dobzhansky consoled Métraux by noting that it was an extremely difficult undertaking to bring all scientists to a uniform position on the race question. There was simply no common denominator between "Darlington, Fisher, and Sturtevant on one side and shall we say, Dunn, Haldane, and Dahlberg on the other." A declaration that could unify such different scientists would really be "a blank sheet of paper."[54]

Finally, in July 1951, UNESCO published the declaration as originally composed and added comments and critiques as an appendix. Even if the declaration could not be seen as a unanimous rejection of political racism by the scientific community because of the limitations in the appendix, it did illustrate the changes that had taken place since 1945. Before 1939 scientists such as Montagu, Dahlberg, and Dunn stood alone in their criticism of race anthropology, but after the Second World War they exerted massive influence on the internal scientific discussion. Their position became ever more dominant based on the altered climate in the scientific community.

The "Descientizing" of Eugenics—And the Formation of Various Specialized Scientific Disciplines

The driving forces of the eugenics movement in Scandinavia, Great Britain, Germany, and the United States after 1945 still held certain basic principles in common with the orthodox eugenicists—belief in a thoroughgoing genetic determinism of human beings and the necessity of improving the human hereditary patrimony. Nevertheless, they more and more distanced themselves from the traditional ideas about race and heredity. The breaking away of the reformers from the classic eugenics à la Davenport, Ploetz, and Darwin was crystal clear in the fact that they belonged to the harshest critics of the racist agenda once laid out by orthodox eugenics. Nachtsheim, the German geneticist involved in the UNESCO declaration, was the most zealous propagandist for a eugenics policy in postwar Germany. Tage Kemp, one of the supporters of the UNESCO declaration, was quite influential in the eugenic orientation of Danish social policy after 1945. Laurence H.

Snyder, one of the leading geneticists in the United States and a critic of orthodox eugenics, at the end the 1940s, was director and later vice president of the American Eugenics Society. Harry L. Shapiro, an anthropologist at the American Museum of Natural History and one of the sharpest critics of a racist oriented eugenics, in 1956 was elected president of the American Eugenics Society.[55] Theodosius Dobzhansky, a coauthor of both UNESCO declarations of 1950 and 1951, allowed himself to be named a director of the American Eugenics Society after the war.

For these eugenicists, it was crucial to set off their policies from the National Socialist crimes. In a fashion quite similar to that of scientists in postwar Germany, they attempted to distinguish between "good eugenics" and "bad," the latter presumably "unscientific" and "misused and perverted" by the National Socialists.[56] For the members of the American and British eugenics societies, the National Socialist "excesses" had been used in an unjustified manner as death knell arguments for all eugenics measures. Blacker in the early 1950s complained that the race doctrines of the National Socialists had severely weakened the position of the eugenicists in Great Britain.[57] Nachtsheim, who tirelessly pleaded for the reintroduction of eugenic measures in Germany, called the "political past" a major "hindrance to eugenics in Germany" and demanded that one should finally be freed of the taboo that had descended on eugenics in Germany since the time of National Socialism.[58]

In light of the discrediting of eugenics by the racist orientation of the early eugenics movement, the reform eugenicists considered it an important step that the eugenics societies should be "descientized" in order to regain respectability in scientific circles. Thus the American eugenicist and human geneticist Bentley Glass emphasized that eugenics was truly based on the science of genetics, but because value judgments could not be avoided, it was not itself true science.[59] Julian Huxley in this sense called eugenics "a form of applied human genetics."[60] Even Carl Bajema, one of the leading postwar American reform geneticists, noted that eugenics was not a science but a social movement with the goal of directing the evolution of humankind in a specific direction.[61] The trend of these arguments was that eugenics should give up any hope of being an academic discipline unto itself.[62]

The artificial division of eugenics as a political program from eugenics as a scientific area for research—"the human genetic patrimony"—demanded from eugenics organizations a readiness to accept other scientific societies in the area of human heredity research, psychiatry, and population science. They had to go even further by noting that the success of eugenics depended at least in part on the successful establishment of these branches of science.[63] Based on this logic, most eugenicists in the postwar period supported what had begun even before 1945, the institutional separation of the new scientific subjects of human genetics and demography from eugenics.

In 1956 for the first time scientists came together for an international congress for human genetics in Copenhagen, a congress which included several eugenics societies as supporters. The American Eugenics Society was a sponsor, as was the Società italiana di genetica ed eugenica. The British Eugenics Society proudly remarked that many of the members of their board of directors were important in the organization of the congress—Aubrey Lewis, J. A. Fraser Roberts, James M. Tanner, and Cedric O. Carter.[64] Leading members of the eugenics movement were among the leadership of the national preparation committee. The German committee included Verschuer, Lenz, and Nachtsheim; the French, Jean Sutter. Corrado Gini took a leading role in the Italian committee, while Torsten Sjögren was chairman of the Swedish committee and Petrus J. Waardenburg of the Dutch committee. Important roles in the five-person British committee were played by Fraser Roberts, a one-time vice president and in the 1950s a fellow of the Eugenics Society, and Eliot T. O. Slater, fellow of the Eugenics Society from 1957 to 1977. The six-person American committee included the eugenicists Franz Kallmann and Sheldon C. Reed; three further members belonged to the American Eugenics Society—the cancer researcher Eldon John Gardner, the human geneticist Arthur G. Steinberg, and the professor of zoology and director of the Dight Institute in Minneapolis, Clarence P. Oliver.[65]

Even though the newly founded scientific discipline of human genetics maintained the procedure of setting itself off against orthodox eugenics, the interests of the human geneticists in the quality of the hereditary patrimony still stood squarely in the tradition of the earlier eugenics movement. With the research into mutations caused by radiation inspired by the development of atomic energy, the human geneticists shared the worry of the earlier eugenicists about the genetic future of mankind. They transferred the old eugenics confession of belief into modern scientific categories of the atomic age.[66] For example, Tage Kemp, president of the First International Congress for Human Genetics, in his welcoming speech brought up the issue that the danger of mutations from radioactive rays and the risk of a declining "genetic virility" was penetrating more and more into the consciousness of the general public. Quite in the tradition of the early eugenics movement, he declared that progress in human genetics gave rise to the hope that in the foreseeable future humans would be able "to control their own biological evolution."[67]

As a result of "descientizing," the eugenics societies also supported the broader establishment of demography on the international level. When some 80 demographers from Europe, the Americas, and Asia got together from August 27 to September 3, 1949, for the first postwar meeting of the International Union for the Scientific Study of Population, a whole series of prominent eugenicists was represented. The five-person British delegation included four leading members of the Eugenics Society, C. P. Blacker and David V. Glass, the statistician Frank Yates, and the economist Julius Issac. Among the eight members of the US delegation, Henry Pratt Fairchild, Frank Lorimer, and Clyde V. Kiser played leading roles in the American Eugenics Society.[68] From France, in addition to the eugenic activist Jean Sutter, there came two demographers quite receptive to eugenics, Adolph Landry and Alfred Sauvy; from Italy, the prewar eugenic activist, the demographer Livio Livi; from Germany, the eugenicist Hans Harmsen.[69]

In its bylaws, the International Union for the Scientific Study of Population (IUSSP) tied into its eugenic roots from the prewar period.[70] The goal of the IUSSP was and continues to be to draw "the attention of governments, international organizations, non-governmental organizations, and the general public to the problems of population."[71] Frank Lorimer in *Eugenical News* pointed out that the Union had the goal of promoting "quantitative and qualitative demography" as a scientific discipline. In a resolution, the IUSSP recognized that "studies on aspects of 'population problems' have often been led into error due to race and class prejudice."

In the words of the IUSSP resolution, Qualitative aspects of the population and their relation to reproduction and migration should still deserve intensive scientific consideration."[72]

It was central to the program that research projects previously integrated into a eugenic science were now divided over a number of disciplines. Of course there were still scientifically based attempts to build bridges among various disciplines, such as psychiatry, human genetics, anthropology, and demography, but scientific careers were ever more strongly oriented now to a *single*, specific discipline. With scientific research broken off into various separate disciplines with their own institutes, journals, and international meetings, this internal differentiation constituted the "descientizing" of eugenics.[73]

"Voluntarism" and "Counseling": The Reorientation of the Eugenics Movement

Thanks to the fact that most eugenicists, and particularly the prominent eugenics societies in the United States and Great Britain, had by now given up their claim to eugenics as an independent scientific discipline, they could henceforth concentrate completely on the development and popularization of their set of eugenic demands. In publicizing their eugenic requests, after the World War eugenicists often ran into a very critical public. The killing and mutilation of hundreds of thousands of people in the name of eugenics under National Socialism and the successes of the social welfare state systems after the war cut the floor out from under the classical eugenic problem set. Even though eugenic sterilization was continued after the war in Sweden, Finland, Norway, and the United States, and even if Japan in 1948 introduced a comprehensive eugenics law on sterilization and abortion, still the climate for the introduction of further eugenic measures was not good, particularly in both America and Europe.[74]

In the face of public distrust of eugenic measures, the American and British Eugenic Societies followed a strategy that Blacker called "hidden eugenics" or "crypto-eugenics."[75] To the family planner Dorothy Brush, Blacker described this strategy as the attempt "to fulfill the aims of eugenics without disclosing what you are really aiming at and without mentioning the word."[76]

Osborn too in his book *The Future of Human Heredity* voiced the opinion that "eugenic goals" are most likely to be reached when one works hard while not using the word "eugenics."[77] Like Blacker, Osborn was convinced that the introduction of effective eugenic abortion laws and sterilization laws would be blocked if one were to put the eugenic arguments up front. In the Galton Lecture of the British Eugenics Society in 1956, Osborn explained that the existence of the great problem that "genetically inferior humans" simply are not ready to accept the idea that the hereditary patrimony from which one's character is formed is inferior and therefore should not be carried over to the next generation. Some people simply would not accept that they are "second-class." That is why eugenicists should stop saying that they have an "inferior genetic quality." Instead, suggested Osborn, eugenicists should support their suggestions with the need for children to grow up in an environment in which they receive affection and responsible care.[78] Because of the critical attitude of the general public to eugenics, birth control should be promoted for other than eugenic grounds "among the least adequate parents in our society." These people would certainly accept the limitations on the number of their births for "entirely other reasons."[79]

The bad reputation that eugenics had because of the racist orientation of the early eugenics movement and because of the National Socialist mass murders led the reform geneticists gradually to give up agitating under the banner of eugenics. The British Eugenics Society replaced its over 50-year-old journal *Eugenics Review* in 1969 with the *Journal of Biosocial Science,* and in 1988 renamed itself the Galton Institute.[80] In 1969 the American Eugenics Society rechristened its journal *Eugenics Quarterly* as *Journal of Social Biology* and in 1972 changed its own name to the Society for the Study of Social Biology.[81] It reassured its supporters that the name changes in no way meant a change in "interests or in policies."[82]

Since the forced measures as practiced by the National Socialists were discredited, eugenic protagonists such as Cook, Osborn, Blacker, Dice, Reed, Nachtsheim, and Kemp limited themselves to the propagation of "voluntary" measures. This notion of "voluntary action" was to be bound up with both positive and negative eugenics in a system of stimuli, indirect pressure, and exerting influence.[83] The social and economic environment was according to Osborn to be so reshaped that human beings would submit to voluntary and largely unconscious selection in a eugenic sense.[84] He thought it possible for physicians, nurses, and social workers to exert such a strong "social and psychological pressure" that the "carriers" of serious genetic defects would absolutely forebear from giving birth to children.[85] Lee R. Dice, director of the first human genetics clinic in the United States and one of the directors of the American Eugenics Society from 1952 to 1971, was convinced of the possibility that every individual can make a eugenically rational decision; he declared that voluntary action was the basis of a modern eugenics. "Persons with superior intellectual abilities"—for example, persons with doctorates from leading American universities—should in Dice's opinion be convinced to have more children through voluntary action by means of economic, social, and psychological support measures.[86]

Eugenic modernization was to slide in under the guise of "population control from below" determined by general political and social conditions. This strategy, discussed as the "eugenic hypothesis," envisaged a way to manipulate social conditions so that "inferior" people would "voluntarily" renounce having children. For people with "superior genetic potential," marked by physical and intellectual health, above average intelligence, psychological stability, and feelings of responsibility, social conditions would be so improved that they would feel encouraged to increase reproduction.[87]

In the opinion of the large majority of eugenicists after the Second World War, a qualitative population policy would be best implemented by internalizing specific eugenic values. The "consumption" of birth control methods as part of a set of social values was to them best suited to a democratic and social welfare state society.[88] Such a eugenics based on voluntary action and "individual decision" in no way meant a departure from the goal of the genetic distinction of the population. The shift of eugenics code word "individualization" referred to the method, not to the content.[89] The goals were to be realized in a subtler fashion than that pursued by the eugenicists at the beginning of the twentieth century.

Based on this logic, the postwar eugenics societies promoted efforts to build up offers of consultation for the broad population based on new knowledge in medical genetics.[90] These counseling offers to couples seeking advice would be supported in the context of the "medicalization" of human genetics. Instead of vague notions of quality characteristic of the early eugenics movement, the categories set up by medicine as "health" and "illness" were to be used as selection criteria. In the mid-1960s Osborn wrote that medicine had taken over research and the prevention of "mental and physical defects."[91] The concept of "medical genetics" was seen to have replaced "negative eugenics."[92]

As part of medical eugenics, which had made enormous strides forward after the Second World War, genetic diseases could be pinpointed.[93] The discovery of the hereditary basis of the metabolic illness phenylketonuria (PKU) was quickly followed by the discovery of other genetic ailments, such as the deadly Tay-Sachs nerve disease and, especially frequent among Africans, sickle cell anemia, which leads to bone and joint damage. The foundation for research into chromosomal diseases lay in new knowledge regarding the structure of the chromosomes, numbering 46 for humans\ as defined at the Copenhagen Congress. Other discoveries followed: testicular malformation in men described in 1942; the Klinefelter syndrome; and the hereditary bases of Turner syndrome, restricted growth of women, and Down syndrome (previously derogatorily referred as "mongoloid idiocy"). By the beginning of the 1970s, some 1,000 diseases could be traced back to a defect in a single gene or to a chromosomal anomaly.[94]

As a result of research on genetically related diseases, human genetic counseling offices became increasingly important. These locations had been initiated in the 1940s overwhelmingly by eugenicists. It is true that the consultants frequently were able to give only quite vague advice about possible hereditary defects in a child, but at least they offered affected couples the possibility of informing themselves of the risks of a genetic defect in their progeny. Tage Kemp, who had set up one of the first counseling offices at his institute in Copenhagen, kept expanding its scope.[95] In the United States, human geneticists and eugenicists such as James Neel, Sheldon C. Reed, and Lee R. Dice took the lead in developing human genetic counseling locations. In Great Britain the eugenicists Cedric Carter and Fraser Roberts, both longtime fellows and vice presidents of the Eugenics Society, were closely involved in the buildup of such institutions.

This counseling was presented as being necessary in the interest of the hereditary health of future generations. The goal was to pose the question to all people before getting married or before having children whether there was an increased risk to the hereditary health of the child.[96] The eugenic perspective, so the reform eugenicists hoped, would implicitly determine the way the consultation was practiced. The physician involved in the genetic counseling, though concerned with the quality of future generations, would simply give advice about a decision of a particular couple. Couples "at risk," as determined by human geneticists for the risk of hereditary defects, would then voluntarily renounce any possibly handicapped children because of concern with their own happiness.

While at first the hereditary counseling for couples was often referred to as "eugenic counseling," the concept was increasingly replaced by the words "genetic counseling." Reed was convinced that the concept of genetic counseling that he had invented and the freeing of counseling offices from all too obvious eugenic rhetoric had contributed to a boom in human genetic counseling in the 1960s and 1970s. Looking back, he opined that genetic counseling would have been rejected if it had been too strongly presented as a eugenic measure.[97] Osborn too declared at the end of the 1960s that he was satisfied that the establishment of heredity clinics and genetic counseling was the "first eugenic suggestion" that had been accepted by public opinion. He contended that this change was related to the fact that the word "eugenics" was no longer associated with heredity clinics.[98]

The Eugenics Movement and the Discussion of "Overpopulation"

At the beginning of the Cold War, the question of overpopulation in the developing countries took on an entirely new dimension. It is true that certain population professionals in the 1930s had drawn attention to the rise of population in Africa, Asia, South America, but the notion of overpopulation as a fundamental threat in the so-called developing countries first became an object of general attention after the Second World War.[99] Due to the "susceptibility" to Socialist thinking by economically backward countries, and due to the growing influence of the Soviet Union in the Third World, demographers became experts in questions of international relations.[100]

Many demographers and family planners no longer called colonialism the basic cause of "underdevelopment" and "backwardness" of the Third World; now they pointed to population growth.[101] At the beginning of the 1950s a broad consensus existed among demographers that rapid population growth was the basic cause of problems such as poverty, criminality, deforestation, erosion, malnutrition, illiteracy, limitation of democratic freedoms, environmental pollution, urbanization, and social tensions.[102] This list of explanations by demographers—illness, war, and communist uprisings—were seen to be preventable by a drastic reduction of population growth in the Third World.[103]

The scientists who at the end of the 1940s took up the struggle against overpopulation were practically all connected to one another through the American and European eugenics movements. Around 1950 a group of prominent American demographers, economists, and health experts laid the scientific foundation for a comprehensive population program in the Third World. This group, meeting under the auspices of the Milbank Memorial Fund and working closely with the United Nations, included the leading ranks of the American Eugenics Society. The meetings were organized by Frank Boudreau of the Milbank Fund and by Clyde Kiser, longtime director and later chairman of the American Eugenics Society. Frank Notestein, the first director of the population department of the secretariat of the United Nations (department started in 1946), chairman of the Population Association of America and a director of the American Eugenics Society, and Pascal Whelpton, Notestein's successor as director of the UN population department and later one of the directors of the American Eugenics Society, were the driving forces in this group.[104] Three other longtime directors of the American Eugenics Society were members of the Milbank Memorial Fund group—Kingsley Davis, professor of sociology at the University of California at Berkeley and later representative to the population committee of the United Nations; Warren Thompson, director of the Scripps Foundation from 1940 to 1945; and Frank Lorimer, professor of demography at American University in Washington, DC.[105]

With the struggle against overpopulation in the Third World, the eugenics movement found a new area of activity after the Second World War. The British and American Eugenics Societies, by their engagement against the population explosion, succeeded in being included in a discourse accepted in broad circles. Being in the vanguard in a discussion of overpopulation brought a certain amount of attention to the eugenics movement in the postwar period, a kind of respect that had been lost by their earlier racist orientation and the National Socialist mass murders.

Eugenicists were concerned about the discrepancy that existed between the decline in the birthrate in the industrial states and the "exploding" population in developing countries. In this vein, C. P. Blacker drew attention to the "population problem" at an international meeting of activists in the birth control movement. On the one hand, in some "developed countries" the "danger of depopulation" due to a declining birth rate existed, while on the other hand, the "threat" of overpopulation in the developing countries existed.[106] His colleague Hans Nachtsheim, who had consistently sounded the alarm about the "population explosion," in a lecture on the "Overpopulation Problem and the Race Profile of Future Humanity" expressed his concern that the result would be "an increase in the colored races [and the] disappearance of the white race."[107]

The way that the eugenicists linked "depopulation" in the industrial countries with the supposed problem of population explosion in the developing countries opened up their line of argument to racist implications. By extending their focus to the developing countries, the reform eugenicists, who wished to emancipate their eugenic policies from race and class prejudices, ran the danger of implicitly building racism into their line of argument. Since there was no generally recognized criterion for a national "population optimum," merely by postulating overpopulation in a developing country meant that a certain value judgment was being made. Why should a developing country, rich in natural resources and with a low population density per square kilometer, be overpopulated, while industrial countries with up to 300 persons per square kilometer, should be "normally populated or even underpopulated"?[108]

For the reform eugenicists, the quantitative and qualitative aspects of population growth were two sides of one and the same basic problem. Again and again they emphasized that "the quantity of people in an area cannot be separated from problems of human quality."[109] For example, Hans Nachtsheim was basing his argument on the work of the Israeli biologist Fritz S. Bodenheimer when he claimed that "population increase with a simultaneous degeneration of the hereditary patrimony is the key question of human existence." The quantitative and qualitative population problems were seen to be even worse than the threat of the hydrogen bomb.[110]

In the last analysis, for the eugenicists the causes of the primarily quantitative and the primarily qualitative population problems were identical ones. Just as medicine had set aside the natural laws of selection in the industrial states and had led to increased reproduction of "intellectually inferior people," in the Third World development aid and improved healthcare would prevent natural shrinking of the total population. The dilemma was similar—on moral grounds one could not deprive specific population groups of the "blessings" of modern medicine, but, as Nachtsheim expressed it, the "natural balance between human increase and human decrease" was thereby disturbed.[111]

The struggle against the population explosion and against a further spread of the "hereditarily sick" was for the eugenicists the proper biological response to the dominant "crisis of human fertility."[112] The eugenicists felt that an all-encompassing eugenic strategy was needed to resolve the two aspects of the worldwide population problem. The two questions of "how many people" and "what kind of people" could only be solved in tandem or not solved at all.[113] Osborn pointed out that birth control measures always had two goals: first, to remove the "dreadful burden of overpopulation" and second, to "raise the level of all our people to the level of the best human beings we know at present."[114]

The concrete solutions suggested by the eugenicists for the various facets of the population problem all came to the same conclusion—rationally managed birth control.[115] Quantitative birth control was thereby both a basic precondition for and a steppingstone to comprehensive qualitative birth control. Fritz Lenz remarked in 1956 that "eugenic viewpoints" were absolutely necessary in the required "drastic limitation of reproduction" in the developing countries.[116] Sheldon C. Reed, director of the American Eugenics Society from 1957 to 1977, fellow of the British Eugenics Society and long-time director of Dight Institute at the University of Minnesota, openly said that "intelligent control of man's evolution" would be possible only if a considerable proportion of the population were to limit family size. To Reed, "Other techniques already are available or could be developed quickly" and formed the key to the "genetic future."[117]

During the 1950s, eugenicists exerted a major influence on all the leading organizations of family planning. In its early years, the International Planned Parenthood Federation (IPPF) was so closely intertwined with the eugenics movements in Europe and America that Günther Repp of the German Society for Hereditary Health Care observed that with the IPPF "the British and North American eugenicists" stood on the "very front line" in the reorientation of eugenics in the postwar period.[118]

The IPPF had been founded in 1952 by Margaret Sanger and her co-agitators Marie Stopes in Great Britain, Rama Rau in India, and Elise Ottesen-Jensen in Sweden.[119] For Sanger and her colleagues, the perceived danger of overpopulation was the major reason for implementing birth control and family planning. The attempt to liberate women from the "constraint of pregnancy" by distributing contraceptives and the goal of improving the economic situation of socially deprived groups by family planning took a back seat to the motivation of stopping population growth through widely implemented birth control measures.[120]

The IPPF, today the umbrella federation of family planning organizations in over 140 countries and the largest nongovernmental organization in the area of family planning, consisted at first of a small international network of activists from the population and birth control movement. In contrast to today, where only the

Pope and a few Islamic fundamentalists oppose contraceptives, in those days after the Second World War activists in the birth control movement ran up against bitter opposition from many religious and political groups. They were far from being supported by national and international institutions. While currently the IPPF is supported by almost $100 million in subventions from Western European and North American governments and the United Nations, in the early years of its existence it depended on contributions from eugenics organizations.[121]

The British Eugenics Society in the early 1930s acquired large assets in a significant inheritance from the Australian property owner Henry Twitchin and used those funds to serve as godfather for the IPPF and for various other initiatives in population policy. So it financed the Population Investigation Committee, which after the Second World War blossomed into the central coordinating office for demographic research in Great Britain. More than that, the Eugenics Society of the British Family Planning Association—a merger of the British Birth Control Association, the Birth Control International Information Center, the British Birth Control Investigation Committee, and the Society for the Provision of Birth Control Clinics—made office space available for free.[122] The International Committee for Planned Parenthood as well, the forerunner of the IPFF founded in 1949, and the IPPF were able to use office space at no cost on the premises of the Eugenics Society.[123] Together with the Foundation for Race Betterment, founded by the eugenicist Dorothy Brush, the Eugenics Society also covered most of the operating costs of the IPPF and helped determine its strategy.[124]

Eugenicists also took over many key positions within the IPPF. C. P. Blacker became vice chairman of the IPPF in 1953, and in 1959 took over the office of administrative director of the organization. Vera Houghton, in the 1950s member and later vice president of the British Eugenics Society, was the first general secretary of the IPPF. Nancy Rose Raphael, in the 1950s fellow of the British eugenics society and director of the organizational committee of the British family planning organization, was the first general secretary of the European regional office of the IPPF.[125] George W. Cadbury, another member of the British Eugenics Society, in 1960 became one of the directors of the IPPF and was the liaison member of the IPPF to the United Nations.[126] G. A. Whyte, member of the Board of Directors from 1949 to 1957 and treasurer of the British Eugenics Society from 1954 to 1961, became financial administrator of the IPPF.[127] Hans Harmsen, one of the most influential German eugenicists and the first president of Pro Familia Deutschland, the highest representative of the German delegation, was a member of the leading circles of the IPPF. Alan F. Guttmacher, director and vice president of the American Eugenics Society in the 1950s and chairman of the medical committee of the Association for Voluntary Sterilization, was chairman of the medical committee of the IPPF in the 1960s and became president of the Planned Parenthood Federation of America.[128]

Blacker and Houghton coordinated a eugenics orientation of the IPPF.[129] According to the bylaws of the IPPF, one of their central goals is still officially to stimulate research on the biological, demographic, economic, eugenic, psychological, and social questions of human fertility and their regulation and to publicize the results.[130] The important role played by eugenics in the early years of the federation became clear in the publication policy of the *Around the World News of Population and Birth Control*, the official organ of the IPPF. The editorial policy of this publication was determined by Margaret Sanger, president of the IPPF; Abraham Stone, director of the research Bureau and member of the Board of Directors; and William Vogt, best-selling author.[131] In 1954 *News of Population and Birth Control* bemoaned the fact that since the war eugenics had to be handled as a "delicate" topic and posed the rhetorical question of what topic could be more important than the quality of the people who populate the earth.[132] One year later Frederick Osborn claimed in the journal of the IPPF that people know more about the heredity of the standard bluebottle fly than they do about the hereditary patrimony of human beings. He praised the American Eugenics Society as the only organization that was trying to unify the knowledge from demography and human heredity into a broad, scientifically based eugenic program.[133]

On the questions of overpopulation, the IPPF worked closely with the Population Council in Washington, another organization that had been dominated in its early years by eugenicists.[134] The Population Council had emerged from a conference of demographers, biologists, economists, and health scientists, a conference to which John D. Rockefeller III had been invited in the summer of 1952.[135] The Population Council was to act as a "catalyst" in the broad field of population questions and to stimulate, support, and perform projects dealing with the world-wide population problem."[136] The background of this initiative was the conviction that the tempo of population growth both among the poorest nations and among the poorest groups in the United States presented a threat to economic development and political stability.[137]

The board of directors of the Population Council included not only Rockefeller and Frank Boudreau of the Milbank Memorial Fund but also the eugenicist and demographer Frank Notestein. The board designated Frederick Osborn as the first executive officer.[138] Osborn was president of the Eugenics Society from 1946 to 1952, then general secretary, treasurer, or director from 1954 to 1973. Dudley Kirk, the first department director for demography of the Council (1954–1968), came from the ranks of the American Eugenics Society. Prior to this post, Kirk served as intelligence research officer in the State Department and then was a member of the US delegation to the UN population commission; he served from 1956 to 1975 as one of the directors of the American Eugenics Society.[139]

In light of this concentration of eugenicists in the Population Council, it is not surprising that the American Eugenics Society and the Population Council for years kept their business offices in the same building in Park Avenue in New York.[140] The Population Council supported the American eugenics movement with large amounts of money. In 1954 it promoted a research project of the American Eugenics Society, the results of which were to show how the hereditary patrimony affected the quality of the American population.[141] It also supported the efforts of the American Eugenics Society to introduce genetic content into the curricula of medical schools.[142]

The Rockefeller Foundation and the Ford Foundation distributed money to the Population Council for demographic studies in the broadest sense. In this way they wanted to create a scientific basis that had so far been lacking for worldwide population planning.[143] A focus in this regard was support for the development of new types of contraceptives. For example, the Population Council commissioned the National Committee of Maternal Health to gather together information about the effectiveness of various contraceptives. The chairman of this committee was Christopher Tietze, assistant director of the biomedical department of the Population Council and member of the American and British Eugenics Societies.[144] Osborn, who succeeded Rockefeller as president of the Population Council, declared frankly in 1966 that the Population Council made available $3 million a year for research into contraceptives and that this investment was one of the most important eugenic measures that had ever been practiced.[145]

Chapter Eight
The Renaissance of Racist Eugenics

At the beginning of the 1960s, at a time when the dissolution of segregation in the United States was moving ahead full speed, Carleton Putnam, a retired American businessman, published a pamphlet titled *Race and Reason—A Yankee View.* According to the author, politically blinded opponents of segregation were attempting to use a conspiracy of levelers to deny any intellectual differences between the races. Two American generations had already fallen victim to a "pseudo-scientific fraud" of Jews and mulattoes, who were attempting to eliminate segregation in the United States. Putnam warned his "white" American fellow citizens about racial mixing with "Negroes" because that would lead inevitably to the decline of the "white race."[1]

The appearance of this publication, initially titled "Warning to the North,' was made possible by financing from Wickliffe Draper. In exchange for publication of the book, Draper guaranteed to the publishing house the purchase of a large number of copies. These purchased books were then sent out at no charge to distributors, to whom Draper promised support in the struggle against race mixing.[2]

Putnam's pamphlet was the product of the overheated atmosphere in America in the 1960s. The American civil rights movement was gaining ground in the battle against race discrimination, and some Southerners in particular felt themselves increasingly placed on the defensive. Putnam's arguments against mixing Negroes and "Whites" could be interpreted, as often occurred, as a wild racist charge of an amateur anthropologist, except for the fact that a whole series of scientists supported it.[3] The geneticist Ruggles Gates, the psychologist Henry E. Garrett, the anthropologist Robert Gayre, and the biologist Wesley C. George gave scientific absolution to Putnam's racist pamphlet. They explained in the introduction to *Race and Reason* that Putnam's scientific foundation was a solid one. In political decisions, one had to consider that races differ not only in their physical characteristics but also with regard to their sensitivity to the world and their intellectual capacity.[4]

In the 1920s opinions against race mixing had met broad agreement in the American Eugenics Society, but in the 1960s the reaction to Putnam's pamphlet showed that the leadership of the American Eugenics Society now took an entirely different view. American eugenicists criticized the support from Gates, Garrett, Gayre, and George for Putnam's hypotheses. Gordon Allen, for many decades one of the directors of the American Eugenics Society, warned on the pages of the *Eugenics Quarterly* that one should "beware of" Putnam's book. Support by the four scientists for Putnam's publication was reprehensible because with their names as scholars they had given legitimation to a book that was absolutely lacking in scientific precision and scientific basis.[5] Allen's colleague on the board of directors of the American Eugenics Society, the geneticist Bruce Wallace, doubted the scientific basis of Putnam's book and also questioned the scientific reputation of Putnam's supporters.[6] Arnold Kaplan, a geneticist from the Laboratory of Medical Genetics in Cleveland, Ohio, in the *Eugenics Quarterly* regretted the fact that during a discussion about the ending of discrimination in the United States several well-known scientists had spoken out against a stronger integration of Whites and Blacks. Their arguments that Negroes on genetic grounds were intellectually inferior and could not develop and maintain an independent civilization were absolutely without foundation in science.[7]

The 1950s saw a split between the moderates, who maintained reserve in questions of race, and a small group of racist eugenicists. Since the American and British eugenics societies during the 1950s and 1960s were on a collision course with racist scientists, one group of race researchers, supported financially by Wickliffe Draper, at the end of the 1950s formed an international organization for eugenics and ethnology. The International Association for the Advancement of Ethnology and Eugenics (IAAEE) intentionally placed itself in the tradition of orthodox eugenics and propagandized for a close connection between eugenic science and basic research. The eugenicists active in this organization maintained close ties to various Fascist and extreme right-wing organizations but at the same time set up a contrast between their supposedly scientific position and the egalitarian position maintained by politically motivated pseudoscientists.

Orthodox eugenic initiatives that emerged at the end of the 1960s from the extreme Right milieu in France and Germany were closely connected to the IAAEE. The international network of eugenicists and race researchers, who down to the present day are supported primarily by Draper's Pioneer Fund, formed the basis for the renaissance of research in the early 1970s into genetically based differences in intelligence between the races.

The Split of Racist Eugenics from the Eugenics Mainstream

The split by the British and American eugenics societies from orthodox racist eugenics, apparent as early as the 1930s, continued after the war, strengthened by public distrust for any form of race research. The distancing of the eugenic mainstream in the United States and Great Britain from race research went so far in the 1950s and 1960s that leading American eugenic race researchers around Gates, Gayre, and Garrett found themselves in the same corner as the Nazi race theoreticians Hans F. K. Günther and Walter Groß. Frederick Osborn noted that unfortunately the theories of "Hitler's pseudo-scientists" flared up again in the controversial discussion about the end of race segregation in the United States.[8] He called the scientists grouped around Gates, Gayre, and Garrett "groups of the radical right" that prostituted science in order to justify their "propaganda" over the supposed inferiority of the "Negro race."⁹ In Osborn's view, "There is still a strong residual suspicion of eugenics as being race or class inspired, and this suspicion to some extent is kept alive by some of the groups in the radical right, such as the so-called International Association for Advancement of Ethnology and Eugenics. In general our feeling is that we should move cautiously with the eugenics argument in this field, for the important thing is to get the result." There is no scientific basis for ranking any one race genetically superior to another. The geneticist finds that no races are pure.[10]

The criticism of the reform eugenicists of their eugenic colleagues focused on the genetically questionable distinction of races. Dobzhansky wrote in *Eugenics Quarterly* that "there is no scientific basis for ranking any one race genetically superior to another" and that "no races are pure." "Race boundaries" had been so blurred by the ongoing exchange of genes that it was simply a question of "convention and convenience" as to which groups one would call races.[11] The reform eugenicists wanted to remove the race question from eugenics in general. Osborn noted in 1963, "Eugenics is not concerned with color of skin or facial or bodily characteristics unless it is shown that these features of man are related to his genetic capacity for socially valuable qualities such as intelligence or character." He went on, "Eugenics is not particularly involved in the unsolved question as to whether the proportion of hereditary ability is greater in one race than another. Eugenics is concerned with saving the genes for superior ability wherever they are found and increasing their frequency."[12]

This moderate course taken by the American Eugenics Society and, with some exceptions, by the British Eugenics Society as well, met major criticism from eugenicists such as Gates, Garrett, and Fisher. They were enraged that the main line of eugenics did not take a clear position on the "race question" and even promoted the end of racial segregation. In a confidential anonymous report to British eugenicists, the argument was made that the policies of the American Eugenics Society toward racial matters were based on the "new eugenics,' which dissociated itself from "the old genetics and the 'racism' of Madison Grant and others." According to the report, this "negative position of the American Eugenics Society was traceable to the Jewish origin of the majority of their officers."[13]

The eugenicists interested in the race questions in the United States felt themselves essentially isolated. Ruggles Gates was indignant that eugenics in the United States had become a "farce" through the use of the slogan "all men are equal."[14] Gates had moved to the United States after the war because he hoped to take over direction of Davenport's Cold Spring Harbor Laboratory, and he was surprised by the vehement criticism of his genetically oriented race research.[15] To Blacker he complained that the "anti-race propaganda" driven by the Jews in the United States had gone so far as to deny the existence of races. In contrast to the "modern" eugenicists in the United States, Galton still had the courage to speak of "inferior races."[16]

Gates made contact with researchers in America who also were threatened by scientific isolation because of their race research. Henry E. Garrett, professor of psychology at Columbia University in New York and one-time president of the American Psychological Association, in a 1947 *Scientific Monthly* article had declared that even among newborns one could determine a higher intelligence of Whites over Blacks. When he was sharply criticized by his colleagues for these arguments, he asked

Gates to defend him against the attacks.[17] Likewise, the eugenicist T. U. H. Ellinger, who was the object of severe criticism because of an article he published during the war explaining National Socialist–based policies and because of his race research on African Americans, turned to Gates for help.[18] He complained that despite many pieces of evidence, many scientists did not want to acknowledge that Negroes had come no further than "apes" and were hardly educable.[19]

Criticism of the main eugenics orientation after 1945 was shared among several eugenicists interested in the race question. Ronald L. Fisher, who had been vice president of the British Eugenics Society in 1933, by the 1940s was distancing himself from the reform orientation of this society. Fisher absolutely rejected "race crossings" and followed the end of race segregation in the United States with great distrust; he was stymied by the reservation of the British Eugenics Society in race questions.[20] John R. Baker, a student of Julian Huxley and up until the end of the 1950s a fellow of the Eugenics Society, complained that the program of the Eugenics Society had become so mild and flabby that it hardly merited the name eugenics. For Baker, the "old Eugenics Education Society" had been a group of men and women who were not afraid of expressing an opinion. Current societies had tied themselves so closely to belief in "race equality" that differences among the races could not even be talked about.[21] Charles Galton Darwin, nephew of Leonard Darwin and one-time president of the Eugenics Society, and C. D. Darlington, longtime fellow of the Eugenics Society, were very disappointed about the trend toward a moderate course being taken by the Eugenics Society. Darwin and Darlington tried in vain to steer the British Eugenics Society onto a clearer course on questions of race.[22]

The critique by broad public opinion after 1945 with regard to genetic research into race and the distancing of the eugenics societies from orthodox eugenics resulted in a situation where it became ever more difficult for scientists interested in race topics, such as Darlington, Gates, Baker, Ellinger, and Garrett, to achieve academic positions, publication possibilities, and grants for research. In this critical situation for race researchers, Wickliffe Draper and his Pioneer Fund leaped into the breach. Draper, who was extending his activity for eugenic and racist topics that he had begun in the 1930s, after the Second World War more and more developed into the gray eminence of European and North American race research.[23]

In 1952 Draper and his Pioneer Fund had helped the American Eugenics Society with a generous contribution to overcome an acute financial crisis in the postwar period and assured the work of the society for three years.[24] However, when Draper realized that the American Eugenics Society under the leadership of the reform eugenicists was becoming ever more reserved in questions of race, he threatened to cut off financing entirely. During a luncheon meeting with Osborn in October 1954, Draper came up with a tempting offer. He would assure the financing of the American Eugenics Society for another five years if the eugenicists would declare themselves ready to speak out for "race homogeneity" in the United States, that is, for a ban on "race mixing" and restrictions on immigration from non-European countries.[25]

When Osborn refused to accept this condition because of the lack of "scientific basis," Draper immediately cut off payments to the American Eugenics Society. He was determined to give money only to those scientists who shared his basic attitude on race questions. Because of the unbridgeable gap with Draper over the question of race, Osborn resigned his office as president of the Pioneer Fund, and the leadership of the American Eugenics Society dissolved what had been until then a practically symbiotic relationship with Draper's foundation.[26] Francis Walter, in the McCarthy era the chairman of the infamous House of Representatives Committee on Un-American Activities, succeeded Osborn on the executive board of the Pioneer Fund.[27]

In the postwar period, Draper understood his task to be to promote race research in particular in British-speaking countries and to have an effect on policy decisions in questions of race by publishing "scientific facts." So in 1954 he hired Ruggles Gates to research human genetic research institutions in Great Britain, Canada, and the United States in order to find out which of these would be prepared to carry out research projects in his sense. Draper's questionnaires contained not only questions about the academic prestige of the institute, about the orientation of its research, and its reputation but also questions about attitudes to "miscegenation," "immigration quotas," and "improving population quality" through negative and positive eugenics.[28]

Two of Osborn's most trusted coworkers in the American Eugenics Society, the director of the Dight Institute in Minnesota, Sheldon C. Reed, and the director of the Genetic Research Institute at the University of Oklahoma, Laurence H. Snyder, refused even to receive Gates because of the racist implications of these questions. In his report to Draper, Gates complained that in addition to Snyder and Reed, geneticists at McGill University in Montréal and at Ohio State University had denied the harmful effects of race crossings. Only the position of Clarence P. Oliver, professor of zoology at the University of Texas and one of the directors of the American Eugenics Society from 1950 to 1956, was "the correct one." On the basis of the "eugenic" position of Oliver, Gates approved the financing of a research project suggested by Oliver on "three race groups in Texas: Whites, Mexicans, and Negroes."[29]

Gates strongly urged Draper to finance an academic chair in human genetics in London to be taken over by his long-time colleague, Fraser Roberts. In a number of letters to Roberts, Gates made clear that his attitude on questions of eugenics, race crossing, and immigration quotas would be decisive for approval of money by Draper. At the beginning of 1955, when Draper kept adding conditions for funding such a professorship, Gates wrote that Roberts's position against race crossings was important, since Draper apparently supported only those academics who took unambiguous positions against the mixing of races.[30]

The Construction of an International Network of Racist Eugenicists: The International Association for the Advancement of Ethnology and Eugenics

The 1950s saw the clash over racial segregation in the United States come to a head. Many African Americans had fought during the Second World War on the side of the Allies against the racist regime of the National Socialists and were no longer prepared to accept race discrimination in their own country. In the judicial decision

Brown versus Board of Education, the Supreme Court in 1954 took the side of the opponents of race discrimination. It declared as illegal segregation by race in schools and universities. In their rationale, the justices referred to the UNESCO resolutions on the race question, according to which inequality between ethnic groups had social causes, not biological ones.[31]

In the court's decision to forbid racial segregation in American schools, it was clear that the discrediting of racism by the majority of scientists in the United States in particular had had an important influence on political decisions. Despite the vigorous arguments after the publication of the two UNESCO declarations on the race question, these documents had materially contributed to the scientific discrediting of the concept of biological races. To the defenders of racial segregation, coming mainly from the southern states, it was obvious that the clash over political questions of racial segregation had been carried out overwhelmingly in the arena of scholarship.[32]

This insight was the occasion for a group of eugenicists interested in race questions to found a new international eugenics organization—the IAAEE. Through the international character of the organization, the founders wanted to raise the issue that the white race was threatened not only in the United States but also in other areas of the world. In contrast to the American Eugenics Society and the British Eugenics Society, which worked primarily on a national basis, the IAAEE was interested in transnational cooperation of eugenicists for the purpose of improving the race. The international claim also protected the organization against too strong monitoring by American authorities, who had not allowed the licensing of a different nationally limited organization for race research at the end of the 1950s.[33]

The IAAEE was supported financially early on by Draper. Its definition of its central goal was to genetically improve the various "peoples, stocks, races, ethnic and cultural groups" through the application of "findings of eugenics, ethnology, history, prehistory, archaeology, and a host of other sciences." It called for special emphasis to be laid on research into "ethnic and cultural problems" with the help of biology, genetics, ethnology, eugenics, and anthropology.[34] Within the framework of promoting eugenics and ethnology, the organization "would publish and disseminate appropriate writings and would provide assistance 'in any lawful manner' to others with an interest in such problems."[35]

For several decades Draper financed the work of the IAAEE. As Draper's advisers, Garrett and Gates guaranteed the close connections between the IAAEE and the fund. Very much in the style of early twentieth-century anthropology, Draper through the IAAEE promoted research projects for measuring head and body shape and for analyzing the pigmentation of "Anglo-Saxon schoolchildren." He also financed through the IAAEE the writing and distribution of materials on race questions.[36]

Leading European and North American scientists had been represented in the International Federation of Eugenics Organizations in the 1920s and 1930s, but in the 1960s the IAAEE consisted primarily of marginal figures from anthropology, ethnology, sociology, and human genetics. Even the IFEO veterans Corrado Gini and Ruggles Gates, who were active in the new International Association, in the 1960s were increasingly academic outsiders.[37] Frank J. C. McGurk and Henry

Garrett, two well-known psychologists active in the IAAEE, lost a good deal of their academic prestige due to their involvement in race questions.[38]

The hard core of the IAAEE—Robert Gayre, Robert Kuttner, A. James Gregor, and Donald Swan—consisted of a small but very adroit group of scientific outsiders with close contacts with extreme right-wing groups. Robert Gayre, a supporter of the National Socialist race theoretician Hans F. K. Günther, stood for "racial stability" of the various European states. At the end of the 1950s he left a professorship at the University of Saugor in India in order to devote himself completely to the study of race questions in Europe.[39]

The biochemist Robert Kuttner, who taught psychology, race history, and political science at various universities in the United States, in the early 1960s unveiled a concept of "biopolitics," which was to serve for the "maintenance of the white race" and the "salvation of European culture." As Kuttner wrote in the extreme right-wing journal *Nation Europa*, "biopolitics openly emphasized the significance of inheritance by blood" and was based on "the simple truth that every human being owed allegiance to this hereditary line given by fate." "Being true to the race [was] natural patriotism raised by biopolitics to a law."[40] Kuttner in the late 1950s regularly wrote for the racist extreme right-wing magazine *Right*, and in 1966 became coeditor of the *American Mercury*, an anti-Semitic and racist magazine.[41]

At the time of the founding of the IAAEE, A. James Gregor was a doctoral student at Columbia University in New York, and later he worked as a political scientist at the University of Kentucky, the University of Hawaii, and the University of California at Berkeley. In the journal *The European*, edited by the British Fascist Oswald Mosley, Gregor in 1958 called the race ideas of the National Socialist race politician Walter Groß the "nucleus" for a worldview that made human beings into creators and the builders of "future races." For Gregor, here was a philosophy, history, politics, race, eugenics, and humanism all together.[42]

Donald A. Swan, who in the years after the founding of the IAAEE held teaching posts at City University of New York and at the University of Southern Mississippi in Hattiesburg, and who received over $100,000 in financial support from Draper's Pioneer Fund, for a long time directed the Institute for the Study of Man.[43] Swan, calling himself an "American Fascist," was convinced that "the selection effects that were determinative for human development and race formation had produced race differences in psychological characteristics as well as in physical characteristics."[44] According to Swan, "intellectual differences in the races" expressed themselves particularly in intelligence tests, which showed that Negroes proportionally had six times as many "feebleminded children" as did the Whites.[45] Swan's relations to the extreme right-wing scene in the United States became well known when in 1966, in connection with a search of his house because of suspicion of counterfeit letters, Nazi flags and photos of Swan with members of the American Nazi party were found in his possession.[46]

In order to counteract the antiracist climate in public opinion and in the media, Gayre, Gates, Kuttner, Swan, and Gregor put together an "alternative statement on the race question," a response to the UNESCO initiated statements against racism. For academics like Gayre, the condemnation of scientific racism by UNESCO amounted to nothing more than scientifically unjustified "attacks against ethnology," which had to be opposed by enlightened researchers.[47] An "alternative declaration on the rights question" composed by leading race researchers was intended by the IAAEE leadership to present "racial differences" as provable scientifically; one should draw conclusions from "race differences" for the way human beings should live together.[48]

In the attempt to discredit the UNESCO position papers on the race question, the IAAEE cooperated closely with the International Institute of Sociology (IIS). This institute, founded in 1893, after Second World War became the organizational retreat for German sociologists such as Karl Valentin Müller, Hans Freyer, and Arnold Gehlen, who had been discredited in the eyes of many of their colleagues because of their activity for National Socialism.[49] The academic policy line of the institute in the late 1950s and early 1960s was set by Müller and the long-time chairman Corrado Gini, who wanted to spread their "anthrosociological, biologistical" approach to research through this close connection.[50]

Gregor, the IIS regional official and one of the coeditors of the *Revue Internationale de Sociologie* at the beginning of the 1960s, put forth the idea that the IIS should be brought into the campaign against the UNESCO resolution. Gregor wrote in a confidential letter to Gates that the institute president, Corrado Gini, did not mince words and spoke out specifically against the UNESCO resolutions. For Gregor, the institute would consciously take a leading role in the development of sociological theory on race questions. He hoped that all the members of the IAAEE would join the IIS in order to jointly expand the race question from the sociological and biological points of view.[51]

For scientific support of the "alternative declaration on the race question," the idea of the IAAEE and IIS leadership was to have "authorities" in the race question put out a collection of essays—by Henry Garret, Ruggles Gates, Eugen Fischer, Ronald Fisher, C. D. Darlington, Ilse Schwidetzky, Friedrich Keiter, Fritz Lenz, Walter Scheidt, Corrado Gini, and Egon von Eickstedt. This book was to resemble in format and chapter headings the collection edited by UNESCO, *The Race Concept in Modem Science*. Contributions from Dunn, Shapiro, Comas, Lévi-Strauss, and Klineberg on the issues of the connection between race and biology, genetics, culture, history, and psychology, were to contradict in detail the UNESCO collection of articles.[52]

With this project directed against the antiracist efforts of UNESCO, the IAAEE hoped to unite scholars from various countries and disciplines into joint action. The leaders of the organization were convinced that such a publication would be much weightier than the work of an individual person. The book was to be financed with contributions from Draper's Pioneer Fund. Swan proposed to Draper's attorney Harry F. Weyher a one-time subvention of $4,000 in order to bring together, with the help of a university, the articles from various authors in the United States and Europe. A second subvention from Draper would then be used to convince a large publishing house to publish the book.[53]

The book appeared in 1967 under the title *Race and Modern Science*. The editor Robert Kuttner referred in his introduction to such "outstanding" scientists as Eugen Fischer and Fritz Lenz, and declared that through the book the "evolutionary value of race prejudices" would be presented in detail.[54] There were 16 articles by well-known race researchers, who had been chosen for their attitude on the race question and not because of their research work, according to L. C. Dunn, the reviewer for the *Eugenics Quarterly*.⁵⁵ The German psychologist Friedrich Keiter wrote about the connection between race and psychology; C. D. Darlington analyzed the impact of genetics on human society; the South African geneticist J. D. J. Hofmeyr dealt with population genetics; the Italian geneticist Luigi Gedda, who in 1961 published a study on the children of Italian women and African American soldiers, examined the significance of race crossing; and Ilse Schwidetzky, a German race anthropologist active as far back as the National Socialists, described the biological history of populations.⁵⁶

The IAAEE consciously was following the traditions of the early eugenics movement.⁵⁷ In contrast to the "descientization" of eugenics as proposed by the US and British Eugenics Societies, the IAAEE activists took a stand against any artificial separation of eugenic science from eugenic policies. In their opinion, the eugenics and race research represented by the IAAEE met scientific standards. Political demands obviously and incontrovertibly flowed from there.

The IAAEE declared as its goal the reprinting of outstanding scientific studies by race researchers and eugenicists. Swan and his colleagues were here thinking of books by the British race researcher Arthur Keith, works of the American anthropologist Carleton Coon, and the chapter regarding the origin of race and race biology from the British translation of the textbook by Baur, Fischer, and Lenz. In addition various German, French, and Italian studies by European scientists were to be translated into English and made available to a broad public in Great Britain and the United States. Some of the scholars mentioned were Eickstedt, Schwidetzky, Fischer, Lenz, Scheidt, Baur, Weinert, Keiter, Günther, Lapouge, Gini, Mjöen, and Reche.⁵⁸

The IAAEE saw the need to pay special attention to the South in the United States, where in their opinion the imminent end of racial segregation made their involvement especially necessary. In 1963 the IAAEE organized a conference in Atlanta, Georgia, where Swan, Garrett, Gregor, and Putnam were active in putting together a case of complaint against "integration of the races." The psychologist Clariette Armstrong, a former member of the Eugenics Research Association, testified to the court for the IAAEE how great a hardship it would be for the Negro child to be in school with brighter, younger Whites, leading to greater delinquency, and so on.⁵⁹ The IAAEE published and distributed arguments by Henry E. Garrett and Ernest van den Haag for racial segregation in the United States.⁶⁰

Swan succeeded in convincing the superintendent of public schools in Virginia, Davis Y. Paschall, to offer a course in all high schools on race science. In this course one was to teach about the bases of "race biology, race psychology, and race history" from the point of view that every "major race" had to retain "its ethnic identity."⁶¹ The IAAEE also planned to have European scientists teach at colleges and universities in the American South. They were thinking of Fisher, Darlington, Keiter, Gini, and Schwidetzky.⁶²

The IAAEE also was interested in an institute for race research, preferably at a university in the South, where research would be carried out on the differences between the "major races" and the various "sub-races" with regard to physical characteristics, intelligence, and criminal behavior. Gayre picked up on this idea and in the beginning of the 1960s attempted to found an "International Institute for Advanced Race Research." Instead of the American Southern states, he thought rather of Edinburgh as the headquarters for the institute since this "center of multi-racism" seemed particularly well-suited as a starting point for a "counterattack."[63]

Financing by public money was hardly to be expected, but once again Wickliffe Draper jumped in and made between 40,000 and 100,000 pounds a year available for such an enterprise. He sent his New York attorney Harry F. Weyher to test the ground in Great Britain. Weyher, who coordinated a large portion of Draper's activities and later took over chairmanship of the Pioneer Fund, attempted to include, besides Gayre and Gates, the general secretary of the Eugenics Society, Colin Bertram, in the planning for the race research institute.[64] In 1958 Bartram, partly in opposition to the restraint of the eugenics societies in race questions, had spoken out in the name of the Eugenics Society for a limitation of immigration from the West Indies and had thrown out the question as to whether race mixing might possibly be harmful.[65]

Setting up an institute for race research did not pan out and neither did the exchange of European and North American scientists. Still, the IAAEE did have some success in what had long been a central goal, the "establishment of the scientific journal for distributing material on race and race problems."[66] The very reserved attitude of the American and British eugenics societies on questions of race and the rising influence of social anthropologists in the anthropological societies had resulted in a situation where academics were hardly able to place their work in the English language journals on the defective intelligence of Blacks, on genetically related increased criminality of Afro-Americans, and on the dangers of race mixing.

Gayre and Gates claimed that an "anti-racist hysteria, [fanned by] liberals and Jews," dominated the atmosphere. Just as in Germany before Hitler, the Jews were said to hold the universities in their grasp. "Jewish voices" blared out from broadcasting stations; they beat the drum against race theories and perverted the "true facts of anthropology." Anyone who questioned the "doctrines of equality," wrote Gayre to Gates, found it almost impossible to publish articles.[67]

The growing scientific isolation led a group of IAAEE members, supported by Northern League activists, to revive the *Mankind Quarterly,* which was to be an international journal on race questions in the areas of ethnology, ethno- and human genetics, ethnopsychology, race history, demography, and anthropogeography.[68] The Northern League had been founded at the end of the 1950s on the initiative of the anthropologist Roger Pearson as a point of meeting for prominent Fascist and racist individuals. Members included the race theoretician Hans F. K. Günther; Ernest Sevier Cox, the head of the American Ku Klux Klan; the British Fascist Oliver Gilbert; Martin Webster from the British National Front; and the one-time SS officer Wilhelm Kusserow.[69]

The league called itself a "pan-Nordic culture society," whose task was to safeguard the biological and cultural patrimony of the Northern European peoples in all parts of the earth.[70] It understood its immediate goal to be the struggle against "external threats" from "enemy populations" supported by Communism and against "internal dangers" such as the "collapse of the cultural and biological patrimony" caused by the immigration of "alien peoples."[71] Race and eugenics, politics, and science were merged into one and the same thing in the ideology of the Northern League, quite in the tradition of the early race-oriented eugenics. In the words of Pearson's *Northlander,* the goals of "eugenicists" and "racists" fit together. Since a race is simply a group of people with the same hereditary patrimony, the efforts of the racists coincided with the goal of eugenicists for "racial improvement."[72]

In the view of the three editors, Gates, Gayre, and Garrett, *Mankind Quarterly* was to take up the neglected "race aspects of human heredity" that had been neglected during the past twenty years.[73] Garrett declared it as "highly desirable" that it should be made clear through *Mankind Quarterly* that not everyone who believed in race differences "burn crosses and go around in bed sheets" like members of the Ku Klux Klan.[74] *Mankind Quarterly* was to be a scientific journal that would take an explicit position on questions of race. Its authors had to establish in the heads of "intelligent educated people" the conviction that "race exists" and that "race, physical, intellectual, and spiritual qualities" were transferred genetically. These "facts" were once believed, but "constant propaganda" had concealed them. Through *Mankind Quarterly* the "truth of these facts" could be reestablished.[75]

Gayre and Gates agreed that a "genetic and ethnological concept of race" absolutely legitimized the position of the apartheid regimes in the American South and in South Africa. Gayre maintained that the antiracist "hysteria" in the United States and Europe had brought about the state of affairs that one could not too openly show political support, and initially one had to behave "Jesuitically." Only gradually would the "problem of the coloreds" be addressed more clearly in *Mankind Quarterly.*[76]

Once again it was Wickliffe Draper who took over financing of *Mankind Quarterly*. It was important for Draper that his financial support should be kept confidential. Through his attorney Weyher, he initially had money sent to Gayre for each issue of *Mankind Quarterly*, and then in 1962 he guaranteed financing for another four years under the condition that the journal would maintain its present "objective attitude on racial problems."[77]

The makeup of the editorial board, which was to secure the scientific reputation of *Mankind Quarterly*, reflected the entire group of leading race researchers in the 1960s. With Torsten Sjögren and Corrado Gini on board as consultants for *Mankind Quarterly*, they had two activists from the dissolved International Federation of Eugenic Organizations. In addition to Gini, two other Italian scientists appeared, the geneticist Luigi Gedda and the anthropologist Sergio Sergi, at whose institute Gayre had given a number of lectures in the 1950s.[78] Other members included Clarence P. Oliver and Sir Charles Galton Darwin, two prominent eugenicists whose extreme positions had made them increasingly isolated in the British and American Eugenics Societies. Henry Garrett made sure that an academic involved in research on the intelligence of African Americans was included by bringing on his well-known student Audrey M. Shuey, financially supported by the Pioneer Fund.[79]

The editors of *Mankind Quarterly,* supported in their work by Swan, Kuttner, and Gregor, were very cautious about letting their political orientation be known immediately. As Gayre expressed it, to nip in the bud beforehand any criticism about a racist orientation of the editorial board, one non-European, the Indian P. C. Biswas, was also taken onto the board. Later on they added the Japanese eugenicist Taku Komai.[80] When Gates considered taking Putnam onto the editorial board in 1962, Putnam refused after consulting with Garrett and Weyher. He felt that it might harm the journal if a nonscientist who was also quite well known for his speaking out in favor of segregation of the races would openly become part of *Mankind Quarterly*.[81]

Accepting German scientists onto the advisory board turned out to be difficult. The National Socialist race theoretician Hans Günther, who had close working relationships with Gayre, even before the appearance of the first edition had given the editors a long list of names of German academics who were finding it difficult to publish in Germany because of their connection with National Socialism.[82] In order not to publicize the close relationships with the Nazi race theoreticians, Gayre and Gates decided to bring in the German scientists onto the editorial board only gradually, and initially only occasionally to print articles by German race theoreticians.[83] And so after the psychiatrist Hans Burkhardt and the anthropologist Ilse Schwidetzky in 1961 and Walter Scheidt in 1965 were the first Germans to be published, only in 1968 were Otmar von Verschuer and the one time SS officer Heinrich Schade accepted onto the advisory board.

The close connection between scientific racism and race policies in the 1960s, strengthened by the isolation of these race scientists in the scholarly world, resulted in a situation where *Mankind Quarterly* essentially found readers in extreme right-wing circles. Gayre knew very well that very few members of scientific societies would subscribe to the journal, and for that reason he moved to the extreme to expand the readership circle for the journal. He used the channels of the Northern League, South Africa House, the North Rhodesia Society, and the racist magazine put out by Willis A. Carto, *Right*.[84] The editorial board of *Mankind Quarterly* attempted to keep secret its close connections to these organizations on the extreme Right. If the relations between Gates, Gayre, Garrett, Gini, or Swan and the extreme Right organizations were to become known, critics would have an ideal point of attack and where possible would destroy the image of scientific respectability that the journal was striving for.

Strategic machinations, however, could not prevent formal attacks in several scientific publications immediately after the appearance of the first edition for being a racist propaganda publication. *Man*, the journal of the British anthropologists, expressed the hope that *Mankind Quarterly* would halt publication as soon as possible before it further "discredited anthropology" and caused "even more harm to humanity."[85] In *Current Anthropology*, the Mexican anthropologist Juan Comas, supported by geneticists such as Dobzhansky, Haldane, and Nachtsheim, basically questioned the scientific seriousness of *Mankind Quarterly*. Comas called the racism clothed in scientific garb in the journal an insult to all of anthropology.[86] Similar cares were expressed by L. C. Dunn, pointing out that *Mankind Quarterly* was attaching itself to the "racist attitudes of an earlier period."[87]

The editors found especially uncomfortable criticism from two academics that had originally been associated with the magazine. The anthropologist Umar

Rolf Ehrenfels complained in *Current Anthropology* that the editors of *Mankind Quarterly* had censored an article that he had published in the first edition. Two paragraphs in which he had criticized the apartheid policies of the South African and Rhodesian governments had been deleted without his consent.[88] Bŏzo Skerlj, for a long time editor of the Journal *Eugenika* and professor of anthropology at the University of Ljubljana, had been nominated by Gayre to the scientific advisory council because the acceptance of an academic from a Communist country was meant to show how "tolerant" *Mankind Quarterly* was.[89] However, right after the first edition Sklerj resigned his position on the council and said that *Mankind Quarterly* was so distorted by racial prejudice that he could not reconcile his cooperation with his "conscience as a researcher."[90]

Any criticism of *Mankind Quarterly* was rejected by the editors as politically and not scientifically motivated. Gayre was convinced that all of social science and a large part of natural science were dominated by "cryptocommunists."[91] Garrett complained that any form of race research was being blocked by a coalition of Communists, Jewish organizations, and scientists in the Boas tradition. In his view, the "egalitarian dogma" threatened to become the scientific falsification of the century.[92]

The staff of *Mankind Quarterly* used the tactic of presenting themselves as objective, value-free scientists, while the critics as a group were defamed as politically blind Lamarckians and Marxists. Clariette Armstong assumed that the critical scientists supported a "communist movement," which desired to destroy the "white population" by "Negro riots" and the like.[93] Gates called Comas's criticism of *Mankind Quarterly* "the most extreme example I have seen of political propaganda." In the words of Gates, a man who "lowers himself to such perversions of scientific truth" loses the right to call himself a scientist.[94] According to the editors' simplistic view of the world, even criticism from colleagues who were not materially interested in race research was written off as influenced by Marxists. In a letter to Draper, Gates reduced the criticism of Sklerj and Ehrenfels to the common denominator that they were both communists.[95]

The scholars involved in IAAEE and *Mankind Quarterly* in the 1960s developed a clear-cut bunker mentality. In that way they attached themselves to the early eugenics movement and understood their eugenic and racist hypotheses as pure scholarship, and as a result any criticism would have to be politically motivated.

The Futile Connection of Race Research with Established Scientific Endeavors: The Balancing Act between White Extremism and Accepted Science

The structure of the network of racist eugenicists after the Second World War bore a certain similarity to the structure of the eugenics movement in its initial phase. Just as in the early days of the eugenics movement, the network around the IAAEE and *Mankind Quarterly* brought together persons from various disciplines, such as psychology, biology, anthropology, political science, and sociology. And just as at the beginning of the twentieth century, the field of eugenicists interested in race research after 1945 included, in addition to academics who initially had a good reputation in their disciplines, laypersons who wanted to see the scientific legitimation of their program of race politics. The group also included politically motivated amateur scientists, who quite often at a late point in their lives were entering what was for them the new field of genetics, long after they had made significant careers in another scientific field.[96]

The main problem of the small international network of race researchers was that the members were unable to get connected either to scientific or current political discussions. At that time the European and North American economies were experiencing tremendous growth. There was a widely shared hope that the social misery in the industrial countries in the Third World could be reduced through policies of social and political development.[97] Most academic studies from the milieu of race scientists were ignored in both political and academic circles. Researchers such as Corrado Gini, Raymond Cattell, Henry Garrett, Luigi Gedda, and Torsten Sjögren were not successful in finding a place on the agenda of the day for genetic differences between the races for their respective disciplines of psychology, psychiatry, human genetics, anthropology, and demography.[98] One simply could not make any headway in these disciplines with race-specific research questions at this time of rapidly diversifying academic disciplines.

The situation changed a bit starting in the 1970s as a result of a broad biological countermovement to what was considered "overemphasis on and environmental influences" in the social sciences. In the opinion of some scientists, the student movement had gained a monopoly of opinion over the central political and scientific questions of the day and had placed a taboo on topics such as "genetically related race differences" and "influence of genetics on human behavior." Emboldened by the economic crisis of the early 1970s, a conservative movement arose against the supposedly scientifically questionable and politically imposed egalitarianism.[99]

In the United States this movement was strengthened by the fact that the movement for civil rights for blacks at the end of the 1960s was losing some of its moral authority. As long as the movement was involved in access for all students to schools, universities, restaurants, buses, and places of public accommodation, broad sections of the population stood behind it. However, when movement activists started to demand preferred treatment of African Americans as reparations for injustices suffered, they ran into opposition in particular from the white American middle class. Easier access for African Americans to places at universities, to jobs, and to public support under the umbrella of what were called affirmative action programs threatened the privileges of the white middle class already weakened by the economic crisis. The anxieties of the white middle class in the face of social and economic decline offered an ideal breeding ground for the revival of race theories in the United States.[100]

The UNESCO declarations on the race question had been primarily the result of political attitudes formed by the rejection of National Socialist race policies; they could not prevent a new debate. In the 1960s discussion had been limited to marginal academic figures of the extreme political Right, but in the 1970s that debate began to expand. The exchanges on race research, taking place primarily in the United States, centered on the question that had aroused conflict in the discussion of the first UNESCO declaration, namely, the supposedly psychological differences between races, particularly in the area of intelligence.[101] The revival of race theories this time did not come from human genetics, anthropology, or demography, but primarily from the field of psychology.[102]

Arthur Jensen, a psychologist teaching at the University of California at Berkeley, started the ball rolling with a rather long article on the failure in school of the less intelligent. Jensen's thesis was that differences in individuals' intelligence were 80 percent determined by genetic factors and therefore could be raised only slightly by better environmental conditions. He gave this well-known genetic-deterministic outlook a racist twist by claiming that the IQs of Negroes were an average of 15 points below those of the "whites." Jensen explicitly attacked attempts to make up for the poor showing of African Americans in education by instituting special educational programs.[103] He was sharply criticized because of his equating IQ test results with the imprecise concept of intelligence, for his artificial division of environmental influences from genetic factors, for his very vague concepts of race, and for the political implications of his hypotheses, but he was supported by a number of colleagues.[104]

Hans J. Eysenck, with whom Jensen had studied in London, supported his one-time student by books and articles written for the general public. The London psychology professor, who until 1970 had never done any race studies on his own, relied primarily on Audrey Shuey's work on "Negro intelligence,' but he declared that the difference in intelligence between Blacks and Whites was scientifically proven. Since one could assume that the white race was intellectually superior to the other "races,' this was for Eysenck sufficient to explain the progress of civilization in European countries as opposed to those outside of Europe.[105]

Richard Herrnstein, a psychologist at Harvard University, picked up on Jensen's and Eysenck's hypotheses. He explained that the more society gave equal opportunity to all people, the more crucial would be in-born intelligence for one's position in society. If social origins were to decline in significance, then one's social position would be primarily determined by one's genetic makeup. In this way Herrnstein, who for some time hesitated to express any opinions about differences in the races, gave social strata in society not only a biological explanation, but he also gave scientific legitimation to these differences. As the historian of science Federico Di Trocchio pointed out very clearly, Herrnstein's arguments said simply that the poor are poor because their parents were poor and stupid, and the reverse, that the rich are rich because they overwhelmingly have intelligent parents.[106]

A central weak point common to the arguments of Jensen, Eysenck, and Herrnstein was that they had to continue the notion of distinguishing among races by reaching back to a contested concept of race typing. Leading human geneticists such as Theodosius Dobzhansky, Walter F. Bodmer, and Luigi Luca Cavalli-Sforza stated that it was scarcely possible to define a hereditary patrimony specific to a race, and this was something that made the distinction among "blacks," "whites," and "yellows" in the 1970s seem so purely arbitrary to many scholars.[107] The Oxford scholar John R. Baker tried to make up for this weakness in the racist set of arguments by studying the biological, anatomical, physiological, and evolutionary differences among the races. Baker was convinced that all races differed in their intellectual abilities and was of the opinion that the concept of "superior" and "inferior" ethnic groups was completely justified.[108] In the end, outside of the network around *Mankind Quarterly*, these considerations essentially found no resonance in scientific work.

Only a group of younger psychologists, anthropologists, sociologists, and political scientists supported Jensen, Eysenck, Herrnstein, and Baker. These social scientists worked almost exclusively with correlation studies similar to those of Francis Galton and Karl Pearson and not with the new molecular biology methods, and in the 1970s and 1980s they went even further with research on intelligence. In this way they expanded the spectrum of explanations for race specific inferences to intellectual abilities. In the tradition of orthodox eugenics, they tried to use statistical calculations to show the supposed differences in talent and intelligence of the various "races" with relation to the rates of criminality.

For example, in building on the work of the Italian criminologist Cesare Lombroso, Nathaniel Weyl asserted that criminality was most widespread among the "races, peoples, and classes" who were the least intelligent and the least creative. For example, he claimed that the "American Negro" tended to have a higher rate of criminality because of his genetically determined lower intelligence.[109] Robert A. Gordon, a professor of sociology at the Johns Hopkins University in Baltimore, steered his criminology research to race and intelligence and traced the high rate of criminality of African Americans to their intellectual inferiority.[110]

J. Philippe Rushton, a Canadian professor of psychology, not only received extensive support from the Pioneer Fund but after the turn of the century also took over leadership of the fund. He pointed to various other social phenomena to be studied by race. Rushton gave the labels "Mongoloid" and "Negroid" to the two opposite poles on criteria such as intelligence, brain weight, size of penis, sexual maturity, frequency of sexual relations, aggression, friendliness, and law-abiding character. He numbered "Negroids" among the human groups with what he called "r-reproduction strategies," meaning that they had lower intelligence and a lower grade of social organization but a higher frequency of intercourse and reproduction. In contrast to "Mongoloids" and "Caucasoids," who very intensively cared for their children, the Negroids compensated for their biological disadvantage with a higher rate of reproduction. The resulting higher frequency of sexual intercourse explained for Rushton why AIDS appeared especially frequently among African Americans.[111]

When Jensen, Herrnstein, and Eysenck and their supporters first published their hypotheses among the broad public, a wave of criticism descended on them. The widespread and vehement criticism of psychological research on intelligence made clear that using biology to explain social inequalities now met a decided denial from a large number of scholars, as opposed to the way they had greeted eugenics in the 1920s. A whole series of geneticists, psychologists, and sociologists voiced the suspicion that race researchers were merely trying to find genetic legitimation for race discrimination. They reproached Jensen, Herrnstein, and Eysenck for being pseudoscientists and pointed to the fact that science, just as under National Socialism, could be misused for political purposes.[112] Groups such as the Society for the Psychological Study of Social Issues, the American Anthropological Association, the American Sociological Society, and the Genetics Society of America criticized the attempt to base the poor results of African

Americans on IQ tests in genetics by noting that it was scientifically not convincing and politically dangerous.[113]

Leading members of the American Eugenics Society joined in with this criticism, being very concerned about the renaissance of "racist eugenics" at the end of the 1960s. Walter F. Bodmer, for a long time one of the directors of the American Eugenics Society, together with his colleague Cavalli-Sforza, lamented that racism and eugenics "frequently seem to go hand in hand."[114] Frederick Osborn observed that there was no scientific basis for the hypothesis that "white people are on the average superior to Negroes." The studies on which the modern race researchers based themselves could properly be dismissed as "fatuous to the point of being childish."[115] Osborn's colleague at the American Eugenics Society, Bruce Wallace, called the ideas of the intelligence psychologists intellectually questionable and called for a complete stop to research on comparative intelligence among the races.[116]

The psychologists under attack answered with the reproach that there was a Marxist inspired conspiracy to repress the necessary race research. A politically motivated "egalitarianism" would lead to a situation where basic scientific knowledge about differences among the races would not be accepted.[117] Quite in the tradition of Gates, Gayre, and Garrett, the new race researchers Jensen, Eysenck, and Rushton defended themselves against their critics with a simple polarity—they viewed themselves as value-free and objective scholars. In their opinion their politically motivated opponents were blinded by Marxism. Jensen countered his critics by noting that it was not possible to prove that environmental factors affected the differences in IQ and that therefore theories built on milieu or education were "nonscientific."[118]

Eysenck claimed in *Mankind Quarterly* that practically every criticism of him and of Jensen was factually incorrect, publically motivated, and useless from the standpoint of the objective scholar.[119] In Eysenck's words, a "new Fascist left wing" was attempting forcibly to prevent the spread of scientific truth.[120] The Nobel Prize winner William Shockley, one of the chief promoters of a racially oriented eugenics in the 1970s, complained in *Playboy Magazine* that a "dogmatism" fit for the dark ages was blocking any "objective information" about the problem of human qualities.[121] Roger Pearson, basing himself on access to the personal files of Jensen, Shockley, and Rushton, put out a comprehensive study regarding the opposition of "Marxists and other left wingers" against research with race implications, and painted a picture of scientifically based research threatened by left-leaning "political ideologues." Unprejudiced race researchers were seen by Pearson as being prevented by left-wing extremists from using their research to contribute to the prevention of the "imminent catastrophe" of the economy and of civilization.[122]

Student-led protests, sometimes violent in nature, against public appearances by Jensen, Eysenck, and Shockley produced a situation wherein they perceived a dichotomy between politically blinded critics and "pure" scholars, who were subject to psychoterror, attempts at intimidation, and death threats.[123] Eysenck, Jensen, and Herrnstein, in conjunction with ten other American and British academics, succeeded in initiating a resolution defending them against "personal and professional disparagement."[124] Here they placed the criticism that they received for their theory of inheritance in the same tradition as the persecution of "well-known scientist victims, includ[ing],Galileo, in orthodox Italy; Darwin, in Victorian England; Einstein, in Hitler's Germany; and Mendelian biologists, in Stalin's Russia."[125] Many of the attacks against the heredity researchers were seen to be made by "non-scientists" or "outspoken enemies of a scientific attitude." Attacks were also deemed to come from "academics who had determined that any explanation of differences among humans could be traced to the environmental theory." In the light of the widespread "environment oriented orthodoxy," there was practically a "witch hunt" against those who held ideas based on heredity or a wish to develop further research work in the area of the biological bases of behavior. In light of the strong resistance against university instructors, researchers, and scientists who followed biological lines of thought, the initiators determined that "hereditary influences" had a strong effect on the qualifications and mode of behavior in human beings, and demanded further "study of the biological bases of behavior." "We deplore the evasion of hereditary reasoning in current textbooks, and the failure to give responsible weight to heredity in disciplines such as sociology, social psychology, social anthropology, educational psychology, psychological measurement and many others."[126]

Since the racist implications of research on heredity by Jensen and Eysenck were discussed only indirectly in the resolution, the initiators persuaded another 64 North American and European scientists to sign, including three Nobel Prize winners. Even some leading members of the British and American Eugenics societies were found among the signers, such as Otis Dudley Duncan, vice president of the American Eugenics Society from 1969 to 1972; Bruce K. Eckland, director of the American Eugenics Society from 1968 to 1982; Garrett Hardin, one of the directors of the American Eugenics Society from 1971 to 1974; and Eliot Slater, one of the fellows of the British Eugenics Society for over 20 years. In Germany the signers who answered the call were not only those known for their eugenic and racist outlook, such as Heinrich Schade, Ilse Schwidetzky, Hans Wilhelm Jürgens, and Friedrich Panse, but also well-known human geneticists such as Friedrich Vogel and Georg Wendt.[127]

The simple dichotomy made in the declaration between the value-free heredity researchers as opposed to their politically blinded critics overlooked the fact of how strongly Jensen, Eysenck, Baker, and Shockley were tightly connected to the extreme right-wing scene. The analysis of an exchange of letters that recently has become available in the archives shows a yoke of three journals with very close connections to the IAAEE binding together race researchers and the extreme political Right: the English language *Mankind Quarterly,* the French language *Nouvelle Ecole,* and the German language *Neue Anthropologie.*

Under the impact of worldwide student unrest, in 1969 several intellectuals from the extreme Right in France founded the Groupement de recherche et d'études pour la civilisation européenne (GRECE—Research and Study Group for European Civilization). The theoretician at the head of the group, Alain de Benoist, as well as the first presidents of GRECE, Roger Lamoine and Jean-Claude Valla, came from the ranks of traditional French right-wing extremists.[128] Under the pseudonym Fabrice Laroche, Benoist in the 1960s wrote for such publications as the magazine *Western Destiny,* the successor to Roger Pearson's *Northlander.*[129] Although Benoist at first flirted with white terrorist organizations, increasingly he devoted his activities to building up GRECE as a type of "anti-egalitarian" think tank, seeking to influence the public debate through its publications.[130]

GRECE and its closely associated newspapers, *Nouvelle Ecole* and *Éléments*, adopted orthodox eugenics as the image of human beings as being overwhelmingly determined by biological evolution and by race. They attempted to make use of eugenics, race research, and sociobiology in the struggle against the basic egalitarian values of the French Revolution. In opposition to "Liberty, Equality, Fraternity,' the new French Right emphasized belonging to a new national community, one that included the natural inequality of races and the ideology of a self-creating elite.[131] Benoist made use of the usual mixture of studies of identical twins and the differing results for Blacks and Whites on intelligence tests to trace back the supposed differences in intelligence of Blacks to their genetic predisposition and not to unfavorable environmental circumstances.[132]

The French eugenicists from the very beginning were closely connected to the IAAEE. Donald Swan acted as general secretary of *Nouvelle Ecole* in the United States, and the French eugenicists referred regularly to the publications of the IAAEE.[133] The sponsoring committee of *Nouvelle Ecole* included scholars who were connected with *Mankind Quarterly*, such as John R. Baker, Luigi Gedda, Robert Gayre, Henry E. Garrett, Cyril D. Darlington, Robert E. Kuttner, Bertil J. Lundman, Stefan T. Possony, Ralph Scott, and Roger Pearson.[134]

Practically in parallel to GRECE, a number of eugenicists led by the Hamburg attorney and neo-Nazi Jürgen Rieger created the Gesellschaft für biologische Anthropologie, Eugenik und Verhaltensforschung (GbAEV—Society for Biological Anthropology, Eugenics, and Behavioral Research) and the journal *Neue Anthropologie*.[135] The GbAEV and *Neue Anthropologie* were by their own admission "closely allied" with the IAAEE. Rieger later became one of the leading promoters of the extreme Right in the Federal Republic of Germany, calling himself an intellectual disciple of Swan. Even during his university days, when he had been involved in the extreme right-wing organization Bund heimattreuer Jugend [Federation of Patriotic Youth], Rieger received all the materials available from IAAEE, which served as the basic source for his first personal work in race anthropology.[136] In his book *Rasse: Ein Problem auch für uns* [Race. A Problem for Us as Well], Rieger called the history of humanity a "history of race wars." He called for the revival of the "power" of the white race that "was lying dormant under the facade of homogenizing civilization."[137]

Many of those who worked with *Neue Anthropologie* and the Society for Biological Anthropology, Eugenics, and Behavioral Research maintained close ties with the extreme German Right wing. Rieger in the 1960s was involved in the ideological successor party to the NSDAP, the National Democratic Party of Germany (NPD), and later became vice chairman of the party. His close associate Rolf Kosiek represented the NPD in 1968 as a state parliament representative and in 1973 became a member of the NPD national board of directors. Hans Georg Amsel was part of the anti-Semitic newspaper *Mensch und Maß*.[138] For the scientific advisory board of *Neue Anthropologie*, Rieger gathered together some coworkers from the German extreme Right and some former activists from the German race hygiene movement, such as the long-time Rüdin colleague Karl Thums, and key figures from the international eugenic network such as Swan und Benoist.[139]

Like the IAAEE and *Mankind Quarterly*, the Society for Biological Anthropology represented—and still represents—an orthodox, race oriented eugenics. This was confirmed by the GbAEV, using a reference to studies by Robert C. Cook and Hans Wilhelm Jürgens, that "school training works as a sieve for talent and leads to a lack of fertility" if "the schools do not provide a biological basis based on eugenic feelings of responsibility" and so long as "marriages before the completion of university education and an abundance of children are not made possible by generous monetary support and tax breaks." The society went on to paint a terrifying picture of the German "population soon consisting only of a younger generation of anti-social people and a mix of illiterates from Southern Europe, the Mediterranean area, Asia, and Africa." Just like the race hygienists of the 1920s and 1930s, the GbAEV claims that "race mixtures result in the increase of many diseases, such as hip-knee luxation, tuberculosis, schizophrenia, and cancer," and therefore the "hereditary quality" of the population was deteriorating.[140]

The activists of the IAAEE developed a close working relationship in the 1970s among *Mankind Quarterly*, *Nouvelle Ecole*, and *Neue Anthropologie*. The magazines regularly reported on the content of their partner journals and reprinted each other's articles. The three of them tried to maintain their scientific credibility by working closely with established race researchers such as Jensen, Eysenck, and Baker. So Rieger was successful in winning over Cyril D. Darlington and Arthur Jensen to the scientific advisory board of *Neue Anthropologie*, while Hans Eysenck served for a long time on the scientific advisory boards of the other two publications. Right down to the 1990s he trumpeted in their pages the credibility of race research. How close the cooperation was between the established academics and the troika of *Neue Anthropologie*, *Mankind Quarterly*, and *Nouvelle Ecole* is evidenced in the fact that articles by Jensen, Eysenck, and Baker appeared word for word in the three journals. Baker, a member of the scientific advisory board of *Neue Anthropologie* and *Nouvelle Ecole*, published a short abstract of his race book in the Journal of the Society for Biological Anthropology, Eugenics, and Behavioral Research.[141] At the initiative of Putnam, translations then appeared in both *Mankind Quarterly* and in *Nouvelle Ecole*.[142]

The English-speaking part of the international network of race researchers continued to be funded primarily by the Pioneer Fund, which after the death of Draper in 1972 had foundation assets of $5 million. The Pioneer Fund, based on its foundation assets and its bylaws as a foundation, acted as a kind of financial permanent financier of race research. Interest income generated by the foundation assets was devoted to projects submitted from the network of *Mankind Quarterly* without subjecting them to the comprehensive scientific expert opinions that are customary for foundations. Since the purpose of the foundation is determined by the bylaws, and since the board of directors of the foundation can itself recruit new members for the board upon the death or resignation of board members, it is guaranteed that the funds available for research can be expended for items that contribute to "improvement of the race."[143]

For many years a large part of the project funds of the Pioneer Fund went for a research project on twins at the University of Minnesota, but the fund has also been involved in almost every piece of research that is intended to determine the psychological differences between the races.[144] Arthur Jensen alone between 1971 and 1992 received over a million dollars. J. Phillippe Rushton in the same period was able to receive over $770,000 in grant funds from Pioneer. William Shockley, who promoted breeding of the highly gifted and to this end made his own sperm available, received about $200,000 for his various activities in the area of race research. Other race researchers as well, such as the professor of education and psychologist R. Travis Osborne, the psychologist Richard Lynn, the sociologist and professor of education Linda Gottfredson, the sociologist Robert Gordon, and the philosopher Michael Levin, received generous amounts from the Pioneer Fund.[145]

Several associations that were funded by Pioneer served race researchers as a means to distribute their publications. The Foundation for Research and Education on Eugenics and Dysgenics (FREED), a group directed by Shockley and advised by IAAEE activist R. Travis Osborne, was supported by the fund to print and distribute pamphlets, articles, and press releases from race researchers.[146] The Foundation for Human Understanding (FHU) was the beneficiary of over $200,000, and it has marketed books and articles about race questions by Jensen, Herrnstein, and Osborne.[147] The Institute for the Study of Man, whose work was increasingly intertwined with that of the IAAEE, received significant amounts of money from the Pioneer Fund for its publications.[148]

The man emerging at the pivot of the far right-wing scene and at the center of activities for North American and European race researchers in the 1970s and 1980s was Roger Pearson. In the 1970s Pearson, backed financially by the Pioneer Fund, took over direction of the Institute for the Study of Man as well as that of *Mankind Quarterly*.[149] After taking over direction of the *Mankind Quarterly,* he attempted to reinforce international visibility in particular by taking European and American scientists onto the editorial board. He made sure that the eugenic orientation of the journal stayed true by expanding the editorial board with such persons as the Italian anthropologist Bruno Chiarelli, the psychologist Raymond B. Catell, the West German demographer Hans W. Jürgens, the East German genealogist Volkmar Weiss, and the American professor of education Seymour Itzkoff.[150]

Using the Institute for the Study of Man and the journal *Mankind Quarterly*, both closely associated with the IAAEE and both recipients of funds from the Pioneer Fund since the beginning of the 1970s, Pearson coordinated various initiatives in the area of race research.[151] In 1978 he was one of the organizers of the eleventh Conference of the World Anti-Communist League (WACL) in Washington, a gathering in which various Fascist, racist, and anti-Semitic organizations took part. After Pearson was forced to resign his position in the WACL in the 1980s because of his very close ties to Nazis, he tried to gain respect in circles of the new American Right, who had become dominant in important governmental decision-making centers upon the election of Ronald Reagan as president of the United States.[152]

Pearson started the Council for Social and Economic Studies and established contact with Jesse Helms, a man of the far Right and later chairman of the Senate Foreign Relations Committee. He also worked for quite some time in conservative

Republican think tanks such as the Heritage Foundation and the Foreign Policy Research Institute. How close his contacts to conservative American leadership circles were became clear when President Reagan in April 1982 praised Pearson's Council for Social and Economic Studies for its considerable contribution to publicizing and maintaining common "ideals and principles."[153]

During the 1970s and 1980s, in the internal correspondence of the recipients of money from the Pioneer Fund, sympathy reigned for the race policies of the National Socialists and an anti-Semitic undertone often was displayed, but later the network around the Pioneer Fund increasingly tried to distance itself from National Socialist race policies.[154] The fact that the Nazis had been engaged in a eugenics program was presented by the authors of an article in the *Mankind Quarterly* as a pure myth. In the words of Seymour Itzkoff, the Holocaust had been "a totally dysgenic program," with which one wanted to liberate a Europe dominated by Christianity from "superior intelligent challengers."[155] According to the literary scholar John Glad, Hitler, though partly under the influence of the eugenics movement, had been driven to discrimination and later extermination of the Jews by his worry that the Jews might be considered a genetically equal or even superior race.[156]

The Controversy about Racist Eugenics in the Mass Media

It would be an exaggeration to attribute to the international network of these new organizations the same significance as to the eugenics movement in the 1920s and 1930s. The eugenics organizations of the postwar period—the *Mankind Quarterly, Neue Anthropologie, Nouvelle Ecole*, the IAAEE, the Institute for the Study of Man, GRECE, the Society for Biological Anthropology and Eugenics, the Foundation for Human Understanding, and the Pioneer Fund—formed a more or less effective connective tissue between "respectable" academics in psychology such as Herrnstein, Jensen, and Eysenck, and the extreme Right scene, but their influence on public discussion was quite small.[157] That the race researchers in the 1980s and 1990 felt themselves compelled to find their support among right-wing extremists shows the comparatively minor influence that racist eugenics had at the end of the twentieth century.

Even if the influence of eugenically interested lobbying groups on national legislatures concerning race issues has dwindled, it should be noted that the biological mode of argumentation of racist eugenics has remained of interest not only to right-wing extremists but also to many right-wing conservative politicians. Especially in view of worsening economic circumstances in many industrialized countries, the growing burden of the social welfare budget, and the expanding tensions between ethnic and religious groups, the explanations offered by racist eugenicists have become attractive to some politicians.

In the mid-1990s, a growing conservative mood in the United States formed the groundwork for an attempt to inject the claims of the dysgenic effect of social welfare into the public debate. In their book *The Bell Curve*, the psychologist Richard J. Herrnstein and the political scientist-eugenicist Charles Murray adjusted the traditional eugenic arguments to current discussions in the United States and in a compact form delivered the material basis for racist and biological legitimation of conservative policies. Quite in the tradition of orthodox eugenics, they determined that the declining intelligence of the American population was responsible for increasing criminality, impoverishment, and the creation of slums in the cities. Thus they used the supposed deficient genetic status of Afro-Americans to explain the prevalence of poverty, violence, and unemployment.[158]

In Germany, in a key reminiscent of Herrnstein and Murray, a similar approach was taken by Thilo Sarrazin, a one-time senator of the city of Berlin and member of the executive board of Deutsche Bank.[159] While the Americans were interested primarily in the genetically rooted socioeconomic differences among "Blacks, Hispanics, and Whites,' Sarrazin focused on the differences between "native-born" and "Muslim immigrants." Still, just like his American counterparts, Sarrazin is concerned with a dysgenic development. Quite in the tradition of the early eugenicists, Sarrazin sees here "signs of collapse." According to him, families with an above-average number of children occur among the "societal layers removed from education,' all of whom have a lower intelligence.[160]

For Herrnstein and Murray as well as for Sarrazin, the scientific references for the eugenic conclusions were drawn in large part from data gathered by researchers in the entourage of the Pioneer Fund and from the journals *Mankind Quarterly, Neue Anthropologie,* and *Nouvelle Ecole.*[161] Herrnstein and Murray were closely tied into the network—all 17 academics from whom the two eugenicists drew legitimation for their theses belonged to the authors and editors of *Mankind Quarterly, Neue Anthropologie,* and *Nouvelle Ecole,* and no fewer than 13 of the researchers quoted by these two were regular recipients of financial support from the Pioneer Fund. However, Sarrazin gave the impression that he was totally unaware of the scientific sources of the problem area into which his book was quickly thrown.[162]

Herrnstein and Murray very consciously drew on references from Richard Lynn, board member of the Pioneer Fund and member of the editorial board of *Mankind Quarterly,* but Sarrazin, in his hastily put together book, seems to have been unclear on the problematic nature of his supposedly scientific sources. Richard Lynn, a member of the executive board of the Pioneer Fund and member of the editorial board of Mankind Quarterly, was quite conscious of his ideas as being drawn from Herrnstein and Murray. For Sarrazin, however, the connection to the Pioneer Fund network of Lynn and other authors that he cites seems not to have been clear. Sarrazin appears to have become aware only after the appearance of his book of the connections of Volkmar Weiss, whose ideas Sarrazin had extensively drawn on. Weiss, who provided sources to Sarrazin in the year before publication, not only is part of the network around *Mankind Quarterly* and *Nouvelle Ecole* but was also nominated by the neo-Nazi NPD party as an expert for an inquiry commission on demographic development.[163]

The style of these new popular scientific eugenic bestsellers from Murray and Herrnstein in the United States and Sarrazin in German is not significantly different from the popular scientific classics on race theory from the Americans Madison Grant and Alfred E. Wiggam, the Englishman Houston Stewart Chamberlain, the Frenchman Alexis Carrel, and the German Hans F. K. Günther.[164] In this new era, the promoters of eugenic ideas have not been recognized academics (except for Herrnstein) but persons who have made their reputation in politics. Their ideas are not being published in articles in scientific journals but as books in editions grown very large because of scandals they have engendered. As a rule no new researches are presented, only their own political positions with references to supporting research results. Even if some gross scientific errors occur in the text, the appearance of scientific status is bestowed by a large number of tables and graphics, a comprehensive footnote apparatus, and references to supposedly broadly accepted studies.[165] It is however curious that these popular scientific pleadings for a eugenic renaissance have taken on a national coloring, and the debates have remained confined within national borders. While the books by Herrnstein and Murray in the United States and by Sarrazin in Germany have occasioned vigorous debates, the controversies have not extended beyond these countries. Such attempts as Richard Lynn's book *The Global Bell Curve* failed to initiate an international discussion because this book had its sensational effect only in particular national mass media.

Even if attempts keep repeating—and repeat in the future—to put a race-based eugenics on the political agenda, this should not be interpreted as the renaissance of a race-based eugenics. Unlike the beginning of the twentieth century, when eugenic claims met broad acceptance and were discussed even in academic circles, the debates now live on primarily through nationally defined outrage in the mass media about the ideas of the neo-eugenicists interested in race questions.

CHAPTER NINE
THE DISSOLUTION OF THE EUGENICS
MOVEMENT: WILL THERE BE EUGENICS
WITHOUT EUGENICISTS?

If one were to be on the lookout in the twentieth century for active eugenicists, it would not be easy to find them.[1] We can hardly talk about an "International of Race Improvers." The vehemence with which critical scholars and politicians turn up as critics when there is only the slightest hint of eugenic approach is a clear sign of the extent to which active eugenicists have been pushed to the margins. The vigorous protests against the director of the German Federal Institute for Population Research, Charlotte Höhn, when she said in an interview that she regretted that one can no longer say that "the average intelligence of Africans is lower than that of other human beings," shows that in particular, racist eugenics of the sort practiced by Ploetz, Mjöen, and Davenport has been totally discredited in scholarship and in politics in Europe.[2] It is true that the vigorous argument about the *Bell Curve* by Herrnstein and Murray is certainly a sign that racist hypotheses at least in the United States are still open to discussion. The vigorous criticism also shows how much these representatives of an orthodox eugenics are scientifically isolated.

A number of indices show that the international network of racist eugenicists appears to be at least partially in a state of dissolution. The International Association for the Advancement of Ethnology and Eugenics has degenerated into an insignificant distribution center for racist and eugenics publications. *Neue Anthropologie* stopped publishing for a few years because Rieger was much too busy in court defending right-wing radicals who had been charged with playing down the Holocaust or charged with fire-bomb attacks against homes for political refugees. The Pioneer Fund was directed for a long time by a small clique of 70-year-olds, who essentially used up all the funds of the foundation. It is true that a younger successor, J. Philippe Rushton, was found to be director of the Pioneer Fund, but it is doubtful that the ongoing fund for race research will be able to continue beyond some point. So it is understandable that the race researchers promoted by the Pioneer Fund speak of themselves as the "last eugenicists."[3]

Even the Osborn-Blacker type of eugenics, distancing itself from race prejudice, has hardly any active supporters. At the end of the twentieth century, it was practically impossible to find an active reform eugenics movement in Europe or America. The reform-oriented eugenics societies that had existed after the war in particular in the United States and Great Britain have completely stopped their work and no longer play a meaningful role in either science or in politics. In Europe, at any rate, eugenics has become taboo, to the point where it would be public suicide for politicians or scholars to openly come out in favor of a state strategy for improvement of the race.

Up to this point, it is possible to speak of a decline or even of a complete disappearance of the eugenics movement. The eugenicists as consciously political and scholarly actors were primarily a phenomenon of the first half of the twentieth century and continued to hang on up until the 1970s, but by the beginning of the twenty-first century, the movement had more or less completely disappeared in Europe and America.

But should one conclude from this development that eugenics as a political or scientific concept is dead? Does eugenics cease to exist with the disappearance of the eugenics movement? A book on the international eugenics movement can obviously not satisfactorily answer such a question. "Eugenics without eugenicists" cannot be explained by a book that has intentionally placed organizations, journals, and conferences in the spotlight.

Nevertheless, the reconstruction of the international eugenics movement can offer a basis for understanding such a "eugenics without eugenicists" not only as a more or less undesired product of the new possibilities of gene technology, but at least partly can hearken back to certain strategic decisions of the reform eugenicists. In particular, the decisions of the reform eugenicists to distinguish between the research areas of human genetics and demography, which had grown almost symbiotically, and to support the turning over of the responsibility for eugenic reproduction to the individual, seem to have played a role in the rise of "eugenics without eugenicists."

A glance at the international eugenics movement gives the first hints that in the second half of the twentieth century, independent eugenic approaches emerged in whose train specific goals of the eugenicists lost their original sense; other goals, in a diluted form, became part of a broad social consensus.[4] This development was labeled with such terms as "neo-eugenics," "new eugenics," "newgenics," "postmodern eugenics," and "liberal eugenics."[5]

A decline in the influence of the eugenicists on international population policy starting in the 1960s was connected directly with this success of the overpopulation paradigm and the forming of a distinct field of population science (demography). Within a mere 20 years, the International Planned Parenthood Federation (IPPF), the Population Council, and various groups of demographers succeeded in establishing as a real concept the notion of "the danger of overpopulation" in the heads of the people of North America and Western Europe. What some critics called the "myth of overpopulation"—a direct connection between the size of the population and poverty, environmental pollution, and political instability—became well established in Western European thought.[6]

Relatively quickly after the Second World War, demographers succeeded in making a name for themselves as experts in the United Nations, the International Labor Office, United Nations Educational, Scientific, and Cultural Organization (UNESCO), and the World Health Organization. Moreover, in the 1960s, international organizations like the United Nations, the World Health Organization, the International Monetary Fund, and the United Nations Fund for Population Activities (UNFPA) accepted the idea of a worldwide "population problem." The large international organizations began to support massive programs for birth control in developing countries.[7] Since the 1960s, an international "family planning industry" has arisen, for which making available means of contraception is the be-all-and-end-all for reducing birthrates in the developing countries.[8]

This development was quite paradoxical for the eugenics movements in Western Europe and North America, the same ones that had set off the discussion of overpopulation. In the 1950s, when the eugenicists still had a strong influence on the discussion regarding population policy, they had to accept that measures to limit the population would be accepted only very slowly in the developing countries and the industrial states. Through the relatively open network organizations involved in population questions—the International Planned Parenthood Federation, the International Union for the Scientific Study of Population, the Population Council, the Population Department of the United Nations, and various national organizations—a direct influence of eugenicists on the discussion was assured. However, the more state and international authorities took on the struggle against the "population explosion," the less important became the direct influence of the eugenics movement. In the 1970s and 1980s, when the topic of "overpopulation" increasingly determined the policies of many governments, eugenicists were hardly able to have included any "points of view regarding quality" in the discussion of population questions. The success of the overpopulation paradigm was then at least one of the reasons for the decline of the influence of the eugenics movement on international population policy.

Similarly, a paradox for the eugenics movement appeared in the area of human genetics. Up until the end of the 1960s, eugenics societies had occupied a major position of influence in the development of human genetic counseling offices, but in this period, the few existing counseling offices had little to offer. To the small number of couples who turned to the counseling offices because of concern about the birth of a sick child, the advisers could really only suggest that one renounce reproduction based on what were frequently very vague estimates of risk.[9]

A fundamental change occurred in the 1970s with the spread of prenatal diagnostic tests developed by the human geneticists and with the introduction of abortion laws in various European and American countries. New molecular biology methods for determining any hereditary diseases prior to birth expanded the strategy of counseling offices by using hereditary prognoses to influence the reproductive behavior of groups at risk. With the help of amniocentesis, it could be determined whether a genetic defect is present in a fetus. Thanks to legalized abortion in many countries, a presumably genetically unsatisfactory birth could be prevented in the case of such a finding.

The simultaneous widespread introduction of amniocentesis for the prenatal discovery of genetic diseases along with the legalization of abortion led to an explosive growth in the number of genetic counseling offices. While the directors of the human genetic counseling offices in the 1940s and 1950s came almost exclusively from the ranks of the eugenics movement, the eugenic societies could not now directly profit from the boom in human genetic counseling. The scientific discipline of human genetics, increasingly differentiating itself and setting itself free of eugenics, took over leadership in the establishment of new counseling offices.

A basic change in the attitude of the broad public to questions of reproduction paralleled the boom in human genetic counseling sites and the successful establishment of an international population policy oriented overwhelmingly to quantitative restrictions, but this was a change that the eugenics movement was hardly able to profit from, despite its strong influence in the early phase. This meant that until the 1960s, the state was by definition given the right to influence the reproductive decisions of its citizens, but now it started to pull back. The individuals deciding for themselves in large measure regarding contraception, abortion, and the number of children was an expression of this retreat by European and American governments in questions of reproduction. Reproduction changed in a few decades from a public to a predominantly private matter.[10]

While such a development would have meant a way out for the interventionist policy of the orthodox eugenicists of the early twentieth century, it was the renunciation by the state of direct management of the reproductive behavior of its citizens that was compatible with the strategy of the reform eugenicists, who wished to turn responsibility for eugenic behavior back to the individual. With the direct management of reproductive behavior by the state, it was no longer to be expected that a government would force its citizens into eugenic reproductive behavior. More likely, a trend emerged in which free citizens under specific social conditions by their own decision would be able to act in the sense of reform eugenics by basing themselves on the new technological possibilities in the arena of reproduction technologies.

The new possibilities in the areas of human genetic diagnostic tests and of contraception, which were partially developed with the support of the eugenicists, made the classic instruments of negative eugenics partially superfluous—health certificate before marriage, placement in asylums, sterilization, destruction of "unworthy lives." Since the ideal of Galton and Ploetz on shifting selection to the level of the nucleus was technically not possible at the beginning of the twentieth century, the eugenicists at that time looked to placement in asylums, sterilization, and destruction of "unworthy lives" as rough means of reproduction management. Nearly 100 years after the rise of the international eugenics movement, Ploetz's and Galton's dream selection on the level of the nucleus could now increasingly be realized. With the technical knowledge regarding prenatal diagnostic tests and treatment of the nuclear cells, today genuinely useful instruments are available that would make state-managed reproduction easily available.

Human geneticists, who understand their role to be scientists independent of eugenics and sometimes even oppose eugenic policies, now make available in a value-free manner the genetic manipulation instruments for an individual decision on reproduction. The technological offering of the human geneticists for prenatal diagnostic tests and in the future possibly even gene therapy are now rarely legitimized with an argument for the genetic quality of the population. Rather, in the arguments for human genetic consultation offices, prenatal diagnostic tests, and in vitro treatment, the deciding role is played by the individual suffering of a handicapped child and its parents.[11]

The question is difficult to answer as to whether the decisions made on the basis of social conditions by couples are made in a eugenic sense. Today, most people are completely indifferent to the genetic level of the German, British, or US populations as a whole. It is also dubious as to whether the genetic patrimony of the population is greatly affected by parents not bringing into the world children who might later be handicapped or whether the increased abortion of defective fetuses is so significant in this regard.

Regardless of whether or not the new genetic modification instruments and societal conditions will lead to "reproduction planning from below," it is quite clear that by the end of the twentieth century, eugenic perspectives had been put aside. Eugenics in the twenty-first century will no longer flow from an international eugenics movement influencing state policies, but rather has become a eugenics unconsciously practiced by every person in the name of "self-determination and freedom of choice."[12]

Afterword

The first edition of this book appeared in German in the mid-1990s. Many times, my colleagues asked me why I did not publish it in English, as I did with my more narrowly focused book, *The Nazi Connection. Eugenics, American Racism and German National Socialism* (Kühl 1994). At the time, I assumed that most scholars, in addition to reading English, would at least have read the German and French literature. However, in reviewing the material on eugenics that has appeared in the twenty-first century, it has become clear to me that academics researching the history of eugenics outside of Europe often have difficulties with a book in German. As a result, a number of problems in understanding my ideas on the international eugenics movement have arisen.[1]

The publication of a significantly expanded and revised German edition of the book gave me the opportunity to prepare an English version as well. In this revision of the first edition, it became clear that I did not have to modify my theses. On the contrary, the studies that have been published since the appearance of the first German edition have extensively confirmed my hypotheses. For example, in the meantime, it has become clear that international networking played an important role in the establishment of national eugenics societies.[2] Even if eugenicists attempted to implement their eugenics programs through policies and individual countries, their eugenics programs, particularly in the first half of the twentieth century, had a strong international orientation, be it in the form of a "blonde international" or a "Race Confederation of European peoples."[3] In connection with my considerations on an international eugenics peace policy, recent studies for individual countries have shown that the eugenicists active there were very concerned with the dysgenic effects of the First World War. After the war, particular eugenic initiatives received significant support.[4] This development is also clearly seen in the way that human genetics and demography had split off from eugenics even before the Second World War, at least institutionally if not always in content. These two disciplines now pursued scientific goals, while eugenics became a political program.[5] National Socialist Germany emerged as an exception to this development in the 1930s, since, as I show, the political and scientific programs of eugenics were fused by active German race hygienists for the benefit of the Nazi state. Despite all international criticism, the representatives of German eugenics were successful at receiving the support of an established international eugenic network in the form of the International Federation of Eugenic Organizations.[6] In addition, a number of recent studies have shown that there was a direct line from the support network of National Socialist race policies to the renaissance of racist eugenics at the end of the twentieth century, not least through the support of the Pioneer Fund.[7]

In addition to correcting a number of minor errors, in the revised version, I have expanded in particular sections informed by the new knowledge gained from research during the last two decades. In describing the Oficina Central Panamericana de Eugenesia y Homicultura, the Fédération Internationale Latine des Sociétés d'Eugénique, the International Group for Human Heredity, and the International Union for the Scientific Investigation of Population Problems, I limit myself to information that explains how these organizations grew as part of the dominant International Federation of Eugenic Organizations, while at the same time partially distancing themselves from the International Federation. Monographs based on original sources regarding these organizations do not exist, and it is my hope that my book on the international movement for eugenics and race hygiene might stimulate research on them.

The focus of my revision has been to sharpen the comprehensive argument on the relationship between science and politics. Even more clearly than in the first edition, I show how the combination of racism, internationalism, and scientism initially gave strong impetus to the movement, while the close connection between science and politics became ever more problematic.

The attempt to present an analysis of the international eugenics movement in the twentieth century was a great challenge to me. The context of the international movement was so varied, the background of the individual eugenicists so complex, and the source material so difficult to get a handle on, that I hesitated for quite some time to take on such a project. Without the encouragement of many colleagues to at least try to describe and explain eugenics and race hygiene in an international context, this study would never have come about.

Gisela Bock of the Free University of Berlin in particular for more than ten years has supported my studies in eugenics and race hygiene. From my first modest research attempts at the beginning of my study, through my publications on the relationships of American eugenicists and German race hygienist under National Socialism, and down to this more comprehensive study regarding the international eugenics movement, she was at all times an irreplaceable support, from conception to completion of the final manuscript.

Peter Weingart of the University of Bielefeld was another important source of support. Precisely because we held partially opposing views regarding the development of eugenics, I had a chance to clarify my own arguments. He also allowed me to continue my research under the aegis of the Institute for the Study of Science and Technology of the University of Bielefeld.

I could not have done my work on this topic if I had not been accepted for extended periods as a visiting scholar by institutes run by Vernon Lidtke at Johns Hopkins University in Baltimore, Paul Weindling at the University of Oxford, and Jean-Paul Gaudillière at Université Paris-7-Jussieu. Among all the colleagues who helped me with my manuscript with discussion and comments, I would like to thank in particular Michael Schwartz in Potsdam, Lene Koch in Copenhagen, Nils Roll-Hansen in Oslo, Benoit Massin in Paris, and Barry Mehler in Grand Rapids, Michigan.

During my work on a history of eugenics and race hygiene, I was supported by various institutions with fellowships and research grants. For this support, I would like to thank the Friedrich-Ebert-Stiftung in Bonn, the Deutscher Akademischer Auslandsdienst in Bielefeld, the Westfälisch-Lippische Universitätsgesellschaft in Bielefeld, the Department of Sociology of the University of Bielefeld, the German Marshall Fund of the United States in Berlin, the German Historical Institute in London, and the Wiener Library in London.

Notes

Introduction

1. In accordance with the custom in the international eugenics movement, in this book, "eugenics" and "race hygiene" will be used synonymously. Although at various times there were discussions in some national eugenics organizations as to whether the two concepts were identical, in the realm of international cooperation of eugenicists, both were generally accepted as having the same meaning. On this, see Turda 2007: 1997 and 2010b: 64ff.
2. For this perspective in the early historical studies of eugenics, see Mosse 1964, 1978; Altner 1968; Gasman 1971; Mühlen 1977. Splitting eugenics away from National Socialism has in the meantime become the established standard in historical writing. On this, see most recently Stern 2005: 2 and Turda 2010b: 1.
3. See the early studies by Adams (1990a) on the Soviet Union, Dikötter (1989) on China, Stepan (1991) on Brazil and other South American countries, and Suzuki (1975) on Japan.
4. See for example the early studies by Schwartz (1995a, 1995b) and Weindling (1987) on socialist and liberal eugenicists, the book by Cleminson (2000) on anarchist eugenicists, and the works by Allen (1988, 1991) on feminist eugenicists.
5. See, for example, Rosen (2004), particularly for the eugenics policies of the Protestants and Catholics in the United States; Leon (2004) in the American Eugenics Society; Richter (2001) on Catholicism and eugenics in Germany; Löscher (2009) on "Catholic eugenics" in Austria; and Falk (2006 and 2010), and Lipphardt (2009) on "Jewish eugenics." John Glad's book (2011: 112ff), which should be read not as an historical monograph but as a contemporary plea for a "Jewish eugenics," has a collection of positive statements by Jews about eugenics.
6. See, for example, the presentation of scientifically important eugenicists in Kevles 1985; Weingart et al. 1988; Weindling 1989a; Mazumdar 1992.
7. This point of view has in the meantime been reinforced. One need only look at the comprehensive *History of Eugenics* (Levine and Bashford 2010), which attempts to show the variety of expressions of eugenics.
8. The concentration on the international forms of cooperation that came about in the twentieth century among eugenicists, race hygienists, demographers, and human geneticists necessarily carries certain limitations with it. From my perspective of the history of international organizations and meetings, I stand in a tradition that presents the organizational history of eugenics and the history of ideas. I cannot add very much to the very praiseworthy attempts to write a political and social history of eugenics, attempts that are being made particularly in Germany and the United States. While it is relatively easy to delineate the scientific, political, economic, and social background conditions of national eugenics movements, the international context is extraordinarily

complex. In my focus on the international movement, I must limit myself to the general political and scientific context (world wars, world economic crisis, worldwide migration, National Socialism, Stalinism, the movement for citizens rights for blacks, the student movement, correlation calculations, Mendelianism, Weismannism, molecular biology).
9. Cf. for example, Semmel 1958: 113; Marten 1983: 174; Kevles 1985: 23; Weiss 1987: 82; Niemann-Findeisen 2004, 12ff. Weindling (1989a: 64) calls attention to the connection between the internationalist outlook and the German expansionist fixation of the race hygienist Alfred Ploetz.
10. Cf. for example, Ludmerer 1972: 148; Kevles 1985: 164–175; Roll-Hansen 1989a, and recently Ekberg 2007, 590ff. Naturally, Germany is considered an exception.
11. On this, see two shorter works that summarize the argument in a compact fashion for a broad public (Kühl 1999).
12. Cf. for example, Gasman 1971; Schmuhl 1987; Tucker 1995; Quine 1996: 12. At this point, I will not go into the problems regarding the concept of Social Darwinism Bannister 1979; Brautigam 1990. See also Carol (1995: 145), who summarily refers to National Socialist eugenics as nationalistic and warmongering.
13. Cf. for example, Ludmerer 1972; Kevles 1985; Weingart et al. 1988; Weindling 1989a; Schneider 1990a. The works that attempt even cursorily to describe the development of eugenics down to the 1960s are based on comparatively meager archival studies. These works can therefore analyze the role of the eugenics societies after 1945 only in a preliminary fashion.
14. This point of view has become widely accepted since the appearance of the first German language edition of this book. See Jackson 2005.
15. The concepts of "race eugenics" or "racist international" are not meant as value judgments, but rather are used to permit an analytical division of eugenicists who primarily were concerned with the "differentiation" of a specific racially defined group from those eugenicists who wished to improve the hereditary patrimony independent of ethnic distinctions.
16. On the determination of values, see Luhmann 1972: 88ff. It has become increasingly emphasized in the historiography of eugenics that eugenics cannot be defined as having a single, unambiguous goal, but is better understood as a value (e.g., Koch 2006: 308).
17. Here I dissent from Luhmann's very narrow definition of the movement as only one of protest (1991: 135ff; 1997: 852).
18. Neidhardt 1985: 194ff. See also Neidhart and Rucht 2001: 541.
19. On the concept of racism, see the summaries by Miles 1989; Fredrickson 2004; and Bogner 2003. Both concepts, racism and egalitarianism, are here presented as sociological categories, and not, as customary among eugenicists, as political campaign goals.
20. See Bock 1986: 16 on the classification of eugenic and ethnic racism. Voegelin 1948 and Olson 2008 on the emergence of the scientist approach.
21. The conception of "science" used in this book is based on the assumption that science as something searching for the "objective," an abstract truth, does not exist. It is therefore also not the goal of this study to discuss whether or not the information obtained by race researchers was "true" or "false."
22. The relationship to other functional systems cannot be discussed in this book. In the meantime, however, a whole series of good studies have appeared that describe the role of eugenics in the mass media and in religion (e.g., Pernick 1996 and Cogdell 2004).
23. On the various logics of politics and science, see the differentiation theory of Niklas Luhmann. On politics, see in particular Luhmann 2005a: 195ff; 2002: 69ff. On science, Luhmann 2005b: 293ff; 1992: 271ff and 630.
24. Despite these differing logics—or better, perhaps because of them—politics and science seek to influence each other in their own particular ways. The politicization of

science—meaning the attempt to use politics to bring about the production of particular scientific knowledge—is just as much a reality as is the "scientification" of politics, the effect that science has on political decision processes.
25. On the spatial, temporal, social, and content dimensions of universality, see the illuminating Stichweh 2003 and 2005.
26. Obviously, this does not rule out the reality that eugenics for a time played a very strong role in the emergence of national self-definition. For example, the role of eugenics in Southeastern Europe (Cf. Promitzer et al. 2011: 17) and in the United States (Cf. Ordover 2003: 3ff).
27. To my knowledge, this understanding comes from Frank Dikötter (1998: 467), who summarizes very well the emerging dominant view of historians.
28. See Stichweh's observations (1984: 13) that the internal differentiation into disciplines was not the result of increasing differentiation in science, but was a decisive precondition for such change in academic pursuits.
29. On this point, see Ash 2001: 117.

1 The Dream of the Genetic Improvement of Mankind—The Formation of the International Eugenics Movement

1. Gilman 1914: 277.
2. Galton, "Eugenics, Its Definition, Scope, and Aims," lecture on May 16, 1904. FGA London, 138–139.
3. See the extensive information on Galton in Gilham 2001 and Brookes 2004. Louçã 2008 has studied the early convergence of eugenics and statistics in the work of Francis Galton and Karl Pearson.
4. A brief view of amateur science—Stichweh 1984: 74ff.
5. From a flyer of the Eugenics Education Society: "The phrase 'National Eugenic,' as defined by Mr. Galton, is the study of agencies under social control that may improve or impair the racial qualities of future generations either physically or mentally." KPA London, 685/3. This definition was used by Galton in an announcement of a research grant for eugenics. Cf. Blacker 1952: 109.
6. Galton 1865; Cf. also Galton 1908: 310; see also Kingsland 1988: 185.
7. Galton 1869.
8. Galton 1908: 323.
9. Galton 1883: 334ff; Cf. Blacker 1952: 96.
10. See Carnegie 1889: 655ff; Ghent 1902: 29. Cf. Hofstadter 1955: 45; Ebbinghaus 1984: 25; Penselin and Strob 1988: 37ff; Mertens 1991: 39–43. I have adopted the concept of evolutionary Social Darwinism from Schmuhl (1984: 49–58) because he very aptly delimits these specific variants of Social Darwinism from other political applications of Darwin's theory of evolution. For a discussion of the concept of Social Darwinism, see Jones 1980; Clark 1984; Weikart 2004.
11. Wagner (1995) speaks in this connection of the "first crisis of the modern age" at the end of the nineteenth century.
12. For example, Ploetz 1895: 196; Gruber 1909: 1993.
13. On the concept of civilization as a chain of Pyrrhic victories, Cf. Zmarzlik 1963; Aldrich 1975: 34; Schmuhl 1987: 357; Roger 1989: 127ff.
14. Cf. Sieferle 1989: 127; Schwartz 1995b: 405.
15. On the basic governmental and economic changes around 1900, see Weinstein 1968; Wiebe 1976; Noble 1977.
16. Cf. Kranz 1984: 242; Mitchell 1985: 154–157; Allen 1986: 225; Wichterich 1994: 11.

17. The concept of "rationalization of sexual life" comes from the interwar period. This was the way it was discussed by the Committee for Family and Population Questions of the Evangelical Protestant Church on September 5, 1930, under the title "Rationalization of sexual life," "conscious separation of the sexual act and the creative act," and the problems of the "regulation of birth as a moral, ethical, psychological, sociological, racial [völkisch], and human problem." Archiv des Diakonischen Werkes (ADW) Berlin CA/G 1700/1. The word was used in the early 1930s in a scientific context by Julius Wolf (1933: 85) and Friedrich Burgdörfer (1933: 352). The social hygienist Alfred Grotjahn utilized the concept "Rationalization of human propagation"; Cf. Weingart et al. 1988: 35; Labisch 1990: 342.
18. Cf. Jones 1988: 61; Mazumdar 1992: 3; Thomson and Weindling 1993: 137.
19. Cf. Schallmayer 1903: 245; see also the description of Schallmayer's eugenic notions in Märten 1983: 172; Weiss 1987: 86; Weindling 1989a: 320.
20. Ploetz 1895: 226–239; Cf. Graham 1977: 1137; and Weiss 1990: 18.
21. Ploetz 1895: 2.
22. Cf. Ploetz 1895: 16; see Heuer 1989: 13; Weiss 1990: 17; Reyer 1991: 16. On the debate over Ploetz's hypotheses in the context of the Deutsche Gesellschaft für Soziologie, see Schmuhl 1991 and Schleiff 2009.
23. Speech to the thirty-fifth meeting of the Deutscher Verein für öffentliche Gesundheitspflege in Elberfeld on October 16, 1910. Manuscript of the speech, APA Herrsching. Cf. Ploetz 1895: 5; 1911a; 1911b. On Ploetz's approach, see Weindling 2011a: 27ff.
24. Memorandum on the founding of the Internationale Gesellschaft für Rassenhygiene 1907, APA Herrsching
25. Cf. *American Breeders Magazine* (1910): 235. On the perspective of breeders in Germany, see Krüger 1998: 383ff.
26. Cf. Allen 1986.
27. Rüdin discusses the problems in a letter to Ploetz, March 14, 1907, APA Herrsching. Cf. Weber 1993: 74. On the recruitment of new members, see APA Herrsching, Jahresbericht 1907 (for the period from December 20, 1906, to December 31, 1907).
28. Cf. Weingart et al. 1988: 327.
29. Mjöen 1914: 346. Cf. Roll-Hansen 1980: 274ff; 1989a: 337; 1989b: 891. See also Lorimer 1932: 336ff.
30. See Weindling 2011a: 45 on the integration of southeastern European eugenicists.
31. Fourth general meeting in Berlin, APA Herrsching; see also Weindling 1989a: 150
32. Members of the International Society for Race Hygiene 1912, ESA London, SA/Eug. L. 33; Cf. Weindling 1989a: 151.
33. Cf. Fischer 1930: 2; see also Kröner 1980: 74ff; Roll-Hansen 1989a: 337; Weindling 1989a; 150ff: Weber 1993: 74; Broberg 1996; Tydén 1996:83; Björkman and Widmalm 2010: 381.
34. Ploetz to Pearson, March 21, 1909, APA Herrsching. Ploetz's hope of getting access to the British Eugenics Society through Karl Pearson was overblown. For his entire life, Pearson refused to become a member of the Eugenics Education Society. He strongly resisted any form of organizational affiliation. Pearson himself, and later the Eugenics Research Institute at University College in London, which he directed, were in constant struggle with the board of directors of the Eugenics Education Society. Still, until Galton's death in 1911, he retained a certain influence over the policies of the Eugenics Education Society.
35. Diary June 26–July 8 1910, APA Herrsching; Cf. Pearson to Ploetz, APA Herrsching, June 29, 1910; see also Muckermann 1930a: 265. On the programmatic agreement of the two societies, see *Eugenics Review* (1910): 91–93 and Schuster 1912b: 52ff.
36. Cf MacKenzie 1981: 84; Searle 1971, 1976.

37. Georgina Chambers, "Notes on the Early Days of the Eugenics Education Society," January 17, 1950, ESA London, SA/Eug B 11; Cf. Jones 1986, 19–24, 43–65; Mazumdar 1992: 34ff.
38. Cf. Gotto to Davenport, April 15, 1910 CDA, Philadelphia: Gotto; Darwin to Ploetz, November 23, 1911, APA Herrsching.
39. ESA, SA/Eug, L/I/: "1. Persistently to set forth the national importance of eugenics in order to modify public opinion and create a sense of responsibility in the respect of bringing all matters pertaining to human parenthood under the domination of eugenic ideals. 2. To spread the knowledge of the laws of heredity so far as they are surely known and so far as that knowledge might effect improvement of race. 3. To further eugenic teaching at home and in the schools." Translated into German by Ploetz 1909: 280.
40. Darwin 1912: 26–38; Cf. *Eugenics Review* 1 (1909): 5.
41. Session of the Council Meeting of the Eugenics Education Society, October 2, 1911, ESA London SA/Eug. I.2.: "1. That the...aims set forth by the International Society for Race Hygiene correctly express the objects which Eugenics Societies hold in view. 2. That an international Eugenic Council be formed of representatives of such Eugenic Societies, as defined above, as may be nominated at the International Congress of 1912. 3. It shall be the duty of this Council to decide when and where international congresses shall be held, and to frame rules with regard to their own constitution and duties which shall be submitted for the confirmation of the international congress, the general aim to further the progress of Eugenic reform." See also the letter from Gotto to Ploetz, October 5, 1911, APA Herrsching.
42. On Ploetz's views, see Ploetz's Diary, August 4, 1911, APA Herrsching. On the views of the British geneticists, see the Fourth Report of the Eugenics Education Society 1911–1912, ESA London, Eug/SA A5; *Eugenics Review* 3 (1911): 279; Cf. Gotto to Ploetz, June 16, 1911, and Pearson to Ploetz, June 29, 1919, APA Herrsching; see also Muckermann 1930a, 265.
43. Eugenics Education Society 1911: 2.
44. *Eugenics Review* 4 (1912); Cf. Laughlin 1934: 1ff.
45. Schuster 1912b: 49; Pearl (1912b: 395) talks of 836 registered participants.
46. "Schönheit und Politik," Vossiche Zeitung Berlin, August 8, 1912. This same Roberto Michels later was immensely influential on research into protest movements with his hypotheses on the iron law of oligarchy.
47. "The First International Congress for Racial Improvement" and "British and Germans at the London Congress for Racial Improvement," *Berliner* Tageblatt, August 1, 1912, and August 2, 1912.
48. Le Matin, August 9, 1912.
49. "What the Eugenicists Say," Public Opinion, August 2, 1912.
50. Mendel 1865. Cf. the popular scientific presentation by Bäumer 1990: 33–39 and Weß 1989: 25ff.
51. Weismann 1892.
52. While individual references to neo-Lamarckian or biometric heredity research appeared in the conference presentations, the international meeting was spared any basic confrontations among the heredity researchers. The most prominent Neo-Lamarckians, the Austrian Paul Kammerer and the Briton Ernst W. MacBride, stayed away from the congress despite their basic eugenic conviction, as did the anti-Mendelian Karl Pearson. Only a small group of new eugenicists at the congress defended a neo-Lamarckian position and sounded warnings, as did the Parisian biologist Frederic Houssay, of "exaggerated Weismannism." Cf. Eugenics Education Society 1913: 34.
53. Pearl 1912a: 47–57.
54. Punnet 1912: 137–138; Cf. *Lancet*, August 3, 1912, and Eugenics Education Society 1913: 17–19.

55. Weeks 1912: 78. The lecture was delivered by Wagenen; Cf. Eugenics Education Society 1913: 20.
56. For eugenic exhibitions, see the informative Rydell et al. 2006: 361ff.
57. Cf. Davenport 1911; see also Scheinfeld 1958: 146; Mazumdar 1992: 92ff.
58. Eugenics Education Society 1912: 2–45.
59. Based on the *Eugenics Review* 24 (1932): 3.
60. Eugenics Education Society 1912: 52ff; Cf. Mazumdar 1992: 89.
61. These arguments were also occasionally found before 1912. See Reyer 1991: 25ff for an extensive presentation.
62. France was an exception, where the birthrate decline started even earlier. Cf. Quine 1996: 52–89.
63. Hoffmann 1912: 337. Cf. Eugenics Education Society 1913: 22.
64. Eugenics Education Society 1913: 23.
65. Wagenen 1912: 462.
66. Wagenen 1912 and Eugenics Education Society 1913: 30–32.
67. Eugenics Education Society 1913: 51; Cf. *Lancet*, August 3, 1912, and The Nation, August 3, 1912; on Kropotkin's Darwinism, see Todes 1989: 123–142.
68. Smith 1912: 485; *Lancet*, August 3, 1912.
69. This paragraph is based on the research of John Boli and George M. Thomas (1999a). Their analysis of data from the Union of International Associations has permitted sorting out the founding of the International Eugenics Society. See also Barrett and Kurzmann (2004) for use of the data records to analyze the eugenics movement.
70. Gruber to Ploetz, July 7, 1912, APA Herrsching.
71. Cf. Ploetz's Diary Notes of July 26 to August 2, 1912, APA Herrsching.
72. Cf. Mann 1980: 338.
73. Eugenics Education Society 1913: 45.
74. Cf. Edmond Perrier, La Société française d'eugénique, EPA Paris, Ms. 2229 and Eugénique 1 (1913): 41–44. Extensive descriptions of the influence of the first eugenics congress on the founding of the French Society are found in Apert 1940; Schneider 1982; Leonard 1983; Schneider 1990a, 1990b; Carol 1995: 77–84; and Fogarty and Osborne 2010.
75. Cf. Fifth Annual Report of the Eugenics Education Society 1911–1912, ESA London Eug/SA, A5. On Italian eugenicists at the congress, see Quine 2010: 378ff, and for extensive information on the Comitato Italiano di Studi Eugenici, Mantovani 2004: 75ff and Cassata 2006b (English translation, Cassata 2011).
76. Cf. Fifth Annual Report of the Eugenics Education Society 1911–1912, ESA London Eug/SA, A5.
77. Cf. *Eugenics Review* 5 (1913): 192. See also Noordman 1989: 93.
78. Cf. Roll-Hansen 1996a: 200.
79. Cf. Fifth Annual Report of the Eugenics Education Society 1911–1912, ESA London Eug/SA, A5. See also Roll-Hansen 1989a: 339; Hansen 1996: 20; Koch 1996: 42ff. The specific reason for both the Norwegian and Danish groupings was that they primarily served to legitimize through the letterhead the individual initiatives of Mjöen and Hansen.
80. Cf. *Eugenics Review* 5 (1913): 286ff; Sixth Annual Report of the Eugenics Education Society, ESA London, SA/Eug A6.
81. See Maiocchi 1999: 10ff.
82. *Eugenics Review* 15 (1924): 645; Hodson 1935: 42.
83. *Eugenics Review* 25 (1933): 144f and Mjöen 1934c: 222; Cf. Roll-Hansen 1980: 275f; 1996a: 183.
84. The program developed by Mjöen was printed in the journal of the Norwegian race hygienists, *Den Nordiske Race*, and then translated into several languages. See, for example, *Eugenics Review* 13 (1922): 551 and Hodson 1935: 42.

85. Mjöen 1914: 140; Cf. Roll-Hansen 1980: 277.
86. Cf. Apert 1913.
87. "The Perfect Race. Mr. Balfour and the Objects of Eugenics," Statesman of August 18, 1912, and "What the Eugenicists Say," *Public Opinion*, August 2, 1912. At least in the view of Beatrice Webb, Balfour was seriously concerned about the supposed racial degeneration and during his period as prime minister pursued an active eugenics policy. Cf. Wright 1990: 89.
88. Cf. Gotto to Davenport, January 22, 1913, CDA Philadelphia: Gotto; *Eugénique* 1 (1913): 149; Laughlin 1934: 24.
89. Between 1900 and 1914 alone, over 300 international scientific, political, cultural federations were founded.
90. Cf. Crawford 1990.
91. Sixth Annual Report of the Eugenics Education Society, ESA London, SA/Eug. A6.
92. "Les échanges de vues qui se sont produits au cours de ces réunions témoignent du besoin de systématisation, de coordination, qui apparait dans tout ordre nouveau de la connaissance, dès que les travaux particuliers se multiplient." *Eugénique* 1 (1913): 149.
93. *Eugénique* 1 (1913): 157–159.
94. At this point, I cannot discuss whether internationalization was an indispensable condition for turning eugenics into a science. The significance of internationalization should be placed in context. In Germany, despite Ploetz's life federation idea ("Lebensbundidee"), after the First World War, a professional scientific organization of German race hygienists did in fact emerge.
95. Cf. Gotto to Ploetz, March 2, 1914, ESA Herrsching and Laughlin 1934: 2.
96. Fischer 1930: 3.

2 The First World War and Its Effect on International Eugenics

1. Irving Fisher: Lecture on Health and Religion. Lecture to the Unitarian Church, Portland, Oregon, 10/21/1917, p. 19f. IFA New Haven, p. III, Box 24, Fol. 376; Cf. Fisher 1915: 65; Aldrich 1975.
2. Irving Fisher: Lecture on Health and Religion. Lecture to the Unitarian Church, Portland, Oregon, 10/21/1917, pp. 18–21. IFA New Haven, p. III, Box 24, Fol. 376.
3. See Dawson 1995 and Weikart 2003.
4. Ammon 1893: 245f. Cf. Kröner 1980: 45; Chickering 1984: 240–245.
5. Pearson 1905: 26–27; Cf. Semmel 1958: 111–125; Stepan 1982: 129; Tucker 1995: 29. See the presentation in Brie (1927: 15f), the source of the citation on the higher and lower races.
6. Melville 1910: 54; Cf. Searl 1976: 37; Stepan 1987: 140. Melville used the concept of "virility." I consider Bowler's conclusion (1989: 275) that eugenics was a product of the age of imperialism to be too simplistic. Eugenicists represented their own variant of a Western drive to hegemony, which, as we shall see, differed basically from the militaristic imperialism of the early twentieth century.
7. Macfie 1917b: 442.
8. Ibid., 442; Cf. Macfie 1917a: 132–137; *Eugenical News* 2 (1917): 89; *Eugenics Review* 9 (1917): 269–270.
9. The eugenic militarists ignored the fact that Darwin (1890: 134) himself in *The Descent of Man* had presented a different position: "In every country in which a large standing army is kept up, the finest young men are taken by the conscription or are enlisted. They are thus exposed to early death during war, are often tempted into vice and prevented from marrying during the prime of life. On the other hand, the shorter and feebler men, with poor constitutions, are left at home, and consequently have a much better chance of marrying and propagating their kind." Cf. Johnson and

Popenoe 1933: 209; Stepan 1987: 140. For the support of this position by French and Italian eugenicists, see Kellogg 1912: 222.
10. See Kühl 2001: 188ff.
11. Kellogg 1913: 99–108, here 108.
12. Jordan 1910: 95, Cf. Jordan 1913a, 1913b.
13. Smith 1914: 2–3 and 5. A list of supporters of Smith's plea for a "eugenic peace" may be found on page 21. Cf. Haller 1963: 88.
14. Smith to Ploetz, 3/13/1914, APA Herrsching. Smith had his speech distributed to all senators and representatives. A very similar position had been taken before Smith by August Forel, who called for a "peaceful confederation of white peoples" in order not to be defeated in the "competitive battle of the races" against the much more rapidly increasing "Negroes, Japanese, and Chinese." Cf. Weber 1993: 24. The citation "interest of the culture-bearing race of Western Europe" comes from a critical presentation of Smith's plan by Lenz in the Archiv für Rassen- und Gesellschaftsbiologie (1914: 558).
15. Cf. Schuster 1912a: 231f; Inge 1914: 261; Schallmayer 1908: 394; Ploetz 1911b: 183.
16. Lenz 1923: 56; Hunt 1934: 244; Wiggam 1923: 218; Ploetz 1936a: 4.
17. Ploetz 1895: 147; Cf. Lutzhöft 1971: 355 and Kater 1990: 112f. For earlier references by Ploetz to the counterselective effects of wars, see Ploetz 1911b: 183. Cf. *Eugenics Review* 2 (1910): 92.
18. Johnson and Popenoe 1918. The citation is from the third edition (1933: 210).
19. Irving Fisher: Lecture on Health and Religion. Lecture to the Unitarian Church, Portland, Oregon, 10/21/1917, pp. 18–21. IFA New Haven, p. III, Box 24, Fol. 376.
20. Kellogg 1914: 48.
21. Lenz 1923: 52.
22. Thomson 1915: 1.
23. Cf. Brie 1927: 17.
24. Kellogg 1912: 222f.
25. Ibid., 223.
26. Cited from Kellogg 1913: 108; Cf. Eugenics Education Society 1913: 47.
27. Eugenics Education Society 1913: 49; Cf. Searl 1976: 37.
28. Eugenics Education Society 1913: 47–49. Cf. United Service Gazette, 8/8/1912: "A Professor's Stumble Over Army Eugenics"; The Westminster Gazette, 7/9/1912: "Future of the Race"; Schuster 1912a: 240.
29. Johnson 1914: 98.
30. The following section will show as an example the outlooks of eugenicists in Great Britain, the United States, Italy, and Germany. On the position of the Hungarian eugenicists, see Turda 2007: 205ff; on Italian eugenicists, see Cassata 2011: 64ff.
31. *Eugenics Review* 7 (1915): 13.
32. Poulton 1916: 34–49; Darwin 1917: 1–10. Cf. *Eugenical News* 1 (1916): 43–44.
33. Poulton 1916: 34–49.
34. Saleeby 1917: 307–316; Cf. *Eugenics Review* 9 (1918): 362f.
35. Crew 1919; Cf. Stepan 1987: 142.
36. Johnson 1915: 548.
37. Johnson and Popenoe 1918; I cite based on the 3rd ed., 1933: 202.
38. Fisher 1921. I quote from the review in *Archiv für Rassen- und Gesellschaftsbiologie*, 14 (1921): 228.
39. Cf. *Archiv für Rassen- und Gesellschaftsbiologie* 14 (1921): 228; *Eugenics Review* 10 (1918): 113.
40. Sergi 1917; Cf. *Eugenics Review* 10 (1918): 113.
41. Haeckel 1916: 140f; see also Semmel: 1958: 123. Cf. Haeckel 1914.
42. Hoffmann 1916 (added in the literature); cited from Kroll 1983: 183.
43. Schweisheimer 1918/1921: 11; cf. *Archiv für Rassen- und Gesellschaftsbiologie* 13 (1918–1921): 176ff and 14 (1922/23): 216. See also Propping and Heuer 1991: 78–93.

44. Until 1917, Ploetz was noted in the *Eugenics Review* as vice president of the Eugenics Education Society. For the first time in February 1918, the British eugenicists actually blocked the German race hygienists from membership in the Eugenics Education Society. Cf. Council Meeting on 2/12/1918, ESA London, SA/Eug. L4.
45. Cf. Fischer 1930: 3.
46. Lenz 1916: 403; Cf. Kroll 1983: 134; Weindling 1989a: 299; Weiss 1990: 24.
47. Invitation to the First German-Austrian-meeting for race hygiene and population policy on September 3, 1918, Budapest, APA Herrsching. Because of the war, the congress was canceled at the last moment. Cf. Weindling 1989a: 298. On eugenics in Austria, Mayer 2007: 162ff., Wolf 2008: 45ff, and a comprehensive discussion, Mayer 2004. On Eugenics in Hungary, see Turda 2007: 186.
48. Weingart et al. 1988: 220–222; Weindling 1989a: 295–305.
49. Lenz 1914: 555; *Archiv für Rassen- und Gesellschaftsbiologie* 11 (1914): 561; Cf. Kroll 1983: 254. See also Weindling 1989a: 298; Weingart et al. 1988: 229; Schwartz 1995a: 171ff.
50. Ninth Annual Report of the Eugenics Education Society, ESA London, SA/Eug A9.
51. Cf. Eighth Annual Report of the Eugenics Education Society, ESA London, SA/Eug A8.
52. Council Meeting of the Eugenics Education Society on 3/23/1915 and 6/16/1916, London, SA/Eug L.4. On this point, I follow the presentation by Soloway 1982: 169.
53. *Eugenics Review* 7 (1915): 96.
54. Schneider 1990a: 76. See Seventh Annual Report of the Eugenics Education Society, ESA London, SA/Eug A7.
55. Seventh Annual Report of the Eugenics Education Society, ESA London, SA/Eug A7; Laughlin 1934: 12.
56. See Weindling (1989a) and Bock (1996), both of whom refer to the central significance of the First World War and the development of eugenic measures in Germany. The question as to whether the killing of sick people and handicapped people can be understood as the ultimate form of negative eugenics was hotly debated among eugenicists in the 1920s. Even today, the question is often underestimated by historians.
57. Anderson 1921a: 481; Cf. Anderson 1921b.
58. Popenoe 1924: 193; Cf. Popenoe 1918: 262.
59. Lindsay 1918: 144. Cf. the discussion in Great Britain about eugenic reforms during the "social reconstruction."
60. Darwin 1918: 146.
61. Ellis 1919: 120f. He continued: "Now the influences of war have so swollen the old evils that only blindness or stupidity can fail to realise that a readjustment of our ideals and our practices has become absolutely necessary...Never before has it been so urgent a demand on us to do all in our power to prevent the breeding of the less fit members of society."
62. Fisher 1921: 217f; Cf. *Archiv für Rassen- und Gesellschaftsbiologie* 14 (1921): 228f. Cf. Fisher 1919.
63. Siemen (1987: 29f and 1993: 99) calculated 70,000 deaths from famine in Germany between 1914 and 1918. According to Schmuhl (1987: 107), in Prussia alone, 45,000 handicapped persons died from lack of food. The psychiatrist Kankeleit (1929 and 1931: 174) estimated that 30–33 percent of institutionalized Germans were victims of the hunger blockade.
64. Kaup 1922: 15 and 132f.
65. Alfred Ploetz: Die rassenbiologische Bedeutung des Krieges und sein Einfluß auf den deutschen Menschen, p. 37, APA Herrsching.
66. Cf. Weindling 1989a: 125. My notions on eugenic peace policies have in the meantime become widespread; see, for example, Mantovani 2004: 145.
67. Galton 1979: 333; see Blacker 1952: 90.

68. Ploetz 1895: 213 and 230; Cf. Doelecke 1975: 78; Weindling 1989a: 125; Weber 1993: 49.
69. Grotjahn 1925: 28.
70. Edwin G. Conklin: The purposive improvement of the race, 1929, EGCA Princeton, Box 16. Conklin continues that precisely those same breeding principles that were instituted for a successful improvement of horses and chickens should be used on human beings.
71. Lundborg 1926: 3.
72. Darwin to Pearson, 12/171918, KPP London 673/6; *Eugenics Review* 11 (1919): 21–23; *Eugenics Review* 11 (1920): 249f; *Eugenical News* 5 (1920): 12.
73. Cf. Osborn 1974: 116.
74. Osborn 1921: 311.
75. Osborn 1921: 312f; Cf. *Archiv für Rassen- und Gesellschaftsbiologie* 15 (1922): 91ff; see also Chase 1977: 278.
76. See the letter from Laughlin to Davenport, 10/1/1923, in which he noted that the Europeans would acknowledge the leading role of Davenport. CDA Philadelphia: Laughlin. In the meantime, case studies have shown the influence of American eugenics on specific countries. For example, Bucur (2002: 54ff), a study on eugenics in Rumania.
77. A series of case studies on sterilization in individual American states have appeared. For example, Ladd-Taylor 2004 on Minnesota; Dorr 2008 on Virginia; Gallagher 1999 on Vermont; Castles 2002, Schoen 2011, and Begos et al. 2012 on North Carolina; Stern 2002 and Lantzer 2011 on Indiana; Largent 2002 on Oregon. For a good overview, see Largent 2008: 72 and 77ff.
78. Popenoe 1924: 184–196. Extensively on Popenoe—Ladd-Taylor 2001 and Stern 2005: 150ff.
79. Minutes of the General Committee of the Second International Committee of the Second International Eugenics Congress, 10/4/1920, ESA London, SA/Eug, D. 104 and *Eugenical News* 5 (1920): 14; see also Chase 1977: 20, 277.
80. Cf. Haller 1963: 73; Barkan 1992: 67.
81. On the Race Betterment Foundation, see Stern 2005: 27ff; on the Eugenics Record Office, Adams et al. 2005.
82. Little 1922: 511.
83. Minutes of the General Committee of the Second International Committee of the Second International Eugenics Congress, 4/10/1920, ESA London, SA/Eug, D. 104 and *Eugenical News* 5 (1920): 14.
84. Cf. Little 1922: 511 and 517; Cf. Popenoe 1924: 19.
85. Little 1922: 511.
86. The lectures by Morgan and Conklin are not reprinted in the congress report. Little (1922: 519) gives information about the topics of their lectures.
87. Darwin 1923a: 7ff; Cf. *Archiv für Rassen- und Gesellschaftsbiologie* 15 (1922): 92.
88. Committee on Publication (Davenport, Osborn, Laughlin, Wissler) 1923: ix.
89. On the origin of the theory of chromosomes, see Sturtevant 1965: 34–36.
90. Davenport 1921: 397; Cf. Barkan 1992: 71ff.
91. Adam 1922: 174.
92. Muller 1923: Ulf.
93. I cannot understand how Ludmerer (1969) can speak of a phase of "disillusionment" of the geneticists regarding eugenics after the First World War. A systematic critique of geneticists of the orthodox eugenics movement first emerged starting in the middle of the 1920s.
94. Myerson 1923: 224f; Cf. *Mental Hygiene* 8 (1924): 1062. Extensive information on Myerson—Trent 2001 and Largent 2008.

95. *Eugenics Review* 13 (1921): 495f.
96. *Journal of Heredity* 15 (1924): 128–130. Citation on p. 128.
97. Darwin 1923b: 409.
98. Besides Mjöen, those in favor of the application included the chairman of the Norwegian consultant to the committee Dr Wille; Haakon Loken, the governor of Christiana; the former parliamentary delegate Alfred Eriksen; and the University professor Wilhelm Keilhan. ESA Philadelphia: Second International Congress.
99. Cf. *Eugenical News* 8 (1923): 5; *Eugenics Review* 15 (1923): 492 and American Eugenics Society 1931: 3. See also the comprehensive presentation regarding the founding of the American Eugenics Society by Mehler 1988: 57f.
100. See the letter from Cora Hodson to S. Wayne Evans, Secretary of the Popular Education Committee of the American Eugenics Society, 6/11/1930, ESA London, SA/Eug, E. l.
101. AESA Philadelphia, 675.06: Am3.
102. Bedwell 1922: 187.
103. Resolution prepared during a meeting on 9/26/1921 in the American Museum of Natural History, 9/26/1921; ESA London, SA/Eug, E.ll.
104. Little 1922: 518.
105. Cf. AESA Philadelphia, 575.06: Am3; ESA London AS/Eug E.ll and Litde 1922: 523. The German race hygienist Otmar Freiherr von Verschuer (1931: 168) claimed that the Committee had been renamed the Commission by 1919. Other sources do not confirm this claim.
106. *Eugenical News* 6 (1921): 67.
107. All information is based on the minutes of the meeting ESA London, SA/Eug, D. 104. On the position of the American eugenicists, see also Davenport to Gotto, 10/8/1919, CDA Philadelphia: Gotto. It appears that at the last minute, the Austrian and Hungarian race hygienists were allowed to participate in the Congress—Cf. Jennings to Davenport, 6/271921, HSJA Philadelphia: Davenport.
108. Jennings to Morgan, 4/141920; Morgan to Bateson, 4/17/1920, Jennings to Davenport, 5/4/1920; Jennings to Osborn, 6/7/1920, HSJA Philadelphia; see also Barkan 1992: 193.
109. Davenport to Mjöen, 12/5/1921, CDA Philadelphia: Mjöen; see also ESA London SA/Eug, D. 104; ESA London Sa/Eug, E.ll; Reichel 1931: 287.
110. Lundborg to Davenport, 8/29/1921, CDA Philadelphia: Lundborg.
111. Cf. Davenport to Jennings, 4/17/1920, HSJA Philadelphia.
112. Davenport to Bluhm, 8/30/1921 and 12/11/1922; CDA Philadelphia.
113. Ploetz to American Ethnology Smithsonian Institution, 2/25/1922, APA Herrsching.
114. Darwin to Govaerts, 13/12/1921, ESA London, SA/Eug, E.ll.
115. *Eugenical News* 7 (1922): 117 and 120; *Eugenics Review* 13 (1923): 285; Laughlin 1934: 5.
116. Davenport to Lenz, 3/17/1923, CDA Philadelphia. On 3/30/1923, he had sent a letter to Baur.
117. Baur to Davenport, 1/5/1923, CDA Philadelphia: Baur; see also Baur to Darwin, 3/8/1923, CDA Philadelphia: Darwin.
118. Lenz to Davenport, 8/8/1923, CDA Philadelphia: Lenz. He mentions to Mjöen the refusal of the international congress; Cf. Mjöen to Davenport, 6/20/1923; Davenport to Lenz, 7/13/1923, CDA Philadelphia: Mjöen.
119. *Eugenics Review* 15 (1924): 644.
120. *Eugenical News* 8 (1923): 116; Darwin to Ploetz, 7/29/1923, APA Herrsching.
121. *Archiv für Rassen- und Gesellschaftsbiologie* 16 (1924): 458. Cf. Kröner 1980: 81.
122. Ploetz-Tagebuch, 9/20/1927, APA Herrsching. A reaction to the German demands is found in Darwin to Davenport, 8/6/1925, CDA Philadelphia: Darwin.

3   Racism, Internationalism, and Eugenics

1. Hall 1919: 126. See also *Eugenics Review* 11 (1919): 97ff.
2. Confirmation letters from Davenport to Draper, 3/23/1923, CDA Philadelphia: Draper.
3. Cf. the letter from Draper's attorney Malcolm Donald to Osborn, 7/22/1947, and Osborn's letter to Donald, 5/25/1948, FOA Philadelphia: Pioneer Fund.
4. For Davenport's reaction, see his letter to Osborn, 4/10/1929, CDA: Davenport: Osborn.
5. Cf. Davenport to Draper, 4/10/1929, CDA Philadelphia: Draper.
6. Eugenics Research Association: Offer of Prize for Best Study on Causes of Fall in Birthrate, 1929, CDA Philadelphia: Essay Prize; Cf. *Archiv für Rassen-und Gesellschaftsbiologie* 22 (1930): 327.
7. IFEO 1930: 76.
8. Davenport to Draper, 3/6/1931; CDA Philadelphia: Draper; Cf. Schubnell 1967: 95f and Tucker 2002: 32f. Of great interest is the definition of the prize in the self-generated description by the Pioneer Fund (see Lynn 2001a: 12ff).
9. It is not possible here to go into the very interesting changes in race terminology. On this, see the very interesting Conze and Sommer 1984.
10. Darwin 1934: 24.
11. Baur 1933: 3.
12. Ibid., 4–7.
13. Wiggam 1923: 225–230. On Wiggam, see Gossett 1997: 402f.
14. Cf. *Eugenics Quarterly* 15 (1968): 56.
15. Johnson 1934a: 235–239.
16. Harry H. Laughlin: Our Faith in the Federal Idea and World Government. HLA Kirksville, B-5–1 B#5.
17. Harry H. Laughlin: The Common Government of the World. HLA Kirksville, B-5–1 B#2. See also the letter of Laughlin to H. G. Wells, printed in Samaan 2013: 475.
18. Harry H. Laughlin: List of Purposes and Responsibilities. HLA Kirksville, B-5–1 B. To Laughlin's great regret, his ambitious plan met little admiration. The heads of government whom he convinced to read his plan ignored his suggestions. American magazines and the large publishing houses alike refused to publish it. His application for the Edward W. Bok American Peace Award got lost in the pile of the other 22,165 peace plans that were submitted. See Laughlin to American Foundation, 3/24/1924 and 12/15/1926, HLA Kirksville, B-5–1 B#7; see also Hassencahl 1970.
19. Lenz 1923: 56.
20. Lenz 1924: 223.
21. Lenz 1923: 272f. In my presentation, I follow Lutzhöft (1971: 254f).
22. Ploetz 1936b: 619f.
23. Alfred Ploetz: Die rassenbiologische Bedeutung des Krieges und sein Einfluß auf den deutschen Menschen. APA Herrsching.
24. Cf. Mjöen 1929: 77. On Grant, see the extensive material in Spiro 2008.
25. Miles 1989: 366; Cf. Grosch-Obenauer 1986: 30f; Goldberg 1993: 68.
26. Baur 1921: 206.
27. Siegfried 1926: 218.
28. Baur 1921: 207.
29. Lenz 1923: 272; Cf. Lutzhöft 1971: 279.
30. Cf. Goldberg 1993: 79.
31. Ibid., 80.
32. Cf. Davenport to Grant, 5/3/1920, CDA Philadelphia: Grant.
33. "Eugenics Seen as Hope of Race." *Battle Creek Morning Journal*, 7/28/1923. Cf. also Aldrich 1975.

34. On MacBride, see Mazumdar 1992: 65.
35. Gates translated from Billig 1981: 105; see also ESA London, SA/ EugC. 120. For quite some time, Gates even took the position that the races were even distinct species.
36. Cf. *Archiv für Bevölkerungswissenschaft- und Bevölkerungspolitik* 7 (1937): 294.
37. Council Meeting of the Eugenics Society, 7/10/1929, ESA London, SA/Eug, L.8. On the world of Pitt-Rivers, see Dietz 2009.
38. There is more to be learned about Pitt-Rivers from his papers in the Churchill Archives Center, Cambridge. See Hart 2012.
39. Cf. Massin 1993: 400 and 2003: 192ff. See also "Ueber die Vererbung geistiger Eigenschaften." *Frankfurter Zeitung*, 18/10/1938.
40. Ploetz to Fischer, 3/9/1930, APA Herrsching; on Fischer's position, see Fischer 1926.
41. On the position of the Scandinavian eugenicists, see Broberg and Tyden 1996; Hansen 1996; Hietala 1996; Roll-Hansen 1996a, 1996b. Research on eugenics in Scandinavia has recently significantly increased; for an overview, see Drouard 1999.
42. Forel in November 1920 in a brochure of Alfred Mjöen, CDA Philadelphia: Mjöen.
43. Cf. Schlaginhaufen 1921a, 1921b.
44. Cf. Gini to the Eugenics Education Society, July 1918, ESA London, SA/Eug, C123.
45. Fantham 1927: 10f.
46. Cf. Schreiber 1932: 320.
47. Mjöen to Davenport, 5/7/1931, CDA Philadelphia: Mjöen.
48. Eugenics, A Birds'-eye View, flyer of the Indian Eugenics Society, Lahore and Simla, ESA London, SA/Eug, E.8. See the flyers of the Indian Eugenics Society, ESA London, SA/Eug, E.8. Cf. also *Eugenical News* 7 (1922): 2. For the relatively minor significance of eugenics in India, see Asana 1934: 10–11. Comparatively little study has been carried out on the eugenics movement in India. However, see the recent Ahluwalia 2008.
49. Cf. *Eugenical News* 19 (1934): 12f; Johnson 1934a: 194. Information on Chinese eugenics is drawn from the review article by Dikötter 1989 and Chung 2010. The most extensive work in English is Chung 2002.
50. Johnson and Popenoe 1933: 369. The information on Japan is based on Johnson 1934a: 189–193; *Volk und Rasse* 7 (1935): 157. See also Suzuki 1975; Chung 2002; Robertson 2010.
51. Cf. *Eugenical News* 8 (1923): 84; *Eugenics Review* 15 (1924): 645; *Eugenical News* 13 (1928): 18. Cf. also Laughlin, *Zeitschrift für Rassenkunde* 1 (1935): 208f. See also: HSJA Philadelphia: Pan-American Association of Eugenics and Homiculture. On the Association, see also González and Peláez 1999: 231ff and Doménech 2004: 294ff.
52. ESA London, SA/Eug, E. 22: Eugenics Society of South Africa.
53. Initial descriptions of both these organizations have become available. For Kenya, see Campbell 2007: 114ff and Campbell 2010. For Netherlands East Indies, Pols 2010.
54. Cf. Hodson to Davenport, 9/30/1930, CDA Philadelphia: IFEO.
55. Davenport to Hodson, 2/2/1935, CDA Philadelphia: IFEO. As early as 1922, the admission of the Japanese eugenicists to the Permanent International Eugenic Commission was under discussion; see Govaerts to Darwin, 1/23/1922, and Darwin to Govaerts, 2/8/1922, ESA London, SA/Eug, E.ll. See also Otsubo 2005: 225.
56. APA Herrsching, Tagebuch 11/101932; see also Hodson to Davenport, 8/7/1935, CDA Philadelphia: IFEO.
57. This view is found in Ludmerer 1972; Kevles 1985; Roll-Hansen 1989, 1996a, 1996b. See Schmuhl 2005: 148ff on the difficulties of assigning Fischer to the orthodox racist variant of eugenics.
58. Cf. Castle 1930; Klineberg 1931. On Davenport's reaction to Castle's criticism, see Davenport to Jennings, 6/17/1930, HSJA Philadelphia; see also Barkan's (1992) study on the "withdrawal" of scientific racism.

59. Davenport to Hodson, 2/24/1930, CDA Philadelphia: IFEO. In a letter on 2/27/1930 to Davenport, Hodson expressed doubt as to whether the International Federation could actually serve as an "anthropological center."
60. Cf. Davenport: Memorandum on the Resolution Proposing the Establishment of Research Committees on Human Measurement, 1930; CDA Philadelphia: Laughlin.
61. Davenport: Report on Standardization Committee 1936 in Scheveningen, MPIA Munich, GDA 50.
62. Davenport: Do Races Differ in Mental Capacity? 1929, CDA Philadelphia: Lecture: Do Races Differ in Mental Capacity?
63. IFEO 1930: 4f.
64. Ibid., 6f.
65. Hodson to Gonzenbach, 29/9/1930, CDA Philadelphia: IFEO. A subsequently established committee for physiology did not take up this work again and is not discussed here.
66. The international congresses for anthropology were suspended after a conference in 1907 in Cologne. The international congresses for anthropology and prehistoric archaeology were halted after the meeting in Geneva in 1912. The meetings of the Paris Institut International d'Anthropologie that took place starting in 1920 were not attended by the German, Austrian, or Hungarian anthropologists. Cf. Verschuer 1931: 167.
67. Hodson to Spearman, 1/19/1935, CDA Philadelphia: IFEO. See also *Eugenical News* 15 (1930): 162.
68. Davenport: Race Crossing in Man, lecture to the IFEO 1927, CDA Philadelphia: Lecture: Race Crossing in Man; Cf. Davenport 1928: 13.
69. Davenport to Hodson, 3/4/1926, CDA Philadelphia: IFEO.
70. Davenport to Hodson, 2/23/1927, CDA Philadelphia: IFEO. On the founding of the committee, see *Eugenics Review* 19 (1927): 262; *Eugenical News* 12 (1927): 153; *Times*, 1/30/1928.
71. Davenport 1917: 366; Cf. also Provine 1973: 791; 1986: 866f; Glass 1986: 132f; Tucker 1995: 64.
72. Cf. Expenses incurred account of Draper Fund, June 1, 1926–March, 31, 1928, CDA Philadelphia: Draper. On Draper's financing of the Jamaica study, see Kenny 2002: 267ff; Lombardo 2002: 766; and Jackson 2005: 34.
73. Davenport 1928, 1930; Davenport and Steggerda 1929; Cf. also *Eugenics Review* 20 (1928): 186f; see also Davenport Lecture: Is the Crossing of Races Useful?, 1929, CDA Philadelphia.
74. See Weikart 2004: 121ff on Fischer's application of Mendelian laws to races.
75. Fischer 1913. On Fischer in general, see Seidler and Rett 1988: 76; Weingart et al. 1988: 102f; Pinn and Nebelung 1992: 12; Ehmann 1993: 6f; Crips 1993; Weingart 1994; and especially Lösch 1997 and Schmuhl 2005.
76. Cf. Köhn-Behrens 1934: 50; see also Pommerin 1979; Pinn and Nebelung 1992: 17.
77. Fischer at the meeting in Farnham 1930; MPIA Munich: GDA 32.
78. See the minutes of the meeting in Munich, CDA Philadelphia: Corrado Gini
79. Cf. Rodenwaldt 1922: 425–422; see also *Eugenical News* 13 (1928): 75f; Fleming 1930: 257–259. In general on Rodenwaldt—Kiminus 2002. On Rodenwaldt's role in the Netherlands East Indies, see Pols 2010: 347.
80. Rodenwaldt 1934: 375. Cf. Ehmann 1993: 8.
81. Nilsson-Ehle 1928; Cf. *Eugenics Review* 19 (1927): 262; *Eugenics Review* 20 (1928): 187; *Eugenical News* 14 (1929): 156.
82. Lundborg 1931; Lundborg had a lecture read in Munich, "Race Mixing among Humans," in Munich in 1928; Cf. *Eugenics Review* 20 (1928): 186f.
83. *Eugenical News* 14 (1929): 156f; Cf. Aikman 1933: 163f.
84. Mjöen 1923: 59.

85. Cf. Gates 1934: 35f; *Eugenics Review* 26 (1934): 73.
86. Lundborg to Davenport, 25/4/1928, CDA Philadelphia: Lundborg.
87. Davenport to Draper, 25/5/1928, CDA Philadelphia: Draper. See also Davenport to Lundborg, 17/5/1928, CDA Philadelphia: Lundborg.
88. IFEO 1930: 17ff; Cf. Aikman 1933: 161.
89. Davenport to various eugenicists, 11/14/1928, CDA Philadelphia: IFEO Committee on Race Crossing.
90. *Eugenics Review* 20 (1928): 186f; *Eugenical News* 14 (1929): 154.
91. After the lecture read by Fischer in Farnham 1930; IFEO 1930: 23–26.
92. Davenport to Fischer, 20/3/1928, CDA Philadelphia: Fischer.
93. Questionnaire on the examination of race crossings; CDA Philadelphia: Fischer. The rate of return of the questionnaires was quite modest.
94. Based on Pinn and Nebelung 1992: 14.
95. Ernst Rüdin: Zu einem Programm der internationalen Erforschung der Rassenpsychiatrie, MPIA Munich, GDA 32.
96. IFEO 1934: 4f.
97. Davenport to Hansen, 14/10/1925, CDA Philadelphia: Hansen.
98. Cf. Mjöen 1929: 77.
99. Davenport to Grant, 5/3/1920, CDA Philadelphia: Grant.
100. Mjöen 1931: 31.
101. Cf. *Archiv für Rassen- und Gesellschaftsbiologie* 18 (1926): 230–233; *Eugenics Review* 18 (1926): 156f.
102. Cf. *Eugenics Review* 17 (1925): 225.
103. Schneidewind 1933: 32.
104. Siegfried 1926: 221; Cf. also Siegfried 1927.
105. Cf. Lenz 1923: 56.
106. Cf. Chase 1977: 289–301.
107. Hall 1919: 126; see also *Eugenics Review* 11 (1919): 97f. The Immigration Restriction League was founded in 1894, well before there was a significant influence of eugenics on American policy. However, the League very soon adopted a basic eugenics orientation. See Tucker 2002 and Spiro 2008.
108. Report of the Committee on Selective Immigration of the Eugenics Committee of the United States of America. *Eugenical News* 9 (1924): 22; Cf. Hassencahl 1970: 232f.
109. Third Report of the Sub-Committee on Selective Immigration of the Eugenics Committee of the United States of America. *Journal of Heredity* 16 (1925): 298.
110. Cf. Hassencahl 1970: 278. See also Hall 1919: 127; *Eugenical News* 11 (1926): 192; *Eugenical News* 13 (1928): 73–75.
111. Cf. Symonds and Carder 1973: 16.
112. Immigration Resolution of the IFEO Meeting 1928 in Munich; CDA Philadelphia: IFEO-Meeting Munich. Cf. *Archiv für Gesellschafts- und Rassenbiologie* 21 (1929): 321–323.
113. Davenport 1934: 18.
114. Cf. *Eugenical News* 19 (1933): 18–20.
115. Cf. Hodson to Davenport, 6/24/1932, CDA Philadelphia: IFEO.
116. In this book, I concentrate on the dominant orientation within the IFEO. A monograph concentrating on the IFEO would have to give greater attention to the internal disputes within the IFEO.
117. Cf. Weindling 1989a: 248; Quine 1996: 7.
118. Sanger 1919b: 10f; Cf. also Sanger 1919a, 1926; see also Kennedy 1970: 115; Gordon 1974: 269; Maas 1976: 22; Ann Taylor Allen 1988, 1991; Hodgson 1991: 15; and Franks 2005. The family planners were not, as Wichterich claims (1994: 11f), "reshaped and coopted," but rather they came by themselves to a eugenics viewpoint. It is important to note that the eugenically disposed feminists were a minority in the feminist movement.

119. Cf. Paul 1995: chapter 5.
120. Eugenics Education Society 1913: 25; Cf. Drysdale 1912; see also Soloway 1990: 103.
121. Ploetz 1913/14; see also Gruber and Rüdin 1911: 171. Cf. *Eugenics Review* 3 (1911): 279f; Mjöen 1914: 151. On Ploetz's attitude to the women's movement, see the comments by Katti Toker Möller, an influential member of the Norwegian birth-control movement. Cf. Roll-Hansen 1980: 285; 1996a: 190. For a different view, see Weikart 2004: 134f.
122. Darwin to Hodson, 6/241927, ESA London; Cf. Kevles 1985: 88; see also *Eugenics Review* 20 (1928): 173.
123. Davenport to Sanger; Cf. also Grant to Whitney, 4/15/1928, CDA Philadelphia: Grant. See also *Eugenical News* 19 (1934): 143.
124. Osborn 1932: 173–179, 1934: 29–41. See the reaction of the editor of the Birth Control Review, Stella Hanau, 8/30/1932, RPA Philadelphia: Sanger.
125. Cf. Apert 1940; Schneider 1982; Leonard 1983; Schneider 1990a, 1990b; Carol 1995.
126. Hodson to Davenport, 4/9/1929, CDA Philadelphia: IFEO.
127. Darwin to Davenport, 5/25/1927, CDA Philadelphia: Darwin.
128. Hodson to Davenport, 7/12/1929, CDA Philadelphia: IFEO.
129. *Eugenical News*, presumably 1925, CDA Philadelphia: *Eugenical News* 11.
130. Cf. Weindling 1993.
131. Darwin to Davenport, 8/1/1924, CDA Philadelphia: Darwin. See also Mjöen to Darwin, August 1926, copy in CDA Philadelphia: Darwin.
132. Cf. *Revue Anthropologique* 37 (1927): 165.
133. Cf. Paul 1995: chapter 7.
134. On Haldane, see Lünen 2009.
135. Cf. Huxley 1936: 18.
136. Cf. Muller 1936; Haldane 1936; see also Freeden 1979; Paul 1984; Weindling 1987.
137. In this section, I follow the groundbreaking work by Schwartz (1994: 437–470; 1995b: 18–32) on socialist eugenics in Germany.
138. Cf. *Eugenical News* 8 (1923): 33f; *Eugenical News* 9 (1924): 69; *Eugenical News* 10 (1925): 57–60. See also Graham 1977, 1981; Adams 1989, 1990a; Spektorowski and Mizrachi 2004; Krementsov 2010. Cf. also Dobzhansky's assessment (to Muller on 5/17/1949), HJM Bloomington, 5/17/1949 Dobzhansky.
139. Cf. Graham 1977; Adams 1990a.
140. After the Russian eugenicists were excluded from the Second International Eugenics Congress, they began contacts with the British Eugenics Society, the Eugenics Record Office in the United States, and with the Deutsche Gesellschaft für Rassenhygiene. In a letter to Govaerts, Koltzoff urged the directors of the Permanent International Eugenics Committee to accept a Russian representative into the original eugenic organization. The committee met this request, and in 1924, Koltzoff participated in a meeting of the International Eugenic Organization. This was however the last time the Russian eugenicists participated in the international eugenics organization. Cf. Govaerts to Darwin, 1/23/1922, ESA London, SA/Eug, E.ll and *Eugenical News* 9 (1924): 105.
141. Cf. Paul 1995: chapter 7.
142. H. J. Muller, "The Dominance of Economics over Eugenics," Abstract of Paper Presented to Third International Congress of Eugenics, and H. J. Muller, "The Dominance of Economics over Eugenics," Presented to the Third International Congress of Eugenics, LCDA Philadelphia: Muller; see also Muller 1934: 143. On the scientific basis of Muller's criticism, see Carlson 2001: 338ff.
143. Dunn to Muller, 6/23/1932 ("I too have been fed up with the orthodox breed of eugenics"); Muller to Mohr, 22/6/1932, HJMA Bloomington; Muller to Dunn, 6/21/1932, and Davenport to Dunn, 6/24/1932, LCDA Philadelphia: Muller.
144. Cf. Abir-Am 1993: 176.

4 The Crisis of Orthodox Eugenics and the Rise of Human Genetics and Population Science

1. Cf. *Eugenical News* 18 (1933): 14.
2. Davenport 1934: 17ff.
3. Cf. Osborn 1974: 118. Grimm 2011: 99 for the first time has made a quantitative comparison by national origins of the three international eugenics congresses. See also Black 2003: 398, who seems to agree with my evaluation.
4. On the congress in general, see Schreiber 1933.
5. See, for example, the speeches at the meeting in Scheveningen 1936; cf. Hodson 1936.
6. Barkan 1992: 158; see also Gates 1934.
7. Cf. *Eugenical News* 15 (1930): 156; 18 (1933): 16–17.
8. Minutes of the Business Meeting of the IFEO, 9/27/1929, MPIA Munich, GDA 32. Cf. *Eugenical News* 15 (1930): 15; *Archiv für Rassen- und Gesellschaftsbiologie* 22 (1930): 433
9. Cf. *Eugenical News* 8 (1923): 116; 9 (1924): 106; 13 (1928): 129f; 19 (1934): 148; *Eugenics Review* 15 (1924): 644.
10. Cf. Minutes of the IFEO-meeting in Farnham, 1930, CDA Philadelphia: IFEO Meeting Farnham; see also Davenport to Gini, 5/5/1930, CDA Philadelphia: Gini; on the general history of the International Research Council, see Schröder-Gudehus 1966: 25, 102, and 111.
11. Cf. Schimank 1995: 96. See Merton 1973 on the idea of organized skepticism.
12. Pearl 1927: 260; 1928b. Cf. Weingart et al. 1988: 317; Propping and Heuer 1991: 80; Allen 1991: 242; Tucker 1995: 71.
13. L. C. Dunn: Report on the Eugenics Record Office, 7/3/1935, CDA Philadelphia: Dunn.
14. L. C. Dunn was an exception. He was one of the first geneticists to distance himself from eugenics in any form.
15. Cf. Blacker 1962a: 10f. See King 1999: 67ff on Blacker's role in the campaign for sterilization.
16. Cf. Osborn 1937a, 1937b, 1939a, 1939b, 1940; Blacker 1952; see also Searl 1979: 159–169; Kevles 1985: 172ff; Allen 1986; Mehler 1988: 265f; Roll-Hansen 1989b: 335; Barkan 1992: 190.
17. Cf. for example, Ellsworth Huntington: The Present Stage of American Eugenics, Presidential Address, 5/51938, EHA New Haven, S. IV, Box 30, Fol. 309; Lorimer to Huntington, 6/111937, EHA New Haven, S. III, Box 79, Fol. 3134.
18. This change was carefully observed by Osborn's critics. Harry F. Weyher (2001: xxxvii), the long-time president of the Pioneer Fund, quoted from Osborn's first draft of the book *Preface to Eugenics* (Osborn 1940: 78): "It is very important that there should be further scientific studies on the genetic capacities of the different races." He then observed that this paragraph was deleted from Osborn's second version (1951a: 122) after the Second World War.
19. Cf. *A Decade of Progress* (1934): 510f; Plate 1.
20. Osborn 1937a: 107; Cf. *Eugenical News* 22 (1937): 62. On changing his opinion, see Osborn to Huntington, 10/23/1936, EHA New Haven, S. III, Box 77, Fol. 3038.
21. Cf. Blacker's lecture to International Union of Family Organizations, 1965, CPBA Oxford, Box 6, in which he attempts to clarify his ideas on the relationship of eugenics to science.
22. On the meaning of the world economic crisis, see Weindling 1989: 9; Schwartz 1995a: 264; Quine 1996: 108. Even the great acceptance of the German sterilization law of 1933 had a close connection to the takeover of power by the National Socialists; it must be understood against the background of the Great Depression.

23. See King 1999: 64ff on the campaign for sterilization in Great Britain.
24. Cf. Reilly 1991: 97; Kühl 1994: 24; Carol 1995: 174.
25. I would like to distinguish this approach from the still dominant historical picture (Cf., for example, Ludmerer 1972; Kevles 1985; Roll-Hansen 1996a: 217). Because of an inadequate distinction between eugenics as an institutionalized science and eugenics as a political movement, the conventional wisdom in history has not made it sufficiently clear that the crisis of eugenics in the 1930s was a crisis only in eugenics as a scientific discipline.
26. The classic understanding of the deterioration of eugenic ideology is found in Ludmerer 1972: 125; Kröner 1980: 32; Mitchell 1985: 165f; Propping and Heuer 1991: 79. Allen (1991: 256) shows this outlook to be the still dominant historical view.
27. Muller 1928; Vavilov 1928. Cf. *Eugenics Review* 20 (1928): 193. On the history of the previous congress for genetics, cf. Nachtsheim 1927: 989f; Verschuer 1931: 168; *Eugenical News* 15 (1930): 54f.
28. On the history of this blocking, see IICE Archives Paris D.IV.4 History 5. Genetics congress.
29. Pearl 1927: 260; 1928b. Cf. Weingart et al. 1988: 317; Propping and Heuer 1991: 80; Allen 1991: 242; Tucker 1995: 71.
30. Pearl 1927; Conklin 1928; Pearl 1928b; Hogben 1931; cf. Ludmerer 1972: 148; Allen 1984: 16; Glass 1986: 137; Mehler 1988: 265f; Kingsland 1988: 191; Roll-Hansen 1989: 338; Kevles 1992: 8f. That Pearl stayed in despite his criticism of eugenics is shown by his readiness to participate in the IFEO meeting in Amsterdam in 1927. Pearl to Davenport, 1/15/1927, CDA Davenport: Pearl.
31. Davenport to Darwin, 6/6/1928, CDA Philadelphia: Darwin.
32. Human Heredity Committee, CDA Philadelphia: IFEO; Frets to Haldane, 3/23/1939, JBSHA London, Box 15.
33. Business Meeting of IFEO, 27/9/1929, AMPI Munich, GDA 32.
34. *Eugenics Review* 20 (1928): 185f; cf. also Nachtsheim 1933: 193–213; Laughlin 1934: 7; History of the Fifth Congress on Genetics, IICEA Paris, D IV.4.
35. Cf. Clark 1956; Mehler 1988: 393f; Gaudilliere 1994; Gaudilliere and Löwy 1994.
36. Gini 1934a: 26.
37. Zentralstelle für menschliche Vererbungslehre, MPIA Munich, GDA 50. This involved a translation by Ernst Rüdin's wife Resa from British.
38. Cf. report from Hodson, November 1932, CDA Philadelphia: IFEO.
39. Statement by the British National Committee for Human Heredity, ESA London SA /Eug, E.l.
40. Hodson to Blacker, 9/23/1936, ESA London, SA/Eug, C.159: Hodson.
41. Cf. *Eugenics Review* 29 (1937): 128.
42. Cf. "Human Heredity. A London Bureau." *Catholic Herald*, 5/6/1936.
43. Cf. Outline of Scheme to be Put Forward by the British Human Heredity Committee at the International Congress in New York, ESA London, SA/Eug E.12.
44. Medical Research Council (Chairman Haldane), 10/18/1937, JBSHA London, Box 15.
45. Human Heredity Committee, MPIA Munich, GDA 33. Membership list of the International Committee on Human Heredity as of August 1939, IHGA Copenhagen: Correspondence Bureau of Human Heredity 1938–1957 and RGA London, Box 10.
46. Report of the Meeting Held in Edinburg, 28/8/1939, RGA London, Box 10.
47. Cf. Searl 1976: 101; 1979; Maas 1976: 29; Roll-Hansen 1980: 286. A letter from Davenport to Sanger on February 13, 1925, in which he rejected her participation at the International Conference for Birth Control in 1925, clearly shows the distrust of the eugenicists of the first generation toward the birth-control movement. CDA Philadelphia: Sanger. See also: Popenoe 1926; *Archiv für Rassen- und Gesellschaftsbiologie* 19 (1927): 446; *Eugenical News* 19 (1934): 143.

48. Carr-Saunders 1935: Ilf and Fairchild at the Preliminary Conference on a Population Association, 12/15/1930, CDA Davenport: Population Association of America; see also *Eugenical News* 17 (1932): 50; *Eugenics Review* 20 (1928): 153.
49. Based on Allen 1991: 247.
50. On Sanger's central role at the congress, cf. C. V. Drysdale: Some Impressions of the First World Population Conference, MSA Washington: 192, p. 9. See also Connelly 2006: 213.
51. Cf. *Eugenics Review* 19 (1927): 143; *Eugenics Review* 19 (1927): 259; *Eugenics Review* 24 (1932): 23–25; Blacker 1964: 136.
52. Cf. Sanger (Hg.) 1927: 11. On van Herwerden, see her book on human genetics and eugenics (Herwerden 1926). Only for the Swiss William E. Rappard could I not show any eugenic engagement. Otto L. Mohr, Norwegian activist for birth control, refused to participate in the congress because of the strong emphasis on eugenics. Mohr to Huxley, 3/12/1927, MSA Washington: 16,700.
53. Purpose and Possibilities of the World Population Conference. RPA Philadelphia: Sanger; Cf. *Eugenical News* 12 (1927): 58; Martyn 1933: 29.
54. Pearl's press declaration to the World Population Conference, MSA Washington, 192. Cf. Sanger (ed.) 1927: 56; *Eugenics Review* 19(1927): 259f; *Eugenical News* 12 (1927): 133f.
55. *New York Times*, 10/16/1927.
56. Pearl to Rublee, 3/28/1927, RPA Philadelphia: Sanger; Pearl to Sanger, 4/13/1927, MSA Washington: 16,700.
57. Methorst 1927; Edin 1927; see also *Eugenics Review* 19 (1927): 259–260.
58. Cf. Popenoe 1930; Burch 1932; see also Hodgson 1991: 16.
59. Cf. Frederick Osborn: Birth Control and Birth Promotion (Original title: Birth Control will Become Eugenic), 3/12/1935, AESA Philadelphia: Osborn. See also Frederick Osborn to Davenport, 9/12/1935 and 10/7/1935, CDA Philadelphia: Osborn.
60. Two other international organizations with contradictory goals were founded in Geneva, the Comité international pour la vie et la famille and the International Medical Group for the Investigation of Birth Control. Neither the pro-natalist Comité International nor the Medical Group, which was allied with the birth-control movement, achieved the influence or significance of the IUSIPP. A study of the development of these two groups would certainly be worthwhile. Cf. *Eugenics Review* 20 (1928): 32f.
61. Cf. Pearl 1929: 18–21.
62. See Pearl 1925; Cf. Allen 1991: 248. On Pearl, see also Jennings 1942; Glass 1986.
63. Cf. Pearl's Report at the Second Assembly of the IUSIPP, London 6/15/1931, RPA Philadelphia: IUSIPP.
64. Informal working contacts among the American IUSIPP delegates existed since the Paris conference of 1928. Because of strained relations between Pearl and Edwin B. Wilson, president of the Social Science Research Council, the official founding of the American committee of the IUSIPP came only in February 1931. Cf. *Human Biology* 6 (1934): 225. For the background of the clash between Wilson and Pearl, see Kingsland 1984; Lebrun 1985: 10. On the origin of the Population Reference Bureau, see *Bulletin of the IUSIPP* 1 (1930): 34f; American Eugenics Society 1931: 51; Birth Control Review, H. 4/1935.
65. The initially differing orientations of the American IUSIPP Committee and the PAA—Dublin as chairman of the American Committee of the IUSIPP was against spreading birth control, while Fairchild, in the name of the PAA, extensively supported it—continued for 30 years. The national committee of the IUSSIP increasingly went along with the Population Association of America and eventually merged with it. Cf. Kiser 1953: 108; P.P.A. Affairs 2/1983; 3/1985. See also Ramsden 2003: 548ff. Edmund Ramsden seems not to have taken account of the first German version of this book because of language difficulties.

66. Cf. the list of the founding directors of the PAA in Fairchild 1934: 79; see also the list of the PPA members of 7/20/1935 in *Population Literature*, H. 1/1935. Of the slightly over 60 persons invited to the founding meeting, 28 in the 1930s belonged to the Advisory Council of the American Eugenics Society; Cf. the data in Hodgson 1991: 23f and Mehler 1988: 307–309.
67. Cf. *Human Biology* 6 (1934): 225f. PAA still professes this goal today. On the history of the PAA, see Lunde 1981; Lorimer 1981; Notestein 1981; Kiser 1981; Hodgson 1991: 21.
68. Cf. Report of Guy Irving Burch, Secretary American Eugenics Society, 5/15/1934. AESA Philadelphia, 575:06:AM3: Minutes 1925–1935 and Osborn 1974: 118. See also the report on the Conference on Population Studies in Relation to Social Planning, 5/2–4/1935, *Journal of Heredity* 26 (1935): 116.
69. Membership list in the *Bulletin of the IUSIPP* 1 (1930): 23–25. See also the list of August 1931 in RPA Philadelphia: IUSIPP. Mazumdar (1992: 52f) counts "only" 14 members. An example of her error involves Sir William Henry Beveridge, director of the London School of Economics and member of the scientific advisory council of the Eugenics Society, whom she counts as a nonmember.
70. Cf. Council Meeting of the Eugenics Society, 10/10/1928, ESA, SA/Eug L.8. Pearl's founding call from the *Times* 9/18/1929 was reprinted in the *Eugenics Review* 20 (1928): 179–181.
71. Report by Pitt Rivers to the general assembly of the IUSIPP, July 1937, in Paris, RPA Philadelphia: Population Association of America and ESA London, SA/Eug, C273: Pitt-Rivers. Cf. Grebenik 1959: 194f; 1991: 9; Mazumdar 1992: 53 and the interesting analysis by Christopher Langford: British participation in the IUSSP in its early years, IUSSPA Lüttich, I. CU 05.04. The British Population Society merged with the PIC World War II.
72. Cf. *Bulletin of the IUSIPP* 1 (1929): 11–15; 2 (1930): 11–13; Cf. correspondence between Fischer and Rüdin, 12/16/1930, 12/22/1930, MPIA Munich: GDA 52. See also Kroll and Weingart 1989; Höhn 1989.
73. Cf. Nora Federici: Italy's participation in IUSSP during the early period of its foundation (1928–31), IUSSPA Lüttich, I. CU 05.04.
74. Cf. Herwerden 1933: 175.
75. Cf. *Bulletin of the IUSIPP* 1 (1929): 11–15; 1 (1930): 23–25, 30–31.
76. Statutes and Regulation, ESA London: SA/Eug D.110. IUSIPP—Zielsetzung, BA Potsdam, 15.01 REM 3058. The citation is from Harmsen (1936): vi.
77. Laughlin, cited from the minutes of the Preliminary Conference on a Population Association, 112/15/1930, CDA Philadelphia: Population Association of America. See Ramsden 2003: 558ff.
78. Second Conference of the Population Association of America, 7/4/1931, CDA Philadelphia: Population Association of America.
79. Hodson to Davenport, 11/19/1927; Hodson to Mallet, 6/22/1928, CDA Philadelphia: IFEO; Cf. *Eugenics Review* 19 (1927): 262.
80. Gini to Pearl, 10/10/1927, and Pearl to Gini, 10/18/1927, RPA Philadelphia: Gini.
81. Cf. *Eugenics Review* 20 (1928): 185.
82. Cf. Gini to Davenport, 3/26/1929, CDA Philadelphia: Gini.
83. Cf. Mallet to Pearl, 9/21/1928, RPA Philadelphia: Mallet.
84. Cf. Pearl to Mallet, 10/2/1928, RPA Philadelphia: Mallet.
85. Cf. Pearl to Pitt-Rivers, 12/21/1932, RPA Philadelphia: Pitt-Rivers.
86. The minutes of the meeting in Rome are found in MPIA Munich, GDA 32, and in CDA Philadelphia: IFEO Rome 1929. On Davenport's and Gini's motives, see their correspondence of 3/26/1929 and 4/15/1929, CDA Philadelphia: Gini and Davenport's letter to Hodson, 4/15/1929, CDA Philadelphia: IFEO.

87. Cf. *Eugenical News* 13 (1928): 131; *Human Biology* 6 (1934): 223. On the conflict, cf. Cassata 2011: 142.
88. The resolution was renewed at the IUSSIP board meeting on May 29 and 30, 1930, in Paris. Cf. *Archiv für Bevölkerungspolitik, Sexualethik und Familienkunde* 1 (1931): 132. See also Gini to Davenport, 2/11/1929, CDA Philadelphia: Gini.
89. Cf. Gini to Wilson (Social Science Research Council), 14/8/1930, copy in RPA Philadelphia: Gini. Cf. Proposal for Establishing in the United States an Executive Office of the International Union for the Scientific Study of Population, 1948, AJLA Princeton: IUSSP. See also Lorimer 1971: 89. For another interpretation, see Lebrun (1985: 12) and Höhn (1987: 9).
90. Pearl to Gini, 22/1/1931, RPA Philadelphia: Gini. The background to the argument lay in the outstanding financial means of the American foundations. The American Social Science Research Council and the National Research Council, through whom the foundation funds were distributed, were skeptical as to whether the IUSIPP was a potentially successful project, and particularly whether a congress of scientific standards could be held in Fascist Rome. Wilson to Gini, 7/3/1930; 3/4/1931, IUSSPA Lüttich, I. CU 05.04. Cf. also Nora Federici: Italy's Participation in IUSSP during the Early Period of its Foundation (probably 1985), IUSSPA Lüttich, I.CU 05.04. The presentations of the London meeting were edited by Pitt-Rivers (1932).
91. Davenport to Gini, 6/24/1931, CDA Philadelphia: Gini. See also Mantovani 2004: 335ff.
92. Cf. *Archiv für Bevölkerungspolitik, Sexualethik und Familienkunde* 2 (1931): 134.
93. Cf. Gini 1933: xlix.
94. Cf. *Eugenical News* 16 (1931): 72f.
95. On criticism of Pearl in the United States, see Kingsland 1984: 10–16.
96. Cf. Barkan 1992: 292.
97. On the problems of the committees due to the conflicts with the Italian committee, see Pitt-Rivers 1936:25.
98. Gini to Davenport, 23/6/1932, CDA Davenport: Gini; Pearl to Close, 12/7/1932, RPA Philadelphia: IUSIPP. See also Nora Federici. Notes sur les relations entre le CISP et l'Union aux débuts de leur activité. IUSSPA Lüttich, I.CU 05.04.
99. See Pearl's Report at the Second Assembly of the IUSIPP, London 6/15/1931, and Close's letter to Pearl, undated, RPA Philadelphia: IUSIPP.
100. Minutes of the Meetings of the IUSIPP in London, June 27 and 28, 1933, RPA Philadelphia: Pitt-Rivers. Cf. also Pitt-River's report to the General Assembly of the IUSIPP, July 1937, in Paris, RPA Philadelphia: Population Association of America and ESA London, SA/Eug, C273: Pitt-Rivers. See also *Eugenics Review* 27 (1935): 147. The analysis of the withdrawal of the IUSIPP from the IFEO is difficult because at the beginning of the 1930s, Gini also fell out with the leadership of the IFEO. Cf. Hodson to Davenport, 1/15/1932, 2/22/1932, and 8/9/1932, CDA Philadelphia: IFEO.
101. Osborn 1937a: 105.
102. Kemp 1932: 311.

5 National Socialist Germany and the National Eugenics Movement

1. Fischer 1936b: 355f. On Fischer's attitude toward the National Socialists, see Lösch 1997.
2. Groß 1933.
3. Hilpert 1933: 97, 100.
4. Cf. Bock 1991b: 233; see also Bock 1991a.

5. See the classification of eugenics in the discussion of the special way in Crook 2002: 370 and Weiss 2010a: 303ff. Weiss in particular argues that there were no specific special qualities that would justify speaking of a eugenic special way. See also Adams et al. 2005. For political differences between countries, see Hansen and King 2001.
6. On the idea of "political opportunity structures," see Meyer 2004. See also Rucht 1996; Schock 1996; and Watanabe 2000.
7. Lenz 1931a: 300–308; Cf. Seidler and Rett 1988: 70–73; Weingart et al. 1988: 373–375.
8. Hitler 1974. The group around the Pioneer Fund interested in questions of race recently has emphasized that the word "eugenics" does not appear in Hitler's *Mein Kampf* and that it is therefore incorrect to label Hitler as a eugenicist (see Glad 2006: 65ff). This overlooks the fact that Hitler in a number of places adopted the German synonym "race hygiene." On distinguishing the "last eugenicists" from Hitler's race policies, see Lynn 2001b: 240ff. On Glad's position, see also Biley 2007: 325ff and Turda 2008: 112ff.
9. Information comes from Ploetz's diary entries of 1/11/1932 and 2/13/1932. APA Herrsching. Lenz proposed the "Wesenverwandtschaft" in the third edition of the volume II of Baur et al. (Lenz 1931b: 550). Cf. Weingart et al. 1988: 382; Weingart 1994: 7. On Muckermann's position at this time, see Richter 2001: 73ff, 274ff, and 314ff.
10. Ploetz to Hitler, 6/4/1933, BAK RMI 15.01 26243; on Ploetz's joining the NSDAP, see his party records in BA Zehlendorf: Ploetz, Alfred.
11. Fischer 1933: 9; Cf. also "Hereditarily Sick Required to Register," Der Angriff, 7/14/1933.
12. Activity report of the KWI, presumably 1933/34, BA Koblenz RMI 15.01 26244; letter from the Reich Ministry of the Interior to Darre, 5/23/1934, BA Koblenz RMI 15.01 26245.
13. Rüdin 1934c; Cf. also Rüdin 1934b.
14. BA Berlin Zehlendorf: Rüdin, Ernst.
15. Bluhm to Ploetz, 21/2/1935, APA Herrsching. See also Bluhm 2007: 196ff.
16. The bylaws of the Deutsche Gesellschaft für Rassenhygiene from 1934 are printed in Kröner (1980: 181–184).
17. PM 21/8/1945; Cf. Weinreich 1946: 33.
18. Rüdin 1934c. For the course of the meeting, see Ploetz diary of 4/22/1934, APA Herrsching. A good analysis of Rudin's "Program" can be found in Roelcke 2003: 43ff.
19. *Ziel und Weg* (1938): 535.
20. Ploetz diary, 5/19/1933, APA Herrsching.
21. For an analysis of the tensions between German eugenicists and the National Socialist administration, see Weiss 2010a: 69ff.
22. Cf. Weinreich 1946: 34; Lifton 1986: 31; Proctor 1988: 23; Gillette 2007: 13. I have not yet been able to locate the original source.
23. See Harten et al. 2006 on race hygiene as the educational ideal of the National Socialists.
24. Cf. Popenoe 1934b: 260; see also Popenoe 1934a.
25. Berger 1939b: 115. Cf. also Berger 1939a, 1939c.
26. Groß 1939: 7, 22f. Cf. also "Sieg des Rassegedankens in der Welt." *Völkischer Beobachter*, 1/19/1939; "Das Weltecho der deutschen Rassenpolitik." *Deutsche Allgemeine Zeitung*, 1/19/1939.
27. Cf. for example, Groß 1939: 27.
28. In a very extensive study, I discussed the "Nazi connection" of American scientists (Kühl 1994; see also a shorter version in Kühl 1998). Black (2003) repeats my arguments in a well-researched book that is written for a broad public in a rather journalistic style.

29. On Nilsson-Ehle and Lundborg, see Goldschmidt 1960: 191; Broberg and Tyden 1996: 92; see also "Erbgang der Begabung in nordischen Sippen." *Völkischer Beobachter*, 15/12/1936.
30. See the practically word-for-word repeat articles in Mjöen 1933a, 1934a, 1935a, 1935b. Cf. also *Eugenical News* 19 (1934): 60.
31. See Hart 2012 for extensive discussion of the support of British eugenicists for Nazi race policies.
32. See *Whitney's Memoirs*, p. 204, LWA Philadelphia; on Laughlin, see "Amerikanische Anerkennung für das deutsche Sterilisationsgesetz." *Der Angriff*, 7/26/1933. Bruinius 2006: 271—a rather journalistic presentation of Laughlin's ties to the National Socialist regime.
33. Laughlin to Draper, 3/15/1937 and 11/9/1938, HLA Kirksville; Cf. Hassencahl 1970: 355f; Mehler 1989: 21. See also *Eugenical News* 22 (1937): 65f; *Ziel und Weg* (1937): 361. On the goals of the Pioneer Fund, see Certificate of Incorporation, HLA Kirksville. See Tucker 2002: 52ff and Lombardo 2002: 789ff on the role of the Pioneer Fund in supporting National Socialist race policies.
34. Cf. Mjöen 1933a; Schrijver 1936.
35. Mussolini, who had close ties to the Italian Eugenics Society and who had earlier received eugenicists like Mjöen, Davenport, and Laughlin, even attended a meeting of the international eugenics organization. (Cf. Leon Whitney's Autobiography, LWA Philadelphia, S. 198; Cläre Mjöen to Hodson, September 1928, ESA London, SA/Eug, C. 235.) The eugenicists, as Eugen Fischer expressed it, represented for Mussolini the only politicians who could and would "truly carry out eugenic rules," but they were disappointed by the meeting (Fischer to Davenport, 7/19/1929, CDA Philadelphia: Fischer). (*Eugenical News* 14 (1929): 154–156. Cf. the minutes of the business meeting of the IFEO, 9/27/1929, MPIA Munich: GDA 32. See also Chase 1977: 345f; see also Lenz in *Archiv für Rassen- und Gesellschaftsbiologie* 22 (1930): 435). Black (2003: 279) does portray Mussolini's visit, but he says nothing about Mussolini's critical position.
36. Resolution of the IFEO requiring action by the president, 1930–1932, CDA Philadelphia: IFEO—Miscellaneous. Cf. *Eugenical News* 15 (1930): 154.
37. Campbell 1936c; Close's position is presented in *Eugenics Review* 28 (1936): 33f. On invalidating basic rights under National Socialism, see Kershaw 1995. On the criticism of democracy in the eugenics movement before 1933, see Goddard 1919; Osborn 1921; McDougall 1921; Cannon 1922.
38. For criticism in the early years, see the second working assembly of the Race Political Office of the NSDAP, June 2–9, 1935, FZA Munich Ma 1159–17448. Cf. also "Welche Staaten bekämpfen Erbkrankheiten." *Völkischer Beobachter*, 3/9/1938.
39. *Revue Anthropologique* 43 (1933): 388f; Cf. also Schreiber 1935a, 1935b, 1936.
40. Rüdin sent a German translation of Schreiber's article to the Norwegian IFEO delegates Alfred Mjöen, Wilhelm Keilhau, and Klaus Hansen, MPIA Munich GDA 32.
41. Rüdin to Fetscher, 4/26/1934, based on Weindling 1989a: 504.
42. In general on the Deutsche Kongress-Zentrale, see Herren 2009: 74ff. Weiss 2010a: 201ff gives an extensive picture of the influence of the Deutsche Kongress-Zentral on German race hygienists.
43. Rüdin's opening speech in Zurich, MPIA Munich, GDA 35 and IFEO 1934: 4f.
44. IFEO 1934: 62f; Cf. *Journal of Heredity* 26 (1935): 10.
45. See Noordman 1989 and Pols 2010 on eugenics in The Netherlands.
46. IFEO 1934: 67; see also Schreiber 1934, 1935a.
47. IFEO 1934: 78f.
48. Cf. the evaluation by Weinreich 1946: 30. Heinz Kürten (1934: 599), representative of the Bavarian government in Zürich, after his return explained to the readers of Ziel und Weg that the resolution "in the sense of the Congress quite clearly and basically" referred to the German sterilization law.

49. *Neues Volk* H. 9/1934: 13.
50. "Sterilisierung—in der ganzen Welt." *NSK*, 8/8/1935; Cf. also "Die Rassen-hygieniker gegen die Kriegshetze." NSK, 7/31/1934.
51. *Rassenpolitische Auslands-Korrespondenz* 5/1934: 3.
52. *Eugenical News* 19 (1934): 140; Cf. *Volk und Rasse* 7 (1935): 155f; *Rassenpolitische Auslands-Korrespondenz*, H. 7/1935: 6.
53. Cf. Frick 1934; Thomalla 1934; the translations were clearly financed by Wickliffe Draper. Cf. CDA Philadelphia: Draper, Wickliffe. Laughlin praised Frick's contribution in a letter to Grant of 1/13/1934 as a "milepost in statesmanship." HLLA Kirksville.
54. *Eugenical News* 18 (1933): 89–92, here 90.
55. *Eugenical News* 19 (1934): 126; *Eugenical News* 20 (1935): 100; *Eugenical News* 19 (1936): 59f.
56. Cf. *Eugenical News* 16 (1931): 171.
57. Close to Pearl, 1/21/1935, RPA Philadelphia: IUSIPP; Close to Lorimer, 5/15/1935, RPA Philadelphia: Population Association of America. For various reasons, the congress was postponed from 1934 to 1935.
58. Resolution of the IUSIPP board of 6/27/1933; Cf. Pitt-Rivers: Report on 1. Progress, 2. Science of Population for the General Assembly, Paris, July 1937, RPA Philadelphia: IUSIPP and ESA London, SA/Eug C 273 Pitt-Rivers.
59. Pearl to Close, 12/17/1934, RPA Philadelphia: IUSIPP; see also Pearl to Lorimer, 5/15/1935, RPA Philadelphia: Population Association of America.
60. Close to Pearl, 12/11/1934 and 1/21/1935, RPA Philadelphia: IUSIPP. Fischer turned down the offer and asked Close to remain in office for another two years.
61. Close to Pearl, 2/8/1936, RPA Philadelphia: IUSIPP.
62. Groß 1936; Cf. also Groß 1935a, 1935b; Harmsen 1935a: 365.
63. Fischer 1935a.
64. Fischer 1935b: 692f; Cf. also Fischer 1936a, 1936b.
65. Telegram of greeting from Fischer to Hitler, Cf. Harmsen and Lohse 1936: 38.
66. See Hart 2012: 46ff.
67. Blacker 1933, 1934; Cf. Soloway 1990: 304.
68. Cf. *New York Times*, 8/30/1935; see also Campbell's letter to Rüdin, 5/5/1935, MPIA Munich, GDA 132. For Hankin's position, see Hankins 1937: 630.
69. Lombardo 2002: 771ff; Kenny 2002: 270 on Draper's participation in the Berlin Congress.
70. Cf. Glass 1935: 209. Like Pearl, Laughlin refused on personal grounds and had his speech read by Campbell. Cf. Kühl 1994: 33f. On Campbell's toast, see Kevles 1985: 347.
71. Campbell 1936a: 602.
72. Harmsen, *Gesundheitsfürsorge*, H. 12/1935: 348; Cf. also *Mitteilungen der Arbeitsgemeinschaft für Volksgesundung*, H. 29/1935.
73. Cf. Close 1936; Linder 1936; Winkler 1936; Wieth-Knudsen and Asbjörn 1936. Harmsen (1935a: 359–368) and Friese (1936) give extensive summaries of the conference contributions.
74. Cf. Burgdörfer 1936; Keiter 1936; Koller 1936. On Burgdörfer, see Thieme 1988: 73ff and Etzemüller 2007: 53ff.
75. Gütt 1936a: 748.
76. Keiter 1936; quoted from "Ein Amerikaner über rassische Pflichten." *Völkischer Beobachter*, 8/29/1935.
77. Harmsen and Lohse 1936; Cf. Bock 1986: 244.
78. Glass 1935: 209.
79. Dalsace 1936: 706–712. See also the article "Empfang der Reichsregierung für die Teilnehmer am Bevölkerungswissenschaftlichen Kongreß." *Völkischer Beobachter*,

8/31/1935, and "Rassenhygiene und Krieg." *Berliner Tageblatt*, 8/31/1935, and the reports by Ruttke (1935) and Harmsen (1935a: 359).
80. *New York Times*, 8/29/1935; Cf. "Empfang der Reichsregierung für die Teilnehmer am Bevölkerungswissenschaftlichen Kongreß." *Völkischer Beobachter*, 8/31/1935.
81. Fischer 1936b: 928; Cf. Bock 1986: 244.
82. On the international congress on criminal law and its presentation in Nazi propaganda, see for example, "Prof. Simon von der Aa über den Strafrechtskongreß." *Völkischer Beobachter*, 8/27/1935; "Gute Zensur. Strafrechtskongreß und Rassengesetzgebung." *Rote Erde*, 8/27/1935; "Wieder eine ausländische Hetzlüge geplatzt." *Völkischer Beobachter*, 8/29/1935; "Der deutsche Standpunkt in der Sterilisationsfrage international anerkannt." *Rassenpolitische Auslands-Korrespondenz*, H. 5/1935: 2; Cf. also Harmsen 1935b: 1208; Wagner 1943: 121f.
83. "Wegweisende Wissenschaft." *Deutsche Allgemeine Zeitung*, 9/6/1935.
84. Schade 1935: 140. On Schade's role in the SS, see *BA Zehlendorf*: SL 9, p. 124.
85. Schrijver 1936, quoted from Schade 1936: 4.
86. "Deutschlands rassepolitische Maßnahmen Vorbild für das Ausland." *Berliner Börsenzeitung*, 9/5/1935.
87. Campbell 1936a: 28. A German translation was published in Erbarzt (Campbell 1936b). For its use in NS-propaganda, see *Angriff*, 7/12/1936; *Preußische Zeitung*, 8/30/1936; *Völkischer Beobachter*, 8/19/1936; *News Bureau of German Newspaper Publishers*, 9/28/1936; *NSK*, 12/11/1936. Campbell's work was reprinted in 1973 in the extreme right-wing American magazine American Mercury on the grounds that it presented "true facts about the German race policies."
88. Quoted from Schade 1936: 3.
89. Pearl to Close, 2/11/1936, RPA Philadelphia: IUSIPP.
90. Glass 1935: 208f. See Campbell's reaction in *Eugenics Review* 28 (1936): 85.
91. Whelpton to Pearl, 3/20/1935, including a resolution of the national committee of the United States, RPA Philadelphia: Population Association of America; Frets's contribution at the board meeting of the IUSIPP in Paris 1937, RPA Philadelphia: Pitt-Rivers.
92. Cf. Hodson's report of March 1939, MPIA Munich, GDA 50.
93. On Mohr, see Roll-Hansen 1980: 290. Whether Mohr himself had applied for membership or whether some members of the IFEO simply considered offering membership to Mohr can no longer be reconstructed. His disagreements with Mjöen lead one to assume that Mohr did not place any great value on membership in the IFEO.
94. Mjöen to Rüdin, 1/7/1934, MPIA Munich, GDA 33.
95. On the lack of interest of the British reform eugenicists, see the letter from Hodson to Davenport, 12/23/1930, CDA Philadelphia: IFEO, and the letter from Blacker to Hodson, 8/271936, ESA London, C 159: Hodson. For Osborn's lack of knowledge about the IFEO, see Osborn to Bertheau, 6/16/1936, EHA New Haven, p. IV, Box 32, Fol. 324. Frederick Osborn first became a member in 1939. Cf. Hodson's report of March 1939, MPIA Munich, GDA 50.
96. Cf. Hodson to Huntington, 11/27/1934, EHA New Haven, p. IV, Box 31, Fol. 315. See the circulating permission for membership in the Reich Ministry of Education, BA Potsdam 15.01 REM 2935. Cf. also the IFEO list of 1937, in MPIA Munich, GDA 33.
97. Stieve to Thummala, 6/16/1936, BA Potsdam, 49–01 REM 3198.
98. Rüdin to German Central Congress Office, 5/8/1936, MPIA Munich, GDA 36.
99. Based on Weber 1993: 232.
100. Minutes of the Scheveningen meeting, CDA Philadelphia: IFEO.
101. Up until 1945, approximately 400,000 people were sterilized on the basis of the sterilization law in Germany and in the annexed territories. Cf. Bock 1986: 238.
102. Cf. Sanders 1936: 106f.
103. Cf. *Der Erbarzt* 3 (1936): 117.

104. On the similarity of the laws, see Hansen 1996: 56; Broberg and Tyden 1996: 115.
105. Quoted from "Nationalsozialistische Rassenforschung im Urteil des Auslandes." *Westfälische Zeitung*, 8/10/1936; "Deutschland Vorbild." *Westfälische Landeszeitung*, 8/10/1936; Cf. "Die deutsche Vorarbeit." *Berliner Tageblatt*, 8/10/1936.
106. "Deutschland führend in der Vererbungsforschung." *Völkischer Beobachter*, 8/2/1936; "Deutschlands Führung in der Vererbungsforschung." *NSK*, 7/31/1936; "Zum Internationalen Kongreß eugenischer Organisationen in Holland." *Rassenpolitische Auslands-Korrespondenz*, H. 7–8/1936: 11.
107. Fischer to Rüdin, 7/13/1936, MPIA Munich, GDA 36.
108. Cf. minutes of the meeting, MPIA Munich, GDA 37.
109. Sjögren to Rüdin, 9/18/1936, MPIA Munich, GDA 34; Cf. Weber 1993: 232.
110. Cf. the fruitless complaint by Dahlberg over Rüdin's participation in the selection process. Dahlberg to the Swedish king, 11/8/1934, MPIA and Rüdin's memo, 9/9/1935, MPIA Munich, GDA 113; see also Broberg and Tyden 1996: 92.
111. Sjögren to Rüdin, 9/18/1936, MPIA Munich, GDA 34.
112. Sjögren to Rüdin, 10/16/1937, MPIA Munich, GDA 34.
113. Minutes of the twelfth meeting of the IFEO in Scheveningen, MPIA Munich, GDA 37.
114. Sanders 1936: 107. Cf. "Deutschland lädt ein zum internationalen Kongreß für Rassenhygiene." *Nachrichten deutscher Zeitungen*, 8/17/1936. For Rüdin's reaction, see Rüdin to the Reichsausschuß für Volksgesundheitsdienst, 1/29/1937, MPIA Munich, GDA 50.
115. Sjögren to Rüdin, 9/18/1936, MPIA Munich, GDA 34; Sjögren to Hodson, 1/6/1937, MPIA Munich, NLR 1.
116. Sjögren to Hodson, 4/11/1938; Hodson to Rüdin, 4/17/1938; Hodson to all members of the IFEO, 5/17/1938, MPIA Munich, GDA 33 and 34; Cf. also *Eugenical News* 23 (1938): 79.
117. Hodson to Rüdin, 7/25/1936, MPIA Munich, GDA 50.
118. Cf. Hodson to Sjögren, 4/24/1938, MPIA Munich, GDA 33.
119. Cf. Hodson to Rüdin, 3/31/1939; MPIA Munich, GDA 50; Hodson to Sjögren 2/22/1939; Sjögren to Rüdin, 3/12/1939; MPIA Munich, GDA 33.
120. Sjögren to the Minister of the Interior, 1/9/1939, MPIA Munich, GDA 33.
121. Minutes of a pre-discussion for the International Congress for Race Hygiene on 2/15/1939, created by Linden on 2/25/1939, BA Potsdam, 49.01 REM 2839.
122. IV. International Congress for Race Hygiene (Eugenics), BHA Munich, M. Inn 79477 Rassenhygiene. Cf. also *Rassenpolitische Auslands-Korrespondenz*, H. 7/1939: 5.
123. See the address list in MPIA Munich, GDA 34. See also the answer of Harry H. Laughlin, printed in Samaan 2013: 474.
124. Linden to the Minister for Science, Education, and National Formation, 10/11/1940, BA Potsdam 49.01 REM 2839.
125. Blacker to Fisher, 5/5/1939, and Fisher to Blacker, 5/9/1939, ESA London, SA/ EugC. 108.
126. On the dispute between Gini and the IFEO, see Hodson to Davenport, 28/7/1932, CDA Philadelphia: IFEO.
127. Gini 1936: 78; Briand 1938: 309. See Schneider 1990b: 97; Stepan 1991: 189; and Cassata 2011: 143 and 177.
128. For example, Gini to Davenport, 1/11/1931, CDA Philadelphia: Gini.
129. *Fédération Internationale Latine des Sociétés d'Eugénique* 1937: 96. Unfortunately, there is still no broad history of the FILDSE. On "Latin eugenics," see Reggiani 2010: 295ff. A second congress of the Fédération planned for September 25–30, 1939, in Bucharest was cancelled (Cf. Turda 2011: 338). No indication of post–Second World War activity of the Fédération has yet been found.

130. For example, see the work of the American Committee for Displaced Scholars, which numbered among its members the eugenicists Curt Stern, Raymond Pearl, and Laurence Snyder, and the Society for the Protection of Science and Learning, in which the eugenicist William Beveridge was active. Cf. Ludmerer 1972: 129; Barkan 1992: 281.
131. I do not go into detail on Dunn below. For his protests against the National Socialist race policies, see Hassencahl 1970: 335; Ludmerer 1972: 128f–130; Chase 1977: 352.
132. Cf. Barkan 1992: 78.
133. Cf. Proctor 1991: 183; Barkan 1992: 81; Kaufmann 2003: 309ff.
134. Boas to unknown, 10/8/1935, FBA Philadelphia: Boas, Franz.
135. Cf. the comprehensive presentation in Barkan 1992: 280–228.
136. In this connection, the French historian Benoit Massin speaks of the Janus-faced Zollschan. Cf. Doron 1980; Bacharach 1984. On Zollschan, see Weindling 2007: 267 and 2011b: 84ff.
137. Zollschan's Jewish nationalism was a reason that Boas did not support his efforts in 1924 to set up an anthropological Institute for the study of race question in New York. As the American historian Elazar Barkan (1992: 319f) shows, Zollschan saw Jewish nationalism, anti-Semitism, and racism as connected questions that have to be addressed simultaneously.
138. Cf. Ruzicka 1935; Brozek 1935; Matiegka 1935; Weigner 1935.
139. Huxley and Haddon 1935; Cf. Fleure 1936: 319; Huxley 1981: 188; Jones 1988: 74f. See also Schaffer 2008.
140. Barkan 1992: 308f.
141. Zollschan: Report on the preparatory work for the planned study of race, 10/1/1934, FBA Philadelphia: Zollschan; Background paper to the UNESCO Statement on Race, 7/19/1950, UNSECOA Paris 323.12. A 102. Cf. Metraux 1950: 386; Barkan 1992: 318–325.
142. Roemer to Pohlisch, 31/1/1937, MPIA Munich, GDA 40.
143. Rüdin to the Reichserziehungsministerium, 10/11/1937, BA Potsdam, 49.01 REM 2839. It would take a separate Investigation to find out why the two American physicians of Jewish origin, Jacob H. Landman and Abraham Myerson, who were numbered among the highest-profile critics of the German sterilization law, did not join with other scientists in criticizing the National Socialists. Cf. also Landman 1932, 1936; Myerson et al. 1936; see also Chase 1977: 115; Kevles 1985: 117; Reilly 1991: 123.
144. "Du Congrès Universel des Races (1911) au Congrès International de la Population (1937)." *Races et Racisme*, 4/1937: 1–3; Race et Racisme: Le Congrès International de la Population, 1937, FBA Philadelphia: Congrès International de la Population; see also Boas to Dunn, 10/25/1937, LCDA Philadelphia: Boas.
145. Landry to Boas, 5/25/1937 and 6/28/1937, FBA Philadelphia: Landry; Cf. Barkan 1992: 326 and Kröner 1998: 50
146. Boas 1938: 87.
147. Zollschan 1938: 97–102.
148. Beck 1938: 107–109.
149. Quoted from Thums 1937: 534.
150. Pfeil 1937: 300f; on Pfeil's general attitude, Cf. Pfeil 1940. For initial analyses of Pfeil's position, see Schnitzler 2009: 328ff.
151. Thums 1937: 534; Cf. also Geyer 1937.
152. On Landry's comments, see *Le Temps*, 7/30/1937; Cf. *Rassenpolitische Auslands-Korrespondenz*, H. 9/1937: 6.
153. Rüdin to the Reichserziehungsministerium, 10/11/1937, BA Potsdam, 49.01 REM 2839; Cf. *Pfeil* 1937: 299 and the official reaction from Rüdin, MPIA Munich, GDA 40.

154. Minutes of the meeting of the executive committee of the IUSIPP, 7/27/1937, RPA Philadelphia: Pitt-Rivers; Cf. Ruttke's report to the Reichsinnenministerium, 11/28/1937, BA Potsdam 49.01 2760.
155. Minutes of the meeting of the executive committee of the IUSIPP, 7/27/1937, RPA Philadelphia: Pitt-Rivers.
156. A. Bohac: Réponse au rapport du capitaine Pitt-Rivers concernant le Tchécoslavaquie, based on: Alena Subtrova: Contribution tchécoslovaque au travail de l'Union internationale pour l'étude scientifique de la population, IUSSPA Lüttich CU.05.04. See also Sharpe 1937; Burgdörfer 1942: 57.
157. Pitt-Rivers 1938; Cf. Griffiths 1980: 323f. See in particular Hart 2012.
158. Cf. Close to Pearl, 12/5/1937, RPA Philadelphia: IUSIPP.
159. Pitt-Rivers: Report on 1. Progress, 2. Science of Population for the General Assembly, Paris, July 1937, RPA Philadelphia: IUSIPP and ESA London, SA/Eug C273 Pitt-Rivers.
160. Close to Pearl, 5/12/1937, RPA Philadelphia: IUSIPP.
161. Verschuer to Reichserziehungsministerium, 9/1/1937, BA Potsdam, 49.01 REM 4901, and Pfeil 1937: 301.
162. There were various criticisms at the Second International Congress for Anthropology and Ethnology, which took place in Copenhagen from July 31 to September 6, 1938.
163. Cf. Paul 1984: 575–583; Weß 1989: 131–133. See also Haldane 1936; Dahlberg 1942.
164. Cf. Schaxel to Muller, June 1935, HJM Bloomington.
165. Quoted from a letter from Müller (Gestapo) to the Reichserziehungsministerium, 6/16/1936, BA Potsdam, 15.01 REM 2935. See also Schaxel to Landauer, 11/28/1935, LCDA Philadelphia: Seventh International Congress of Genetics, and Dunn to Landauer, 9/30/1935, LCDA Philadelphia: Landauer. Schaxel was the victim of a Stalinist purge in 1937. For more information on Schaxel, see Weindling 1989a: 327–329; Deichmann 1992: 44. See also Gütt 1936b.
166. Landauer to Dunn, 11/20/1935 and 12/5/1935, LCDA Philadelphia: Landauer.
167. Letter from 31 American geneticists to Levit, 4/2/1936, HSJA Philadelphia: Levit. Two other signatories, Laurence H. Snyder and Harold H. Plough, later joined the American Eugenics Society. For evidence of the coordination by Landauer, see the letter from Landauer to Demerec, 2/29/1936, MDA Philadelphia: Landauer.
168. Levit to Jennings, 4/27/1936, HSJA Philadelphia: Seventh International Congress of Genetics; Levit to Landauer, 6/25/1936, LCDA Philadelphia: Seventh International Congress of Genetics; Levit to Huxley, 6/25/1936, ESA London, SA/Eug, D. 130.
169. Cf. the report of the German embassy, 6/22/1936, BA Potsdam, 39.01 2969.
170. Rüdin to Mjöen, Lundborg, Nilsson-Ehle, and Sjögren, 3/16/1936, MPIA Munich, GDA 36.
171. Müller (Gestapo) to the Reichserziehungsministerium, 6/10/1936, BA Potsdam, 15.01 REM 2935.
172. Results of a discussion at the Foreign office on 8/21/1936, BA Potsdam, 15.01 REM 2935.
173. Linden (Reichsinnenministerium) to Rüdin, Fischer to both the Reichserziehungsministerium and the Race Policy Office of the NSDAP, 2/24/1937, BA Potsdam, 15.01 REM 2935. Attached was a copy of a letter from A. I. Muralow to Paula Hertwig of 2/8/1937. Muralow was a well-known geneticist serving as president of the congress.
174. *Journal of Heredity* 28 (1937): 55; Cf. Jones 1988: 3.
175. Cf. Muller to Huxley, 8/2/1939, and Huxley to Muller, 8/3/1939, HJMA Bloomington: Huxley.
176. In my description of the congress, I extensively follow Roth 1986: 11–13; Weß 1989: 155; see also Verschuer's report to the Reichserziehungsministerium, 9/4/1939, BA Potsdam 49–01 Rem 3198.

177. On the story of the origin of the Geneticists' Manifesto, see Muller to Paul Blanshard, 8/20/1951, and Muller to Helen Bowyer, 9/16/1946, HJMA Bloomington.
178. Some of the journals that reprinted the Geneticists' Manifesto were: *Eugenical News* 24 (1939): 63f; *Journal of Heredity* 30 (1939): 371–373; *Nature* 144 (1939): 521f.
179. Cf. Roth 1986: 14. Commentaries on the Geneticists' Manifesto are also to be found in Ludmerer 1972: 129; Paul 1984: 583; Weingart et al. 1988: 542; Weß 1989: 155–157.

6 The Second World War and the Mass Murder of the Sick and Handicapped

1. IFEO 1934: 4; Cf. also "Humane Rassenhygiene." *Völkischer Beobachter*, 8/5/1935.
2. Rüdin 1937: 108.
3. Rüdin 1939: 444f: 235; Cf. also Weber 1993: 235. Generally on Rüdin—Weber 1996 and Weber 2000.
4. Rüdin 1940: 4.
5. Frick 1936: 11; Cf. also "Die nationalsozialistische Weltanschauung ist der Garant des Friedens." *Indie*, 9/3/1935.
6. Groß 1935a: 6. Groß gave a similar lecture to the German colony in London; Cf. "Rassenpolitik ist Friedenspolitik." *NSK*, 7/12/1935.
7. Cited from Frercks 1937: 46. Here I disagree with Bergman's view (1992: 116f) that the Nazis glorified war because they expected positive selection.
8. IFEO 1934: 78. On the position of the British Eugenics Society, see the decision of the Board of Directors in the mid-1930s, *Eugenics Review* 28 (1937): 297. On the position of the American Eugenics Society see American Eugenics Society 1938: 20. Only Gini (1934a) continued to refer to the positive effect of the war on the hereditary patrimony of a people. See also Kühl 2001: 200ff.
9. Groß for example referred explicitly to the Zurich resolution.
10. Ploetz—Tagebuch, entries of 7/3/1935, 7/16/1935, 7/17/1935, and 7/26/1935, APA Herrsching.
11. Ploetz 1936b: 616, 1936a: 3f. Ploetz's lecture also appeared in *Ziel und Weg* (H. 9/1935) and in the *Süddeutsche Monatshefte* (H. 10/1935); see also "Rassenhygiene und Krieg." *Berliner Tageblatt*, 8/30/1935.
12. Cläre Mjöen to Ploetz, 1/17/1936, APA Herrsching; Cf. *Volk und Rasse* 11 (1936): 151; *Rasse* 3 (1936): 151.
13. See the entries in Ploetz's Tagebuch, 6/22/1936 und 11/20/1936, APA Herrsching.
14. For example, see Graßmann 1936; *Völkischer Beobachter* 2/19/1936.
15. Bluhm to Ploetz, 11/26/1936, APA Herrsching; also published in Bluhm 2007: 201.
16. Gesunde Rasse erhält ein Volk. *Fränkische Tageszeitung*, 11/26/1936.
17. For differing appraisals, see Doelecke 1975: 109; Lutzhöft 1971: 355.
18. The Hartheimer statistics for what was called the Action T4 show exactly how many mentally handicapped and mentally ill persons were killed in Germany (Cf. Friedlander 1997: 1990). However, the total number of the ill and handicapped under the Nazi regime in Germany *and* in the occupied areas can only be roughly estimated. If the numbers of ill and handicapped in the occupied territories are included, it can be assumed that the total number lies far over 200,000. See Kaminsky 2008: 274ff.
19. On this point, see the controversy between Michael Schwartz (1996, 1998, 2008), who has spoken out against a close connection between eugenics and euthanasia, and Hans-Walter Schmuhl (1987, 1997), who considers the killing of the handicapped and sick as a logical continuation of eugenics and race hygiene. For the controversy, see also Trus 2002: 247ff. and Kaminsky 2008: 270f. Black (2003: 247ff), like Schmuhl, seeks a continuous line in the United States from eugenics to euthanasia. On the debate in the United States, see Joseph 2006: 171ff.

20. Cf. Weingart, Kroll, and Bayertz 1988: 524; Winau 1989: 163.
21. Lenz 1923: 307; Rüdin in the first draft of a memo, MPIA München, GDA 54; Luxenburger 1931: 753; cited from Schmuhl 1987: 39; Weber 1993: 270; Weingart et al. 1988: 524. This opinion was shared abroad—see *Eugenical News* 20 (1935): 38f.
22. As Maretzki (1989: 1321) shows, the question of motivation even for physicians who participated is not clear. For the race hygienists I suggest reasons to be presented below, which obviously can be seen only in the context of a network of interconnected reasons.
23. *Völkischer Beobachter*, 8/7/1929, Cf. Bock 1986: 24; see also Dörner 1967: 131; Steinbach and Tuchel 1984: 19. In a conversation with his Reich physician director Gerhard Wagner in 1935, he laid out his basic agreement with the killing of the mentally handicapped. Because of the expected church protests, the murder action could not be begun until after the beginning of the war. Cf. Mitscherlich and Mielke 1948: 184; Lifton 1986: 50; Schmuhl 1987: 181.
24. Without including the eugenic "peace policy," the psychiatrist Klaus Dörner (1967) argued in a similar vein in a pathbreaking article regarding National Socialism and the destruction of life.
25. Cf. Popenoe 1923: 275f; Shelven quoted from "Nationalsozialistische Rassenforschung im Urteil des Auslandes." *Westfälische Zeitung*, 8/10/1936.
26. Quoted from Stroothenke 1940: 113; Cf. Dörner 1967: 131.
27. Ruttke 1934: 603.
28. Groß 1935b: 5f.
29. Stähle 1935: 9; Cf. Bock 1996: 326. After Second World War, Stähle justified his participation in the murder of this with the same argument. Cf. Klee 1983: 90. On Stähle, see Burleigh 1994: 135.
30. Ploetz 1936b: 618; see also Platen-Hallermund 1948: 34; Dörner 1967: 131.
31. Gütt, Ruttke, and Rüdin 1936: 72; Cf. Steinwallner 1937: 251.
32. On the concept "Kriegserklärung nach innen," see Hartmut Brodersen in Dörner et al. 1980: 44f.
33. Platen-Hallermund 1948: 21; Cf. Dörner 1967: 149.
34. BA Koblenz, R96 1/2: 1266; Cf. Weingart et al. 1988: 443.
35. Rüdin to Schütz, 10/28/1942, MPIA München, GDA 8; Cf. Weber 1993: 279. That Rüdin was already aware of the euthanasia action by the early winter of 1939–1940 is shown by his letter to Nitsche, 1/18/1940, MPIA München, GDA 131. On Rüdin's role in the killing of the psychologically ill and the mentally handicapped, see Roelcke 2000. 112ff and Schmuhl 2003: 16f
36. Groß 1940: 16.
37. Keiter 1941.
38. Grobig 1943; Cf. Weber 1993.
39. Verschuer 1944: 3; Cf. Weber and Weisemann 1989: 170.
40. Burgdörfer 1942: 29; see also Burgdörfer 1940.
41. Report by Eugen Fischer, 1/9/1940, MPGA Berlin 1.1A No.: 2400.
42. Cf. Lösch 1990. On the planned renaming of the Institute, see the notice in the file of 10/18/1940, MPGA Berlin 1.1A No.: 2399.
43. Reichsführer SS—personal staff of the Reichsgeschäftsführer of Ancestral Patrimony, 6/23/1943, BA Berlin Zehlendorf: Abel, Karl SS HO 4001.
44. Cf. Nyiszli 1960. Nachtsheim's comment in 1961 shows that the staff of the KWI were aware of the source of the organ and blood samples: "I must admit that it was the greatest shock that I ever experienced in the entire Nazi period when one day Mengele sent me the eyes of a Gypsy family executed at the Auschwitz concentration camp. The family had heterochromia of the iris, and one of the researchers at the Institute who was working on heterochromia had previously expressed interest in these eyes." Nachtsheim to Dunn, 2/14/1961, LCDA Philadelphia: Nachtsheim. Karin Magnussen was among the staff. On this, see the extensive description in Hesse 2001 and Schmuhl

2005: 470ff. On Verschuer's role, see in particular Müller-Hill 2000: 190ff and the earlier Müller-Hill 1984: 198ff. The leading role of this geneticist in race hygiene research under National Socialism has not yet been adequately evaluated.
45. See the reports by Stoddard (1940: 11 lf) and Ellinger (1942); Cf. Kühl 1994: 53–64.
46. Cf. Deichmann 1992: 156f; Macrakis 1993: 137–150.
47. Vertraulicher Erlaß des Reichserziehungsministeriums, 10/21/1940, MPGA Berlin 1.1A. No.: 1066.
48. Fischer to the Reichserziehungsministerium, 2/23/1942, MPGA Berlin 1.1 A. No.: 1067.
49. Fischer to the Reichserziehungsministerium, 4/29/1942, MPGA Berlin 1.1 A. No.: 1068.
50. "Les tendances morales et toute l'activité des Juifs bolchéviques décèlent une mentalité si monstrueuse que l'on ne peut plus parler que d'infériorité et d'être d'une autre espèce que la nôtre." Fischer 1942: 106; Cf. Briand 1941: 195f; Schneider 1990b: 1.
51. Fischer to the Reichserziehungsministerium, 11/13/1940 und 12/3/1940, BA Potsdam REM 4901–3038; Cf. Minutes and notes of 11/12/1940 und 1/6/1941 in BA Potsdam 15.01 REM 3058; see also Schneider 1990b: 257.
52. Minutes of a meeting in the Reichserziehungsministerium, 1/17/1941, BA Potsdam 15.01 REM 3195.
53. Lorimer 1971: 90; Lebrun 1985: 13; Höhn 1987: 17.
54. Cropp to the Reichserziehungsministerium, 3/7/1941, BA Potsdam, 15.01 REM 3058.
55. Cropp to the Reichserziehungsministerium, 3/21/1941, BA Potsdam, 15.01 REM 3171.
56. Cf. Roll-Hansen 1996a; Runcis 1998; Tydén 2010. The Norwegian law was reformulated in 1940 under Nazi occupation.
57. Blacker to Darwin, 8/11/1937, CPBA Oxford, Box 1.
58. Huxley to Blacker, 4/29/1933, ESA London, SA/Eug, C. 185.
59. In the first edition of my book I included Roswell Johnson (1934b: 117ff) among the reform eugenicists. The work by Gillette (2007: 75ff, 137ff, and 195) has convinced me that this was wrong.
60. Osborn to Boas, 11/27/1933, FBA Philadelphia: Osborn, Frederick; Osborn's summary of minutes of an AES Conference, 2/24/1937, ASEA Philadelphia 575.06: AM3 Conference on Eugenics in Relation to Nursing 1937; Cf. Allen and Mehler 1977: 12f; Mehler 1987: 14, 1988: 223.
61. Osborn, 5/5/1938, ASEA Philadelphia 575.06: Am3 Frederick Osborn. See also Osborn to Gosney, 6/24/1936, ELA New Haven, p. III, Box 77, Fol. 3038: "Dr. Kopp's report strongly supported the care and intelligence with which sterilization has been carried out in Germany and its beneficial effects." On the founding of the population commission, see Boas to Osborn, 12/20/1937, and Osborn to Boas, 12/23/1937, FBA Philadelphia: Osborn, Frederick; Cf. also Barkan 1988: 180–205; Barkan 1992: 333f.
62. Osborn to Blakeslee, 4/23/1940, CDA Philadelphia: Osborn.
63. Cf. Osborn 1939a: 2; see also "Eugenics for Democracy." *Time*, 10/9.
64. Myrdal 1939: 4f. On Myrdal's hypotheses, see in detail Etzemüller 2010.
65. Cf. Osborn to Huntington, 4/9/1937, EHA New Haven, p. III, Box 79, Fol. 3135; Huntington to Bigelow, 6/15/1944, EHA New Haven, S. III, Box 92, Fol. 3814.

7 On "Good" and "Bad" Eugenics: Refocusing on Human Genetic Counseling and the Struggle Against "Overpopulation"

1. Blacker to Hodson, 7/16/1948, ESA London, SA/Eug C160: Hodson.
2. On Carrel, see Reggiani 2007. The praises of Nazi race policies are found in the German edition of Carrel's "L'homme cet inconnu" (Carrel 1937).

3. Peltier 1949: 13; Cf. also Sauvy 1959: 184f; see also Sutter 1946, Sutter 1950; on Sutter, see *Population* 25 (1970): 749–758; *Population* 26 (1971): 717–720; on INED, see Girard 1986, Girard 1987; Schneider 1990b: 290; Drouard 1992a, 1992b, 1992c; Bachelard-Jobard 2001: 83ff.
4. Gini to Gates, 10/25/1949 and 1/5/1950, RGA London, Box 12; Gini to Kemp, August 14, 1950, IHGA Kopenhagen: Correspondence Bureau of Human Heredity 1938–1957.
5. Kemp to Snyder, 9/1/1950, and Kemp to Gini, 9/23/1950, IHGA Kopenhagen: Correspondence Bureau of Human Heredity 1938–1957.
6. Shapiro 1985: 40.
7. See Verschuer (1961: 356), who emphasized the eugenics movement in the United States, Great Britain, and Scandinavia.
8. Graveside eulogy delivered by Prof. Kurt Pohlisch, Bonn, APA Herrsching.
9. Cf. Lenz to Nachtsheim, 4/2/1946 (?), HNA Berlin, N. 18. Lenz received his professorial appointment on 10/16/1946—see the long discussion in Kröner 1998: 67ff.
10. Cf. Weindling 1989a: 566 and extensively Kröner 1998: 174ff. In particular on Verschuer, see Weiss 2010a.
11. Cf. Weingart, Kroll, and Bayertz 1988: 457; Pinn and Nebelung 1992: 27.
12. Here I am following the studies by Lösch (1990: 225–246) and Massin (1996), who for the first time showed the broad identification of young German scientists with the Nazis.
13. Response to charges, MPIA München, NLR 5 and defense brief regarding denazification, 11/7/1945, MPIA München, NLR 1.
14. Defense brief regarding denazification, 11/7/1945, MPIA München, NLR 1.
15. Rüdin's explanation of his political party stance, MPIA München, NLR 6.
16. Lenz to Gates, 3/18/1946, RGA London, Box 10.
17. Cf. Verschuer's position on an article in the Neue Zeitung of 5/3/1946, MPGA Berlin A 2- II 56.
18. Verschuer to Gates, 8/28/1946, RGA London, Box 10.
19. Verschuer to Muller, 9/30/1946 and 7/31/1947, HMA Bloomington. More generally, see Kröner 1998: 274f.
20. Mohr to Kemp, 5/16/1947, IHGA Kopenhagen. See Koch 1994: 5. Thanks to Lene Koch (1994: 5) for reference to this source. However, see Koch 1996 and Kröner 1998: 274.
21. Cited from the letter from Lenz to Kemp, 5/7/1947, IHGA Kopenhagen: Correspondence 1945–1948; see also RGA London, Box 10. See also Lenz to Verschuer, 2/11/1946; correspondence between Verschuer and Lenz (in private hands), cited according to Kröner 1998: 274.
22. Kemp to Lenz, 6/9/1947, IHGA Kopenhagen: Correspondence 1945–1948; Cf. Koch 1994: 5.
23. Memo from Kemp, 2/22/1949, see also Kemp to Verschuer, 6/9/947, IGHA Kopenhagen: Correspondence 1945–1948. On this period see also Koch 1996: 213ff.
24. Cf. Provine (1973: 796), who was the first to refer to this development based on the discussions of race mixing.
25. Cf. Huxley 1946.
26. Ironically, it was Huxley who was not in agreement with the others on the race question in the first UNESCO declaration. Cf. Huxley to Muller, 3/11/1950, HJMA Bloomington: Huxley. Both the UNESCO resolution and the reactions of the network of racist scientists have been treated in many studies (e.g., Kohn 1996: 40ff.), but there is still no monograph on this topic based on the UNESCO archives.
27. The task was taken up by the fourth general assembly of UNESCO in 1949. Cf. Resolution 4.2 of UNESCO.
28. See Montagu 1972: 1–6. Surprisingly, Montagu does not mention in his book that Muller as well submitted a critique of the original version; see also Muller to Montague, 2/27/1950, LCDA Philadelphia: Muller.

29. UNESCO Official Background Paper 104, Paris, July 1950, UNESCOA Paris 323.12 A 102; Cf. Metraux 1950: 385f.
30. 1. Statement on Race, July 1950, UNESCOA Paris, SS/1; Cf. Kuper 1975: 343–347. See also Lerch (1950: 172–174) for a partial German translation. An extensive rationale is found in Montagu 1972.
31. "No Biological Justification for Race Discrimination, say World Scientists." Paris, 7/18/1950, UNESCOA Paris, 323.12 A 102.
32. "'Race' a Social Myth." *Times*, 7/18/1950; "No Scientific Basis for Race Bias Found by World Panel of Experts." *New York Times*, 7/18/1950; Cf. Barkan 1992: 341.
33. Cf. UNESCO 1969: 494f.
34. *Eugenics Review* 42 (1950): 22; Cf. Jones 1988: 70.
35. Note on the UNESCO Statement on Race by the UK Secretariat of UNESCO, 11/17/1950, UNSECOA Paris, 323.12 A 102.
36. Cf. Montagu 1942.
37. Criticisms of the "Statement on Race" (1950), Question for discussion for the meeting of experts on June 4–8, 1951, UNESCOA Paris 323.12 A 102.
38. UNESCO 1952: 7; Cf. Montagu to Metraux, 4/4/1951, UNESCOA Paris, 323.12 A 102.
39. In a letter to Provine 8/13/1971, Darlington pointed out that he and his colleague Fisher would "obviously" not be accepted onto the committee because of their genetic race convictions. Dunn, in a letter to Provine 11/19/1971, agreed in principle with Darlington's perception. LCDA Philadelphia: Provine.
40. Cf. a UNESCO working paper that works through the differences between the first and second declaration. UNESCOA Paris, 323.12 A 102.
41. Penrose to UNESCO, 11/22/1951, UNESCOA Paris, 323.12 A 102 and LSPA London; Cf. UNESCO 1952: 24f. On Penrose see Kevles 1992: 13. His anti-Semitic attitude is clear in many writings after 1945 (e.g., Penrose 1951, 1963); see also *The Globe* and *Mail*, 11/28/1966
42. This view was presented retrospectively in letters to the historian William B. Provine by Sewall Wright (8/17/1971), Curt Stern (8/18/1971), C. D. Darlington (8/13/1971), and Paul Popenoe (8/10/1971), LCDA Philadelphia: Provine.
43. Muller to Metraux, 4/2/1952, UNESCOA Paris, 323.12 A 102. See also Muller to Darlington, 5/19/1952 and 4/23/1953, HMA Bloomington: Darlington.
44. Darlington to Metraux, 12/7/1951, UNESCOA Paris, 323.12 A 102; Cf. UNESCO 1952: 26f.
45. Internal file notice of UNESCO, UNESCOA Paris, 323.12. A 102; Cf. UNESCO 1952: 27; Jones 1988: 70.
46. Lenz to Nachtsheim, 7/27/1951 and 8/4/1951, HNA Berlin, N 18, M 118; Lenz withdrew his first draft of 7/27/1951 after a comment by Nachtsheim. His second version is found in a letter from Lenz to UNESCO, 9/7/1951, UNESCOA Paris 323.12 A 102; Cf. also Weingart, Kroll, and Bayertz 1988: 609f.
47. Salier to Nachtsheim, 7/24/1951, HNA Berlin, N 18, M 118; copy UNESCOA Paris, 323.12 A 102.
48. Scheidt to Nachtsheim, 7/23/1951, HNA Berlin, N 18, M 118; copy UNESCOA Paris, 323.12 A 102; on Scheidt's anti-Semitism, see his letter to Boas, 4/19/1933, FBA Philadelphia: Scheidt.
49. Fischer to Nachtsheim, 8/28/1951, HNA Berlin, N 18, M 118; copy UNESCOA Paris, 323.12 A 102.
50. Weinert to Nachtsheim, 8/28/1951, HNA Berlin, N 18, M 118; copy UNESCOA Paris, 323.12 A 102. On the position of the German academics, see UNESCO 1952: 30–35; Jones 1988: 71f; Weingart, Kroll, and Bayertz 1988: 605–610.
51. Nachtsheim to Sturtevant, 4/29/1952, UNESCOA Paris, 323.12 A 102; on Nachtsheim's position regarding the commentaries of his colleagues, see the letter to Metraux, 9/25/1951, in the same place.

52. Quoted from the translation of Weingart, Kroll, and Bayertz 1988: 613; Muller to Huxley, Darlington, Gowan, Cook, Sturtevant, Snyder, Boyd, Sonneborn, and Cleland; copy to Nachtsheim, 4/8/1952, HNA Berlin, N 18 M 118.
53. Metraux to Nachtsheim, 3/8/1952 and 3/19/1952, UNESCOA Paris, 323.12 A 102. See also Metraux to Dunn, 4/18/1952, LCDA Philadelphia: UNESCO-Race and Society.
54. Dobzhansky to Metraux, 4/29/1952, UNESCOA Paris, 323.12 A 102.
55. Cf. *Eugenics Quarterly* 3 (1956): 3.
56. See for example Blacker 1952: 144; Shapiro 1959: 4; and G. Allen 1968: 193. For the defense strategy of German race hygienists, see Rüdin's defense brief, MPIA München, NLR 1; Lifton 1986: 131.
57. Blacker 1952: 141; Cf. Blacker 1955: 130f, 1962a: 22; see also Thomson and Weindling 1993: 149.
58. Nachtsheim: Die qualitative Bevölkerungsbewegung: Erbgesundheitspflege—Notwendigkeiten und Möglichkeiten einer Planung; HNA Berlin III 20B:15; see also Nachtsheim 1963a, 1963b.
59. Cf. Glass 1955: 314.
60. Cf. Huxley to L. C. Dunn, 7/5/1961, LCDA Philadelphia: Huxley.
61. See Mehler 1988: 10.
62. As Grimm (2011: 104) has shown, this observation correlates with the declining number of articles with the keyword "eugenics" in Science, Nature, Journal of Heredity, and the American Journal of Public Health for the period 1900–2007 in comparison with articles with the keyword "genetics."
63. See Osborn to Johnson, n.d., FOA Philadelphia: Osborn Concerning Eugenics. This contradicts the notion that eugenics after 1945 was taken over by human genetics. It was rather the case that eugenics as an academic field was taken over by human genetics. Eugenics as a social and political movement continued to exist, often in close connection with human genetics.
64. Cf. Annual Report 1956–1957, ESA London, SA/Eug, A. 49. See Roelcke 2003: 39f on the role of Eliot Slater, Rockefeller fellow at Rüdin's Forschungsanstalt für Psychiatrie starting in 1934, who even after 1945 still reacted positively to the studies of the institute.
65. Membership list of the national committee in Kemp, Hauge, and Harvald 1956: viii–ix; on the German committee, see Koch 1993: 295f. See Cassata 2011: 288ff on the conflicts within the Italian Society for Genetics and Eugenics.
66. Cf. Paul 1989; Weingart 1994: 9.
67. Kemp, Hauge, and Harvald 1956: xii–xiii.
68. From 1939 to 1965 and from 1969 to 1972 Lorimer was director of the American Eugenics Society. He was vice president from 1966 to 1968. Kiser was director of the AES from 1958 to 1971, interrupted by only 4 years in which he served as president of the American Eugenics Society. Fairfield was president from 1929–1931, and director of the AES from 1939 to 1951. On the interesting figure of Fairchild, see Gossett 1997: 384ff.
69. Based on Blacker's lists of the expected participants, ESA London, SA/Eug, D. 11. The list in *Population Index* 16 (1950): 13f varies slightly. A systematic presentation of the overlapping between the IUSSP and the eugenics societies is difficult because I have membership lists only for the British and American Eugenic Societies. A comparison of the two shows that in 1955, of the 66 American members of the IUSSP, 28 were also active in the American Eugenics Society (including Kingsley Davis, Louis I. Dublin, Otis Dudley Duncan, Halbert L. Dunn, Frank H. Hankins, Philip M. Hauser, Dudley Kirk, Frank W. Lorimer, Frank W. Notestein, Frederick H. Osborn, Lowell J. Reed, Norman Burston Ryder, Christopher Tietze, Charles F. Westoff). Of the 15 British members of the IUSSP, 10 belonged to the British Eugenics Society

(including Blacker, Norman H. Carrier, James William Bruce Douglas, Glass, Eugene Grebenik, Richard M. Titmuss, Conrad Hal Waddington). The membership list of the IUSSP is reprinted in *Le Démographe* 1 (1955): 65–137.
70. Notestein 1968: 554.
71. IUSSP 1993: ii.
72. Lorimer 1949: 46f.
73. Barrett and Kurzmann (2004: 499ff) were unable to recognize this important aspect with their quantitative, neo-institutional study due to the fact that they surveyed the number of international conferences only for eugenics, not for demography.
74. For example, in 1947 the board of directors of the American Eugenics Society decided that was not the right time for "aggressive eugenic propaganda." Instead of being an organization recruiting for eugenics, the Eugenics Society should rather be a "forum" for the further development of eugenic ideas. Cf. Osborn 1974: 121; Kevles 1985: 252; Meehan 1993b.
75. Schenk and Parkes 1968: 155.
76. Blacker to Brush, 4/13/1956, CPBA Oxford, Box 7; Cf. also Meehan 1993d: 2.
77. Osborn 1968: 104; Cf. O'Keefe 1993: 24.
78. Osborn 1956: 21f; Cf. Meehan 1993b.
79. Osborn to Barrows, 3/25/1965, 4/8/1965, and 8/25/1965, ASEA Philadelphia: Osborn Letters on Eugenics; Cf. Meehan 1993g.
80. Cf. Hall 1990: 332; a critical assessment of the renaming of the Eugenics Review can be found in New Society, 3/6/1969.
81. Cf. Proctor 1991: 192.
82. Based on Osborn 1974: 126. See also Black 2003: 425ff for several new developments.
83. Cf. the Statement of the Eugenic Position (Gordon Allen, Dudley Kirk, J. P. Scott, Harry L. Shapiro, and Bruce Wallace), 1961, FOA Philadelphia: Concerning Eugenics.
84. Osborn 1983: 1026.
85. *Eugenical News* 38 (1952): 8; thanks to Mary Meehan (1993c) for the reference to Osborn's comment.
86. Dice 1960: 21; Cf. Weingart 1994: 9. See Ekberg 2007: 586ff as an example of how complicated this line of argument can become.
87. Osborn 1951a: 21, 1952: 8; Osborn's definition is also found in *Eugenics Quarterly* 2 (1955): 194f; see also Osborn 1940; G. Allen 1956: 4–8. Criticism of this approach is found in Neel and Schull 1954.
88. Cf. Blacker 1955: 133; see also Weingart 1994: 15.
89. Cf. Schulz 1992: 131.
90. Niklas Luhmann here speaks of a second polar set for the health system, which normally uses healthy/sick. Eugenics had developed a second polar set of "genetically OK/ questionable," which should become part of the medical system through genetics. NLA Bielefeld notebox 7/3513a: 1. See the discussion in Luhmann 2005a: 185.
91. Osborn to Causbie, 2/24/1966, AESA Philadelphia: Osborn—Letters on Eugenics.
92. Osborn to Popenoe, 3/25/1965, AESA Philadelphia: Osborn—Letters on Eugenics; thanks to Mary Meehan for this reference (Cf. 1993c).
93. Cf. McKusick 1975.
94. Cf. Kevles 1985: 255.
95. On Kemp's genetic consulting offices, see *Around the World News of Population and Birth Control*, No. 55/1957.
96. See Wendt 1975; Weß 1986: 13.
97. Reed 1974.
98. Osborn 1968: 91 Heredity clinics are the first eugenic proposals that have been adopted in a practical form and accepted by the public…the word eugenics is not associated with them. See also Osborn to P. S. Barrows, 8/25/1965, ASEA Philadelphia: Osborn Letters on Eugenics.

99. On the warnings in the 1930s, see Symonds and Carder 1973: 9 and 92f.
100. Cf. Szreter 1993: 678.
101. For example, Blacker 1960: 236.
102. For example, see the listings in Hauser 1962.
103. Szreter 1993: 679f.
104. Cf. Heim and Schaz 1993: 4; more generally, see Heim and Schaz 1996.
105. Symonds and Carder 1973: 52. The data on the activity in the AES come from O'Keefe (1994), who put together the membership lists of the AES and the British Eugenics Society. The sociology professor Philip M. Hauser was in the American Eugenics Society. Ronald Freedman, who also participated in group meetings, first appears on the membership list of the eugenic society for 1974.
106. Blacker 1955: 131.
107. Nachtsheim: Das Überbevölkerungsproblem und das Rassenbild der zukünftigen Menschheit, 1968, HNA Berlin, III 20 B 16. Nachtsheim in the 1960s also spoke out against the reduction of the birth rate in the industrial countries.
108. The line of argument in the 1920s and 1930s was often explicitly racist. For example, see a lecture by Davenport in 1929, in which he asked whether the nations with "inferior hereditary patrimony" reproduced more than did those with "superior heredity." The Norman Wait Harris Lectures Chicago, MSA Washington 192.
109. Here for example Osborn in a letter to David Bosanquet, 4/9/1954, FOA Philadelphia.
110. Quoted from Nachtsheim: Die qualitative Bevölkerungsbewegung: Erbgesundheitspflege—Notwendigkeit und Möglichkeiten einer Planung, 1965, HNA Berlin, III 20B: 15; Cf. Pinn and Nebelung 1992: 38f
111. Nachtsheim: Überbevölkerung der Erde—Zentralproblem der Welt, 1966, HNA Berlin, III 20B: 15; Cf. on the issues of development aid Nachtsheim: Überbevölkerung—Weltproblem Nr. 1, 1966, HNA Berlin, III 20B: 15.
112. Cf. Nachtsheim: Die Zukunft des Homo sapiens. Betrachtungen eines Erbbiologen zur quantitativen und zur qualitativen Bevölkerungsbewegung, 1966, HNA Berlin, III 20B: 15
113. Osborn to Dobzhansky, 12/12/1961, TDA Philadelphia.
114. Here cited in a report of the director of the Human Betterment Association of America, n.d. (presumably 1953), RGA London, Box 14.
115. Nachtsheim: Gefahren der gegenwärtigen quantitativen und qualitativen Bevölkerungsbewegung für die Zukunft der Menschheit, 1968, HNA Berlin III 20B: 1.
116. Lenz 1956: 20.
117. Reed 1968: 235 and 242.
118. Repp 1967: 12; Cf. also Repp 1966.
119. On the founding conference in Bombay 1952, see *Around the World News of Population and Birth Control* 8 (1952) and *Around the World News of Population and Birth Control* 13 (1953); see also: Heim and Schaz 1993: 13. According to Meehan (1993d: 2), one half of the British delegates and a good one-third of the American delegates were members of the British or American eugenics societies.
120. Cf. p. 7 of Avabai B. Wadia: Who is IPPF? 1977, IPPFB London.
121. Cf. IPPF 1994: 34; on the history of the IPPF see Blacker 1964; Deverell 1968; Symonds and Carder 1973: 102–105; Suitters 1973; Dennis 1973; Foley 1988; Connelly 2006: 220ff.
122. Annual Report 1939–1940 and 1941–1942, ESA London, SA/Eug, A. 32.
123. Cf. Suitters 1973: 39, 260; *Eugenics Review* 60 (1968): 139 and 141; see also Blacker 1964: 139; Meehan 1993d: 2.
124. The Eugenics Society from 1953 to 1957 bore part of the operating costs of the IPPF and made money available for the IPPF conferences in Bombay, Stockholm, Rome, and Tokyo; ESA London, SA/Eug D. 104; Cf. Around the World News of Population

and Birth Control 131 (1964). Dorothy Brush, one of the directors of the American Eugenics Society and editor of the Journal of the IPPF, directed the Brush Foundation from 1956 to 1953. In total, the Russia Foundation supported the IPPF and its predecessor organization from 1949 to 1966 with about $450,000; Cf. Meehan 1993d: 2. On the role of Brush in the American Eugenics Society, see *Eugenics Quarterly* 3 (1956): 125; on the role of Brush in the IPPF, see *News of Population and Birth Control* 96 (1961).

125. Cf. Suitters 1973: 86; O'Keefe 1993: 251.
126. Cf. Blacker 1964: 140f; Suitters 1973: 194f.
127. On the composition of the first board of directors of the IPPF, see *Eugenical News* 38 (1953): 140 and Blacker's copy of the annual report of the IPPF 1952/53, IPPFB London.
128. On Guttmacher's role in the American Eugenics Society see *Eugenics Quarterly* 3 (1955): 3; Cf. Meehan 1993c; O'Keefe 1994: 12.
129. On the role of Blacker and Houghton in the creation of the constitution, see *Eugenical News* 38 (1953): 139; Blacker 1964: 140; see also Suitters 1973: 39, 54, 57, 187–198.
130. Suitters 1973: 398; see also Foley 1988: 205.
131. They were members of the Advisory Council in April 1954, November 1955; Cf. *Around the World News of Population and Birth Control* 39 (1955).
132. *Around the World News of Population and Birth Control* 24 (1954).
133. Osborn 1955:1.
134. Cf. Dennis 1973: 416.
135. Cf. Notestein 1979; Rockefeller 1979; G. E. Allen 1991: 253; Heim and Schaz 1993: 2f; see also Weissman 1973: 82f; Maas 1976: 37.
136. Cf. Weissmann 1973: 83.
137. Notestein 1968: 553; Cf. Symonds and Carder 1973: 105.
138. Cf. Symonds and Carder 1973: 105; Notestein 1979: 510.
139. Cf. Meehan 1993b, 1994; see also Heim and Schaz 1993: 3, 1994: 132f.
140. After the Pioneer Fund cut off funding for the AES in 1954, the Population Council supported the AES from 1955 to 1958 with $12,000; *Eugenics Quarterly* 2 (1955): 67; Cf. Osborn 1974: 121f; Meehan 1994.
141. *Eugenics Quarterly* 2 (1955): 3; Cf. Heim and Schaz 1993: 5.
142. Cf. Osborn 1974: 123.
143. Osborn to Stratton, 1/12/1966, AESA Philadelphia; Cf. Meehan 1993g.
144. Cf. Notestein 1968: 555; Nachtsheim 1968: 7.
145. Osborn to Stratton, 12/1/1966, ASEA Philadelphia; Cf. Meehan 1993g.

8   The Renaissance of Racist Eugenics

1. Putnam 1961: 23.
2. See long discussions in Tucker 2002: 104ff and Jackson 2005: 94ff.
3. See for example Montague (to Dobzhansky, 1/29/1962, TDA Philadelphia: Montague), where he called Putnam and his colleagues as sick as Hitler.
4. In Putnam 1961: vii–viii.
5. *Eugenics Quarterly* 8 (1961): 105–107; cf. the dispute between Garrett (1961a: 218) and G. Allen (1961: 222); see also Putnam to Gates, 4/11/1962, RGA London, Box 23.
6. Wallace 1962: 161–163.
7. Kaplan 1963: 188; cf. Huxley 1963.
8. Osborn 1963: 104.
9. Osborn to P. S. Barrows, 3/25/1965, and to Shibdas Burman, 9/8/1965, ASEA Philadelphia: Osborn Letters on Eugenics.
10. Osborn 1966: 161; cf. also Pettigrew 1964b: 210, 1964a. See also Osborn 1964.

11. Dobzhansky 1963: 153f. Cf. also Dobzhansky 1964.
12. Osborn 1963: 108.
13. Letter of 1960, RGA London, Box 21. See also Schaffer 2008.
14. Gates to Blacker, 2/28/1949, ESA London, SA/Eug C. 120.
15. For Gates's plans in Cold Spring Harbor see Hodson to Gates, 10/9/1945, RGA London, Box 10.
16. Gates to Blacker, 11/2/1949 and 5/19/1952, ESA London, SA/Eug C. 120.
17. Cf. Jones 1988: 119.
18. Ellinger to Gates, 10/17/1946, RGA London, Box 10.
19. Ellinger to Gates, 1/24/1946, RGA London, Box 12.
20. Fisher to Collier, 6/30/1941, ESA London SA/Eug C. 108; for his position on race questions, see Gates, 8/27/1954, RGA London, Box 15.
21. Baker to Bertram, 12/26/1961, ESA London SA/Eug C. 13; see also his letter to Schenk, 11/9/1968, ibid. He resigned from the society and completely stopped all work in cooperation with the society in 1973. Baker to Blacker, 23/2/1973 and 11/32/1973, CPBA Oxford, Box 17.
22. Cf. *Eugenics Review* 52 (1960): 138; Jones 1988: 128.
23. Draper's position and the work of the Pioneer Fund have in the meantime been well researched. See Winston 1998; Lombardo 2002; and especially Tucker 2002. These descriptions provide an interesting contrast to the self-descriptions of the Pioneer Fund by their representatives Weyher (2001) and Lynn (2001b).
24. Cf. *Eugenics Quarterly* 2 (1955): 3. At the end of the 1958 Osborn adjusted his stance to Draper's position. He declared that he completely agreed with Draper that one should especially improve the genetic qualities of the race group that was still in the majority in the United States. Osborn to Donald, 10/20/1947, FOA Philadelphia: Pioneer Fund.
25. Memo from Frederick Osborn, 12/16/1954, ASEA Philadelphia: 575.06: Pioneer Fund Foundation Grant.
26. Cf. Osborn 1968; see also Osborn to Draper, 6/14/1956, FOA Philadelphia: Draper. Osborn declared in 1958 that he completely agreed with Draper's goal of improving the genetic potential of the American people, but still there were differences on the proper path to follow. Osborn to Draper, 4/28/1958, ASEA Philadelphia: Pioneer Fund. See Tucker 2002: 56ff on the disagreements between Osborn and Draper.
27. Cf. May 1960: 420; Mehler 1984; Jones 1988: 123. I was unable to verify this information. A query to the Pioneer Fund to confirm this information was not answered.
28. Draper to Gates, 8/7/1954, RGA London, Box 15.
29. Gates to Draper, 11/9/1954, RGA London, Box 15.
30. Gates to Roberts, 9/27/1954, 11/30/1954, and 1/15/1955, ESA London, SA/Eug D. 104.
31. On the situation in the United States, see King 1995.
32. See the long discussion in Jackson 2005.
33. On this point cf. Tucker 1995: 172 and Jackson 2005: 103ff.
34. Articles of Incorporation of the IAAEE (4/23/1959), State Tax Commission of Maryland in Baltimore. See Tucker 1995: 172.
35. Cf. Swan to Gates, 4/23/1960, RGA London, Box 21. There is still no comprehensive monograph on the IAAEE. Building in part on my studies, Tucker (2002: 78ff), Jackson (2005: 103ff), and Cassata (2011: 354) offer a first comprehensive overview. See Weyher 2001: lv on the financing of the IAAEE by the Pioneer Fund.
36. Cf. for example the Tax Records of the Pioneer Fund for the years 1976–1981, FC Washington: Pioneer Fund; see also Tucker 1995: 173.
37. Gini and Gates were listed on the letterhead of 1977 as founders of the IAAEE; cf. JBA Oxford, E. 136. On the isolation of Gates within the community of geneticists, see Cook to Journal of Heredity, 1948, HMA Bloomington: Cook.

38. Cf. McGurk 1956; see also Billig 1981: 96–98; Maoläin 1987: 308, Tucker 1995: 152f; Schaffer 2008: 105ff.
39. Gayre 1944: 11; cf. *Mankind Quarterly* 11 (1970): 3f. On Gayre see Tucker 2002: 73ff, 91ff, and 105ff.
40. Kuttner 1962; see also Kuttner 1960, 1963.
41. On Kuttner see Billig 1981: 102; Mintz 1985: 68f; Maoläin 1987: 342; Jackson 2005: 60ff. See also Searchlight 1984b: 3.
42. I quote from Gregor (1982: 20), based on the reprint in the *National Socialist*. On Gregor, see Jackson 2005: 106ff.
43. Miller 1994: 114.
44. Swan 1973: 46; Swan's description of himself as an American Fascist is based on information from the New York Post; cf. May 1960: 422. Thanks to Barry Mehler for a look at the original source. On Swan, see Tucker 2002: 85ff and Jackson 2005: 104ff.
45. Swan 1974: 37; he based himself on research by Shuey (1966).
46. Anderson and van Atta 1989; cf. *Nation Europa* 12 (1974): 40; *Neue Anthropologie* 9 (1981): 100; Swan himself shrugged off these photos as merely a meeting with former university classmates. Swan's presentation is interesting in Lynn's "History of the Pioneer Fund" (2001b: 143ff).
47. Gayre to Gates, 2/4/1957, RGA London, Box 18.
48. Swan to Gates, 7/13/1960, 11/21/1960, RGA London, Box 21.
49. Cf. Pinn and Nebelung 1990: 202; see also Weyer 1984: 80–81, 1986: 280–304.
50. Cf. Stölting 1987: 158. For a long time Müller was general secretary of the IIS; cf. Le Démographe 6 (1957): 67.
51. Gregor in the name of the IIS to Gates, 11/17/1960; cf. Gregor's article in *Mankind Quarterly* (1960: 128); see also Jones 1988: 122. In particular because of the role of the IIS in sociology today, its role in the late 1950s and early 1960s merits more intensive research. In the light of clear evidence of Gini's engagement in the network, it is unclear to me why Quine (2010: 392) calls Gini's post-1945 eugenic position "deracialized." See Cassata (2011: 362ff) for both Gini's clash with the representatives of the IAAEE and the cooperation between the IAAEE and IIS.
52. Outline of Race Symposium Book, 1960, RGA London, Box 21.
53. Swan to Garrett, 6/4/1960, RGA London, Box 21. It is not clear from the documents whether Draper actually made the money available. See also Tucker 2002: 982ff.
54. Kuttner 1967: xv; cf. also Tucker 1995: 178.
55. Cf. *Eugenics Quarterly* 15 (1968): 301. The sources on the preparation of this work in RGA London support this assumption.
56. Cf. Kuttner 1967; on the prior history, see Gregor (IIS) to Gates, 4/10/1962, RGA London, Box 23. On Gedda, see Cassata 2011: 335ff.
57. See Swan to Gates, 9/1/1961, RGA London, Box 22.
58. Pamphlet of the IAAEE, 1960, RGA London, Box 21.
59. Armstrong to Baker, 6/18/1963, JBA Oxford E.88.
60. See the notice in *Mankind Quarterly* 5 (1965); cf. also Bellant 1991: 63.
61. Cf. Swan to Gates, 5/11/1960, RGA London, Box 21.
62. Pamphlet of the IAAEE, 1960, RGA London, Box 21.
63. Gayre to Gates, 6/13/1960, RGA London, Box 21. On this, see Tucker 2002.
64. Memo from Bertram about a visit from Weyher (including Weyher's business card), 2/11/1961, and a letter from Weyher to Bertram, 3/4/1962, ESA London SA/Eug D.104; the delegation of the work from Draper to Weyher is clear in a letter from Draper to Gates, 3/20/1961, RGA London, Box 22.
65. Cf. The *Times*, 9/30/1958; see also positive reactions in *Right. A Journal of Forward-Looking American Nationalism* 42 (1959). The critical reaction of members of the Eugenics Society is clear in *Around the World News of Population and Birth Control* 73 (1959). Bertram was able to get financing for two research projects from Draper

through his pamphlet for the Eugenics Society—one on the fertility of immigrants from the West Indies and Pakistan, and a second one on results of "hybridization" between "Whites and Negroes from West Africa." Cf. Weyher to Bertram, 4/11/1961 and 7/6/1961; Bertram to Weyher, 6/7/1962 and 10/1/1962; Report to date on the use of the Fund, 4/29/1963, ESA London, SA/Eug D. 104.
66. Pamphlet of the IAAEE, RGA London, Box 21.
67. Gayre to Gates 10/18/1958, 12/8/1958, and 4/20/1959, RGA London, Box 20; his colleague Gates felt called on—sponsored by Eugen Fischer—to publish articles in the German Zeitschrift für Morphologie und Anthropologie; cf. Schaeuble to Gates, 6/29/1959, RGA London, Box 20.
68. Donald Swan wrote to Wesley Critz George: "The IAAEE is closely associated with the publication and distribution of the scientific journal, *The Mankind Quarterly*." Swan to George, 12/12/1960, WCG Chapel Hill Box 8: cited from Jackson 2005: 148f. As part of Pearson's worldwide promotional tour for the Northern League, which was to be the beginning of the defense of the white race against its "internal and external enemies," Pearson and Gayre agreed that Gayre would take the lead in organizing a scientific journal. Cf. Gates, 9/12/1959, RGA London, Box 20; see also the *Northlander*, 2 (1959), S. 2.
69. Cf. Anderson and Anderson 1986: 94; Maoläin 1987: 200. For more on the Northern League, see Coogan 1999.
70. Pearson to Gates, 4/17/1958, RGA London, Box 19. See also Pearson 1959.
71. Explanation of the principal worldview (ideology) and aims of the Northern League, written by Hans F. K. Günther in 1959; pamphlet in the possession of the author.
72. The *Northlander* 8 (1959), S. 2; cf. also Pearson 1959 and Kohn 1996: 52ff.
73. Gayre in the introduction to the first edition, *Mankind Quarterly* 1 (1960): 1.
74. Garrett to Gates, 11/28/1959, RGA London, Box 20. See Winston 1998: 179ff on Garrett's role.
75. Gayre to Gates, 10/28/1959, RGA London, Box 20; cf. Jones 1988: 123–125.
76. Gayre to Gates, 4/15/1960 and 4/21/1960, RGA London, Box 21. The spokespersons for apartheid in the Southern states immediately turned to information from *Mankind Quarterly* to justify their position; cf. Kilpatrick 1962: 46f.
77. Weyher to Gates, 2/22/1962, RGA London, Box 23; cf. Garrett to Gates, 6/16/1961; Weyher to Gayre, 6/15/1961, 4/4/1961, and 12/131961, RGA London, Box 22. See also Garrett to Gates, 11/23/1959; Gayre to Gates, RGA London, Box 20; see also Jones 1988: 123.
78. Cf. Gayre to Gates, 12/17/1957, RGA London, Box 20. On Gini, see in particular Cassata 2006a.
79. Cf. Shuey 1966. On the support of Shuey by the Pioneer Fund, see Weyher 2001: xix and Lynn 2001b: 91ff.
80. Gayre to Gates, 2/29/1960, RGA London, Box 21.
81. Putnam to Gates, 2/14/1962, RGA London, Box 23.
82. Cf. Gayre to Gates, 3/14/1960, RGA London, Box 21.
83. Cf. Gayre to Gates, 4/21/1960, RGA London, Box 21.
84. Cf. Gayre to Gates, 3/31/1960, RGA London, Box 21; on Right, see Mintz 1985: 70f.
85. Cf. Billig 1979: 18, 1981: 114.
86. Comas 1961; cf. Jones 1988: 124. On the commentaries by Haldane, Nachtsheim, and Dobzhansky, see *Current Anthropology* 2 (1961): 314–334.
87. Dunn 1962a: 74; cf. Dunn 1962b; see also *Mankind Quarterly* 3 (1962): 48–50.
88. Ehrenfels 1962: 154; cf. Billig 1979: 18, 1981: 115. Ehrenfels at first used the given name Omar, but later he spelled it Umar.
89. Gayre to Gates, 4/13/1960 and 4/22/1960, RGA London, Box 21.
90. Skerlj 1960: 172f; see also Skerlj in *Current Anthropology* 2 (1961): 329; cf. Billig 1981: 115.

91. Gayre to Gates, 4/13/1960, RGA London, Box 20.
92. Garrett 1961c: 480, 1961b: 253–255; cf. also Tucker 1995: 155.
93. Armstrong to Baker, 6/18/1963, JBA Oxford, E. 88.
94. Gates to the editor of *Current Anthropology*, 2/28/1961, RGA London, 11/99.
95. Gates to Draper, 5/3/1961, RGA London, Box 22. In the case of Skerlj, the editors felt compelled to take special steps, since his criticism had sorely damaged the reputation of *Mankind Quarterly*. Gates urged Draper to pay for the charge of slander against Skerlj. Weyher, who took soundings for Draper regarding the chances for success of a lawsuit, succeeded in turning Gates away from a lawsuit because the "thousands of dollars" that a legal controversy would cost would be better spent on scientific research. Weyher to Gates, 4/19, 5/16, and 8/1/1961, RGA London, Box 22.
96. On the heterogeneity of the network, see Lynn's presentation (2001b) of the recipients of the Pioneer Fund, summarized—ignoring the heterogeneity of the source—under the keyword "the Science of Human Diversity."
97. Cf. Billig 1981: 67.
98. Gedda's recruitment for *Mankind Quarterly* at the Second International Congress for Human Genetics went largely without response. Cf. Rome Genetics Conference News, October 1960; see RGA London, Box 22.
99. Cf. Proctor 1991: 194.
100. Cf. Tucker 1995: 180, 207.
101. Cf. Weingart, Kroll, and Bayertz 1988: 618.
102. Lynn 2001b: 199 gives an overview of the funding areas of the Pioneer Fund.
103. Cf. Jensen 1969, 1972, 1973, 1985; see also *Nation Europa* 1 (1974): 41.
104. For criticism of Jensen, see Rose 1976; Gould 1979; Billig 1981: 71–81.
105. Cf. Eysenck 1971, 1976; see also Eysenck's foreword in Galton 1979: v–vi; cf. *Nation Europa* 1 (1991): 38–40.
106. Herrnstein 1971; cf. Di Trocchio 1994: 127.
107. Cf. Dobzhansky 1963, 1964; Bodmer and Cavalli-Sforza 1970, 1976.
108. Baker (1974) used the words "superiority" and "inferiority." I rely on Baker's April 1967 and December 1968 synopses for his book on race, which he used in his communications with publishers. Cf. JBA Oxford, E.21.
109. Weyl 1973b: 45–48.
110. Cf. Gordon 1975, 1980, 1986, 1987.
111. Rushton and Bogaert 1989; cf. Rushton 1988a, 1988b, 1993; see also *Dialyse Intern* 2 (1993): 30. See Tucker 2002: 197ff and 214 on Rushton's role in the Pioneer Fund. Lynn 2001b: 359ff heaps praise on Rushton.
112. Cf. Epps 1973; Ehrlich and Feldman 1977; Gould 1979.
113. Cf. Provine 1986: 878.
114. Bodmer and Cavalli-Sforza 1976: 672. Based on information from Kathy O'Keefe (1994), Cavalli-Sforza appears on the 1974 membership list of the American Eugenics Society.
115. Osborn to Shockley, 2/20/1968, FOA Philadelphia: Shockley. Osborn relied on Shuey's (1966) study.
116. Cf. Wallace 1975.
117. The criticism by the group around the Pioneer Fund became so personal that the scientific standing of their critics was called into question (cf. Weyher 2001: xliii on Barry Mehler). It is curious that my detailed reconstructions of the close ties between the scientific and political motives of the racist eugenicists (1994 and 1996), which appeared in well-respected scientific publications and which were based on an analysis of the internal correspondence of racist eugenicists, have so far been largely ignored in the publications of the Pioneer Fund.
118. Jensen in *Neue Anthropologie* 1 (1972): 39; cf. Rexilius 1980: 113.
119. *Mankind Quarterly* 17(1976): 149; cf. Billig 1981: 93.

120. Cf. Eysenck in Pearson 1991: 17. He uses the concept of the "new Fascist left" in contradistinction to the old left, which had remained open to rational arguments.
121. Shockley 1992b: 241; see also Shockley 1972, 1974, 1992a.
122. Pearson 1991: 13–15. See Winston 1998: 195ff on Pearson.
123. A list of the protests appears in Pearson 1991; see also *Nation Europa* H. 1/1974, S. 41.
124. The ten other scholars were Jack A. Adams, Raymond B. Cattell, Francis H. Crick, C. D. Darlington, Charles W. Eriksen, Eric F. Gardner, Quinn McNemar, Eliot Slater, Robert L. Thorndike, Frederick C. Thorne, and Philip E. Vernon; cf. JBA Oxford, G. 14. The activity was at first coordinated by the psychologist Ellis B. Page of the University of Connecticut; cf. Jensen to Dobzhansky, 9/2/1972, THA Philadelphia: Jensen, Arthur R.
125. John Randal Baker Papers, Bodleian Library Oxford G 14.
126. The resolution appeared for the first time in American Psychologist (1972): 660f; cited from the translation in *Nation Europa* 12 (1972): 21–23.
127. Cf. Pinn and Nebelung 1992: 22f; Vogel in his role as chairman of the German Gesellschaft für Anthropologie und Humangenetik criticized a GfbAEV pamphlet against race mixing. Vogel's (1983) unclear attitude on possible biological-genetic reasons for IQ differences between "blacks and whites in the USA" and between "Ashkenazi Jews and other Europoids" was criticized by Pinn and Nebelung (1992: 25).
128. Cf. Haller 1981: 91.
129. Northlander was renamed Folk in 1963 and renamed again in 1964 as Western Destiny. Cf. Maoläin 1987: 97f and 360.
130. On Benoist, see Böhm 2008.
131. Jaschke 1990; cf. Assheuer and Sarkowicz 1992: 191–195; Fromm and Kernbach 1994: 305; see also Haller 1981: 86f.
132. Cf. Moreau 1983: 148f.
133. Cf. *Mankind Quarterly* 11 (1970): 62; cf. Putnam to Baker, 10/16/1975, JBA Oxford; see also Billig 1979: 24f, who quite early on noted the network made up of the *Mankind Quarterly, Neue Anthropologie,* and *Nouvelle Ecole*.
134. A complete list is found in Duranton-Crabol 1988: 254–258. After 1981 Roger Pearson is no longer found on the list of *Nouvelle Ecole*; cf. Duranton-Crabol 1988: 158.
135. The GbAEV was a successor organization to the Deutsche Gesellschaft für Erbgesundheitspflege.
136. Cf. *Neue Anthropologie* 9 (1981): 100, 15 (1987): 28.
137. Rieger 1969: 45, 50; cf. Billig 1981: 118–120. Since Rieger demanded not only segregation of the races between "great races" but also demanded segregation between subraces, his writing itself was criticized in *Mankind Quarterly* (11 (1970): 62). A defense of Rieger's work (Kiesel 1971) appeared shortly thereafter.
138. Cf. Heidenreich and Wetzel 1989: 157; Lange 1993: 143 and 149.
139. For the scientific advisory council, see *Neue Anthropologie* 8 (1980) and 13 (1985).
140. Gesellschaft für biologische Anthropologie, Eugenik und Verhaltensforschung 1987; Informationsblatt 4, distributed together with *Neue Anthropologie* 2 (1987).
141. Baker 1975; on the connection of Baker to *Neue Anthropologie* see Rieger's letter to Baker, 12/12/1974, JBA Oxford, E. 106/107.
142. Baker to Gayre, 12/3/1975, JBA Oxford, E. 184–187.
143. See the copy of the Certificate of Incorporation in Lynn 2001b: 556. Presumably because of public criticism, the bylaws were slightly changed on June 14, 1985. The goal was changed from "human race betterment" to "race betterment." See Tucker 2002: 7.
144. On the twins research project at the University of Minnesota, in which a broad genetic determination of humans was established, see Bouchard and McGue 1990a; Bouchard et al. 1990b. Bourchard's work in Lynn 2001b: 297ff is revealing. See Joseph 2004: 11ff on the "genetic illusion" of the research on twins.

145. Cf. Miller 1994 and Tucker 2002 ; see also tax returns of the Pioneer Fund in FC Washington: Pioneer Fund. Lynn 2001b in an in-house publication gives an overview of the recipients of funding from the Pioneer Fund.
146. Entries for FREED from 1973 to 1981 in the FC Washington: Pioneer Fund; see also Tucker 1995: 193.
147. Entries for FHU from 1973 to 1992 in the FC Washington: Pioneer Fund.
148. Entries for the Institute for the Study of Man from 1976 to 1993 in the FC Washington: Pioneer Fund.
149. Roger Pearson alone by 1994 had received over $750,000 from the Pioneer Fund over 25 years. Cf. Miller 1994. See Lynn 2001b: 455 on the annual subventions starting in 1973.
150. See Lynn 2001b: 459f and V. Weiss 2012: 99 on the expansion of the editorial board of *Mankind Quarterly* after Pearson took over.
151. See IAAEE, Application for Recognition of Exemption, Internal Revenue Service, Form 1023, submitted by the Institute for the Study of Man, 5/25/1975; cf. Bellant 1991:104.
152. Cf. also Anderson and Anderson: 1986: 92–97; Bellant 1991: 60f; Tucker 1995: 256, 2002: 159ff; Jackson 2005: 45ff, 195ff.
153. Cf. Bellant 1991: 63; Harris 1994: 54f. See also Searchlight 1984b: 3, 1984a: 9.
154. See Weyher 2001: xl, where Barry Mehler, one of the first to call attention to Draper's "Nazi connection,' is called a "Swastika painter."
155. Itzkoff 2006: 8. He writes: "Rather, the Holocaust was a vast dysgenic program to rid Europe of superior intelligent challengers to the existing Christian domination by a numerically and politically miniscule minority."
156. For this argument see Glad 2006: 73 and 2011: 67ff, who overlooks the tight connection between the German eugenicists and the National Socialist regime. The distinction between a "good" American eugenics and a "bad" German race hygiene is possible only because both Glad and Itzkoff ignore information about the "Nazi connection" (Kühl 1994) and the "Nazi nexus" (Black 2009).
157. See the publications of the Marburg professor of psychology Detlef H. Rost (2009) as an example of how strongly psychological intelligence research is based on the research coming from the Pioneer network.
158. See Herrnstein and Murray 1994: 23ff and the long review of *The Bell Curve* by Lynn 2008: 1ff. Sesin 1996: 46ff shows that the cooperation between Charles Murray and Richard Herrnstein was mediated by Robert Gordon.
159. Cf. Sesin 2012: 29 on Sarrazin's references (2010a) to *The Bell Curve*.
160. Sarrazin 2010a: 51ff and 331. See Weingart 2012: 19ff on Sarrazin's basic eugenic approach.
161. See Lane 1994 and Naureckas 1995.
162. Cf. Lane 1994: 15.
163. See Sarrazin 2010b. There has as yet been no reconstruction of the way Sarrazin's book was put together. It would be interesting for Sarrazin to clarify further his research strategies. Sarrazin makes use of V. Weiss's book *Die IQ-Falle*, which relies heavily on the structure and argumentation of Murray and Herrnstein's Bell Curve (V. Weiss 2012: 393). He also extensively uses sources made available to him by V. Weiss (email from V. Weiss to the author, 6/22/2012). The heavy reliance of V. Weiss's *Die IQ-Falle* on *The Bell Curve* by Murray and Herrnstein came about because originally V. Weiss wanted to undertake a German translation of *The Bell Curve*. See Sesin 2012 and Kemper on Sarrazin's English and German sources. The work of Kemper (see also 2011) is particularly interesting because he shows how extensively Sarrazin made use of V. Weiss's book *Die IQ-Falle*, though Sarrazin makes many errors in citations. See the comparison of V. Weiss 2000: 27 with Sarrazin 2010: 96. Of interest as well is V. Weiss's book on psychogenetics that appeared in East Germany in 1982. It shows

that in the Socialist states there was no systematic problem in staying abreast of discussions in the West (V. Weiss 2012: 98ff. on the publication history).
164. See Grant's *Passing of the Great Race, or, The Racial Basis of European History* (1916), Wiggam's New Decalogue of Science (1923), Chamberlain's *Die Grundlagen des neunzehnten Jahrhunderts* (1899), Günther's Rassenkunde des deutschen Volkes (1922), and Carrel's L'homme cet inconnu (1935).
165. See Etzemüller 2012: 158 on the style of popular scientific eugenic books. The only exception is Murray's coauthor Richard Herrnstein, who occupied a professorial chair in psychology at Harvard University and who has made a scientific name for himself with his matching theory. It is especially noticeable in Sarrazin's book (2010a: 175) that errors have crept in because of the rapid transcription of ideas. For example, he writes that children of the lower class inherit "the intellectual abilities of their parents, in conformity with Mendelian laws."

9 The Dissolution of the Eugenics Movement: Will There Be Eugenics without Eugenicists?

1. Largent (2008: 128f) shows the quantitative decline of eugenics on the basis of negative, neutral, and positive mentions in US biology textbooks.
2. Cf. taz [Tageszeitung], 12/21/1994. The verbatim text of Höhn's interview was: Höhn—"It is unfortunately statistically provable. I know that today one can say that. That's really too bad." Question: "What is provable?" Höhn: "For example, there are differences in the distribution of intelligence. You can talk about that perhaps without the words superior or inferior, but even that one can't do today. What I observe...with a certain distress is this type of taboo on ideas that is generally widespread." Question: "What do you mean by taboo on ideas?" Höhn: "For example, that one says that the average intelligence of Africans is lower than that of other people."
3. See Lynn 2001b: 39.
4. Such an approach would go against the tendency of most historians and social scientists to portray the development of nonracist eugenics after 1945 with the omission of the actors. At this point, I cannot discuss my ideas on this matter more comprehensively. Support for historical studies about eugenics after 1945, which hardly exist at all, could help to test the tenability of these hypotheses.
5. See Ekberg 2007: 581 and Raz 2009: 605ff on neologisms in the description of post–Second World War eugenics.
6. Cf. Hartmann 1987: 4 and Frey 2007.
7. Cf. Macura 1986: 20. [OK]
8. Donaldson 1990: 19 and 87; Cf. Szreter 1993: 682; Heim and Schaz 1994: 131.
9. Cf. Paul 1995, Chapter 7.
10. Here I follow the very stimulating ideas of Paul (1995: Chapter 7).
11. On the lively discussion France, see Simonnot 1999: 141ff; Bachelard-Jobard 2001: 89ff.
12. Köbsell 1994: 90. Kerr and Shakespear 2002: 101ff; Duster 2003: 114ff.

Afterword

1. This problem of reading German and other "foreign" languages is apparent even in studies that explicitly deal with internationalization, internationalism, and the cosmopolitanism of eugenics (Cf. Kevles 2004 and Bashford 2010), which are based almost exclusively on English-language primary and secondary sources.
2. See Grimm (2011: 127ff), who has reconstructed the international contact network of important US eugenicists as part of a network analysis; he has shown the

overlapping of memberships in eugenics societies and in the participation in international congresses.
3. See Turda 2010a: 9 and 40ff, who, like me, emphasizes the importance of the Second World War in the development of eugenics.
4. Weikart 2003: 273ff and Engs 2005: 227ff.
5. See in particular Ramsden 2003.
6. Björkman and Widmalm 2010 and Bär 2002 on the support of Swedish eugenicists for the National Socialist race policies. On support by British eugenicists, see Hart 2012: 33ff. On support by US eugenicists, Sheila Weiss (2010a: 265ff), which to a large extent follows my presentation. Additional information is offered in particular by her two valuable chapters on the Munich Institut für Psychiatrie and the Berlin Kaiser-Wilhelm-Institut für Anthropologie, Human Heredity Theory and Eugenics. Much more problematic from an academic point of view is Edward Black's book. He acts as though he were the first to show international support for the National Socialist race politicians, intentionally ignoring my two studies on the *Nazi Connection* and *Die Internationale der Rassisten* as well as the works of other historians. Without any references to secondary literature, in the description of many events, he simply repeats information that had been published previously.
7. In this regard, Tucker 2002; Jackson 2005; and Cassata 2011 are particularly valuable.

Sources and Bibliography

1. Archives

ADW Berlin	Archive of the Diakonisches Werk, Berlin
ASEA Philadelphia	Archive of the American Eugenics Society, American Philosophical Society, Philadelphia
AJLA Princeton	Alfred J. Lotka Archive, Princeton University
APA Herrsching	Alfred Ploetz Archive, Herrsching
BA Koblenz	Bundesarchiv, Koblenz
BA Potsdam	Bundesarchiv, Außenstelle Potsdam
BA Zehlendorf	Bundesarchiv, Außenstelle Berlin-Zehlendorf (formerly Berlin Document Center)
BHA München	Bayerisches Hauptarchiv, München
CDA Philadelphia	Charles Davenport Archive, American Philosophical Society, Philadelphia
CPBA Oxford	C. P. Blacker Archive, Wellcome Unit, Oxford
EGCA Princeton	Edwin G. Conklin Archive, Princeton University
EHA New Haven	Ellsworth Huntington Archive, Yale University, New Haven
EPA Paris	Edmond Perrier Archive, Musée d'histoire naturelle, Paris
ESA London	Eugenics Society Archive, Wellcome Institute, London
FBA Philadelphia	Franz Boas Archive, American Philosophical Society, Philadelphia
FC Washington	Foundation Center, Washington
FGA London	Francis Galton Archive, University College London
FOA Philadelphia	Frederick Osborn Archive, American Philosophical Society, Philadelphia
HHLA Kirksville	Harry H. Laughlin Archive, Missouri State University, Kirksville
HMA Bloomington	Hermann J. Muller Archive, Indiana University, Bloomington
HNA Berlin	Hans Nachtsheim Archive, Archiv der Max-Planck-Gesellschaft, Berlin
HSJA Philadelphia	Herbert Spencer Jennings Archive, American Philosophical Society, Philadelphia
IFA New Haven	Irving Fisher Archive, Yale University, New Haven
IFZA München	Institut für Zeitgeschichte Archive
IHGA Kopenhagen	Institute for Human Genetics Archive, Kopenhagen
IICEA Paris	International Institute of Intellectual Cooperation Archive UNESCO, Paris
IPPFB London	International Planned Parenthood Federation Library, London
IUSSPA Lüttich	International Union for the Scientific Study of Population Archive, Lüttich
JBA Oxford	John Baker Archive, Bodleian Library, Oxford

JBSHA London	J. B. S. Haldane Archive, University College, London
KPA London	Karl Pearson Archive, University College, London
LCDA Philadelpia	L.C. Dunn Archive, American Philosophical Society, Philadelphia
LSPA London	Lionel S. Penrose Archive, University College, London
LWA Philadelphia	Leon Whitney Archive, American Philosophical Society, Philadelphia
MDA Philadelphia	Milislav Demerec Archive, American Philosophical Society, Philadelphia
MPGA Berlin	Max-Planck-Gesellschaft Archive, Berlin
MPIA München	Max-Planck-Institut für Psychiatrie Archive, München
MSA Washington	Margaret Sanger Archive, Library of Congress, Washington
NLA Bielefeld	Niklas Luhmann Archive, University of Bielefeld, Bielefeld
RAC Tarrytown	Rockefeller Archive Center, Tarrytown
RGA London	Ruggles Gates Archive, King's College London
RPA Philadelphia	Raymond Pearl Archive, American Philosophical Society, Philadelphia
TDA Philadelphia	Theodosius Dobzhansky Archive, American Philosophical Society, Philadelphia
UNESCOA Paris	UNESCO Archive, Paris
WCG Chapel Hill	Wesley Critz George Papers, Southern Historical Collection, Chapel Hill.

2. Journals and Periodicals

Acta Genetica
Allgemeines Statistisches Archiv
American Breeders Magazine
American Journal of Sociology
American Mercury
American Sociological Review
Angriff
Archiv für Bevölkerungswissenschaft und Bevölkerungspolitik
Archiv für Rassen-und Gesellschaftsbiologie
Around the World News of Population and Birth Control
Ärzteblatt für Württemberg und Baden
Atlantic Monthly
Battle Creek Morning Journal
Berliner Börsenzeitung
Berliner Tageblatt
Birth Control Review
Bulletin of the IUSIPP
Catholic Herald
Child Study
Current Anthropology
Den Nordiske Race
Deutsche Allgemeine Zeitung
Deutsche Medizinische Wochenschrift
Dight Institute for Human Genetics Bulletin
Erbarzt
Erbe und Verantwortung
Erfelijkheid bij de Mens
Eugenical News

Eugenics Quarterly
Eugenics Review
Eugenika
Eugenique
Europäischer Wissenschafts-Dienst
European
Forschung und Fortschritt
Fortschritte der Erbpathologie, Rassenhygiene und ihrer Grenzgebiete
Frankfurter Zeitung
Fränkische Tageszeitung
Genius
Gesundheitsfürsorge
Globe and Mail
Hamburger Fremdenblatt
Harvard Educational Review
Hereditas
Human Biology
Indie
Journal of Biosocial Science
Journal of Heredity
Journal of State Medicine
Jungdrogist
Kongreß-Korrespondenz
Lancet
Le Démographe
Le Matin
Le Problème Sexuel
Le Temps
L'Humanité
Man
Mankind Quarterly
Mental Hygiene
Milbank Memorial Fund Quarterly
Mitteilungen der Arbeitsgemeinschaft für Volksgesundung
Nachrichten deutscher Zeitungen
Nachrichtendienst der Reichsfrauenführung
Nation
Nation Europa
National Socialist
Nationalsozialistische Parteikorrespondenz (NSK)
Nature
Naturwissenschaften
Neue Anthropologie
Neue Zeitung
Neues Volk
New Society
New Statesman
New York Times
Nieuwe Rotterdamsche Courant
Nord und Süd
Northern World
Northlander
Nouvelle Ecole

N.S. Beamten-Zeitung
Öffentlicher Gesundheitsdienst
Perspectives in Biology and Medicine
Playboy
PM
Population
Population Index
Preußische Zeitung
Public Opinion
Race et Racisme
Rasse
Rassenpolitische Auslands-Korrespondenz
Reich
Revue Anthropologique
Revue Internationale de Sociologie
Revue Politique et Parlementaire
Right
Rote Erde
Royal Society of Health Journal
Schulungsbrief
Science
Science News Letter
Scientific American
Scientific Monthly
Scientific Worker
Siècle Medical
Social Biology
Social Hygiene
South African Journal of Science
Soziale Praxis
Stanford Daily
Statesman
Süddeutsche Monatshefte
Survey Graphic
Time
Times
United Service Gazette
University of Pittsburgh Bulletin
U.S. News and World Report
Völkischer Beobachter
Volk und Rasse
Vossische Zeitung
Western Destiny
Westfälische Landeszeitung
Westfälische Zeitung
Westminster Gazette
Zeitschrift für induktive Abstammungs- und Vererbungslehre
Zeitschrift für Morphologie und Anthropologie
Zeitschrift für psychische Hygiene
Zeitschrift für Rassenkunde
Zeitschrift für Volksaufartung und Erbkunde
Ziel und Weg
Züchtungskunde

Bibliography

3. Literature

Abir-Am, Pnina (1993). "From Multidisciplinary Collaboration to Transnational Objectivity: International Space as Constitutive of Molecular Biology, 1830–1970." In Elisabeth Crawford, Terry Shinn, and Sverker Sörlin, eds. *Denationalizing Science. The Contexts of International Scientific Practice.* Dodrecht; Boston; London: Kluwer Academic, pp. 153–186.

Adams, Mark B. (1989). "The Politics of Human Heredity in the USSR, 1920–1940." *Genome* 31:879–884.

——— (1990a). "Eugenics in Russia 1900–1940." In Mark B. Adams, ed. *The Wellborn Science: Eugenics in Germany, France, Brazil and Russia.* New York: Oxford University Press, pp. 153–216.

——— (1990b). "Toward a Comparative History of Eugenics." In Mark B. Adams, ed. *The Wellborn Science. Eugenics in Germany, France, Brazil and Russia.* New York; Oxford, pp. 217–232.

Adams, Mark B., Garland E. Allen, and Sheila F. Weiss (2005). "Human Heredity and Politics: A Comparative Institutional Study of the Eugenics Record Office at Cold Spring Harbor (United States), the Kaiser Wilhelm Institute for Anthropology, Human Heredity and Eugenics (Germany) and the Maxim Gorky Medical Genetics Institute (USSR)." *Osiris*, Second Series, 20:232–262.

Adam, George (1922). "The True Aristocracy. An Address Contributed to the International Eugenics Congress Held in New York in September, 1921." *Eugenics Review* 14:174–186.

Ahluwalia, Sanjam (2008). *Reproductive Restraints: Birth Control in India, 1977–1947.* Bloomington: University of Indiana Press.

Aikman, K. B. (1933). "Race Mixture." *Eugenics Review* 25:161–166.

Aldrich, Mark (1975). "Capital Theory and Racism: From Laissez-Faire to the Eugenics Movement in the Career of Irving Fisher." *Review of Radical Political Economy*, no. 3/1997 33–42.

Allen, Ann Taylor (1988). "German Radical Feminism and Eugenics, 1900–1918." *German Studies Review* 11:31–56.

——— (1991). "Feminismus und Eugenik im historischen Kontext." *Feministische Studien* 9:46–68.

Allen, Garland E. (1984). "A History of Eugenics." Science for the People. Sociobiology Study Group, *Biology as Destiny: Scientific Fact or Social Bias?* Cambridge: Science for the People, 13–19.

——— (1986). "The Eugenics Record Office at Cold Spring Harbor, 1910 to 1940: An Essay in Institutional History." *Osiris* 2:225–264.

——— (1991). "Old Wine in New Bottles: From Eugenics to Population Control in the Work of Raymond Pearl." In Keith R. Benson, Jane Maienschein, and Ronald Rainger, eds. *The Expansion of American Biology.* New Brunswick; London: Rutgers University Press, pp. 231–261.

Allen, Garland E., and Barry Mehler (1977). "Sources in the Study of Eugenics. Inventory of the American Eugenics Society Papers." *The Mendel Newsletter* 14:9–15.

Allen, Gordon (1956). "To the Editor." *Eugenics Quarterly* 3:4–8.

——— (1961). "Reply to Garrett Concerning Putnam's Race and Reason." *Eugenics Quarterly* 8:221–222.

——— (1968). "Eugenics." *International Encyclopedia of the Social Science.* New York: Macmillan, pp. 193–195.

Altner, Günther (1968). *Weltanschauliche Hintergründe der Rassenlehre des Dritten Reiches. Zum Problem einer umfassenden Anthropologie.* Zürich: EVZ.

American Eugenics Society (1931). *Organized Eugenics.* New Haven: American Eugenics Society.
——— (1938). *Practical Eugenics: Aims and Methods of the American Eugenics Society.* New York: American Eugenics Society.
Ammon, Otto (1895). *Die Gesellschaftsordnung und ihre natürlichen Grundlagen.* Jena: G. Fischer.
Anderson, Hjalmar (1921a). "The Swedish State-Institute for Race-Biological Investigation." *Eugenics Review* 13:480–482.
——— (1921b). *The Swedish State-Institute for Race-Biological Investigation.* Stockholm: No publisher.
Anderson, Jack, and Dale van Atta (1989). "Pioneer Fund's Controversial Projects." *Washington Post*, November 16, 1989.
Anderson, Scott, and Jon Lee Anderson (1986). *Inside the League: The Shocking Expose of How Terrorists, Nazis, and Latin American Death Squads Have Infiltrated the World Anti-Communist League.* New York: Dodd, Mead.
Apert Eugene (1913). "Une science nouvelle, l'eugénique." *L'Hygiène*, no. 42:12–13.
——— (1940). "L'eugénique en France." *Revue Anthropologique* 50:207–216.
Asana, J. J. (1934). "Eugenics in India." *Eugenical News* 19:10–11.
Ash, Mitchell G. (2001). "Wissenschaft und Politik als Ressourcen füreinander. Programmatische Überlegungen am Beispiel Deutschlands." In Jürgen Büschenfeld, Heike Franz, and Frank-Michael Kuhlemann, eds. *Wissenschafts–Geschichte heute: Festschrift für Peter Lundgreen.* Bielefeld: Verlag für Regionalgeschichte, pp. 117–134.
Assheuer, Thomas, and Hans Sarkowicz (1992). *Rechtsradikale in Deutschland. Die alte und die neue Rechte.* Munich: Beck.
Bacharach, Walter Zwi (1984). "Ignaz Zollschans "Rassentheorie." In Walter Grab, ed. *Jüdische Integration und Identität in Deutschland und Österreich 1848 bis 1918.* Tel Aviv: Jahrbuch des Instituts für Deutsche Geschichte. pp. 179–197.
Bachelard–Jobard, Catherine (2001). *L'eugénisme, la science et le droit.* Paris: Presses Universitaires de France.
Baker, John R. (1974). *Race.* New York; Oxford: Oxford University Press.
——— (1975). "Die Rassenwirklichkeit." *Neue Anthropologie* 3:41–44.
Bannister, Robert (1979). *Social Darwinism: Science and Myth in Ango-American Social Thought.* Philadelphia: Temple University Press.
Barkan, Elazar (1988). "Mobilizing Scientists against Nazi Racism." In George Stocking, ed. *Bones, Bodies, Behavior: Essays on Biological Anthropology.* Madison, WI: The University of Wisconsin Press, pp. 180–205.
——— (1992). *The Retreat of Scientific Racism.* New York: Cambridge University Press.
Barrett, Deborah, and Charles Kurzmann (2004). "Globalizing Social Movement Theory: The Case of Eugenics." *Theory and Society* 33:487–527.
Bashford, Alison (2010). "Internationalism, Cosmopolitanism, and Eugenics." In Alison Bashford and Philippa Levine, eds. *The Oxford Handbook of the History of Eugenics.* New York: Oxford University Press, pp. 154–172.
Baur, Erwin (1921). "Die biologische Bedeutung der Auswanderung für Deutschland." *Archiv für Frauenkunde und Eugenik* 7:206–208.
——— (1933). *Der Untergang der Kulturvölker im Lichte der Biologie.* Munich: J. F. Lehmanns Verlag.
Bär, Gesine (2002). "Wir stehen nicht allein: Schwedische Eugenik im Spiegel der deutschen nationalsozialistischen Rassenforschung." *Nordeuropa Forum*, 12, no. 2/2002:25–41.
Bäumer, Anne (1990). *NS-Biologie.* Stuttgart: Hirzel.
Beck, Maximilian (1938). "Unabhängigkeit der geistigen Kultur von der Rasse." Congres International de la Population, Paris 1937. Paris: Hermann et Cie.,Éditeurs, Vol. 8, pp. 106–116.

Bedwell, C. E. A. (1922). "Eugenics in International Affairs." *Eugenics Review* 14:187–189.
Begos, Kevin, Danielle Deaver, John Railey, Scott Sexton, Paul Lombardo (2012). *Against Their Will. North Carolina's Sterilization Program and the Campaign for Reparations.* Apalachicola, FL: Gray Oak Books.
Bellant, Russ (1991). Old Nazis, the New Right, and the Republican Party: Domestic Fascist Networks and their Effects on U.S. Cold War Politics. Boston, MA: Political Research Associates.
Berger, Erich (1939a). "Die Gesetzgebung für Erbkranke." *Hamburger Fremdenblatt*, 1.4.1939.
——— (1939b). "Gibt es eine rassenpolitische Propaganda Deutschlands im Ausland." *Ziel und Weg* 9:113–115.
——— (1939 c). "Das deutsche Beispiel macht Schule." *Ziel und Weg*, 9, 44–47.
Bergman, Jerry (1992). "Eugenics and the Development of Nazi Race Policy." *Perspectives on Science and Christian Faith* 44:109–135.
Biley, Francis C. (2007). "Pondering the Future of Humanity: Reflections on John Glad's 'Future Human Evolution; Eugenics in the Twenty–first Century.'" *Journal of Psychiatric and Mental Health Nursing* 14:325–329.
Billig, Michael (1979). *Psychology, Racism and Fascism*. Birmingham: Searchlight.
——— (1981). *Die rassistische Internationale. Zur Renaissance der Rassenlehre in der modernen Psychologie*. Frankfurt a.M.: Neue Kritik.
Björkman, Maria, and Sven Widmalm (2010). "Selling Eugenics: The Case of Sweden." *Notes & Records of Royal Society* 64:379–400.
Black, Edwin (2003). *The War against the Weak. Eugenics and America's Campaigne to Create a Master Race*. New York: Four Walls Eight Windows.
——— (2009). *Nazi Nexus: America's Corporate Connections to Hitler's Holocaust*. Washington, DC: Dialog Press.
Blacker, C. P. (1933). "Eugenics in Germany." *Eugenics Review* 25:157–159.
——— (1934). *Voluntary Sterilization: The Last Sixty Years & Transitions Throughout the World*. London: The Garden City Press Limited.
——— (1952). *Eugenics: Galton and after*. London, U.K.: Duckworth.
——— (1955). *Family Planning and Eugenic Movements in the Mid-Twentieth Century*. Report of the Proceedings: The Fifth International Conference on Planned Parenthood, 24–29 October, 1955. Tokyo, Japan. London: The International Planned Parenthood Federation: pp. 126–134.
——— (1960). "World Population Trends." *Royal Society of Health Journal* 80:236–238.
——— (1962a). "Voluntary Sterilization: The Last Sixty Years." *Eugenics Review* 54:9–23.
——— (1962b). "Voluntary Sterilization: Transitions Throughout the World." *Eugenics Review* 54:143–162.
——— (1964). "The International Planned Parenthood Federation. Aspects of Its History." *Eugenics Review* 56:135–142.
Bleker, Johanna, and Svenja Ludwig, eds. (2007) *Emanzipation und Eugenik: die Briefe der Frauenrechtlerin, Rassenhygienikerin und Genetikerin Agnes Bluhm an den Studienfreund Alfred Ploetz aus den Jahren 1901–1938*. Husum: Matthiesen.
Boas, Franz (1938). "Heredity and Environment." In Démographie de la France-d'Outremer, ed. *Congrès International de la Population, Paris 1937*. Paris: Hermann et Cie., Éditeurs pp. 83–92.
Bock, Gisela (1986). *Zwangssterilisation im Nationalsozialismus. Studien zur Rassenpolitik und Frauenpolitik*. Opladen: Westdeutscher Verlag.
——— (1991a). "Antinatalism, Maternity and Paternity in National Socialist Racism." In Gisela Bock and Pat Thane, eds. *Maternity and Gender Policies: Women and the Rise of the European Welfare States, 1880s-1950s*. London: Routledge pp. 233–255.
——— (1991b). "Krankenmord, Judenmord und nationalsozialistische Rassenpolitik: Überlegungen zu einigen neueren Forschungshypothesen." In Frank Bajohr,Werner Johe,

and Uwe Lohalm, eds. *Zivilisation und Barbarei. Die widersprüchlichen Potentiale der Moderne*. Hamburg: Christians, pp. 285–306.

——— (1996). "Sterilization and 'Medical' Massacres in National Socialist Germany." In Manfred Berg and Geoffrey Cocks, eds. *Medicine and Modernity: Public Health and Medical Care in 19th and 20th-Century Germany*. Cambridge: Deutsches Historisches Institut Washingston.

Bodmer, Walter F., and Luigi Luca Cavalli-Sforza (1970). "Intelligence and Race." *Scientific American* 43:301–327.

——— (1976). *Genetics, Evolution, and Man*. San Francisco: Freeman.

Bogner, Alexander (2003) "Wissenschaft und Ideologie: Die liberale Bioethik und der Strukturwandel rassistsicher Diskurse." *Wiener Zeitschrift zur Geschichte der Neuzeit* 3:64–86.

Boli, John, and George M. Thomas (1999). "INGOs and the Organization of World Culture." In John Boli and George M. Thomas, eds. *Constructing World Culture: International Nongovernmental Organizations since 1875*. Stanford, Calif.: Stanford University Press, pp. 13–49.

Bouchard, Thomas J., and Matt McGue. (1990a). "Genetic and Environmental Influences on Reli-gious Interests, Attitudes, and Values: A Study of Twins Reared Apart and Together." *Psychological Science* 1:138–142.

Bouchard, Jr., Thomas J., William M. Grove W., Elke D. Eckert, Leonard Heston, Nancy Segal, and David T. Lykken (1990b). "Heritability of Substance Abuse and Antisocial Behavior: A Study of Monozygotic Twins Reared Apart." *Biological Psychiatry* 27:1293–1304.

Bowler, Peter J. (1989). The Mendelian Revolution: The Emergence of Hereditarian Concepts in Modern Science and Society. Baltimore: The Johns Hopkins University Press.

Böhm, Michael (2008). *Alain de Benoist und die Nouvelle Droite*. Münster: Lit Verlag.

Bräutigam, Jeffrey C. (1990). "Sorting Out Social Darwinism." *History and Philosophy of Life Sciences* 12:111–116.

Briand, Henri (1938). "L'eugénique et la conservation des qualités de la race." *Revue Anthropologique* 48:55–68.

——— (1941). "Une Conférence du Professeur Eugen Fischer à Paris." *Revue Anthropologique*, 51:195–196.

Brie, Friedrich (1927). *Der Einfluß der Lehren Darwins auf den britischen Imperialismus*. Rektoratsrede. Freiburg im Br.: Speyer & Kaerner.

Broberg, Gunnar, and Mattias Tyden (1996). "Eugenics in Sweden: Efficient Care." In Gunnar Broberg and Nils Roll-Hansen, eds. *Eugenics and the Welfare State: Sterilization Policy in Denmark, Sweden, Norway, and Finland*. East Lansing: Michigan State University Press, pp. 77–150.

Brookes, Martin (2004). Extreme Measures: The Dark Vision and Bright Ideas of Francis Galton. London: Bloomsbury Publishing PLC.

Brozek, Artur (1935). "Der biologische Begriff der Rasse." In Karel Weigner, ed. *Die Gleichwertigkeit der europäischen Rassen und die Wege zu ihrer Vervollkommnung*. Prag: Verlag der Tschechischen Akademie der Wissenschaften und Künste, pp. 29–48.

Bruinius, Harry (2006). Better for All the World: The Secret History of Forced Sterilization and America's Quest for Racial Purity. New York: Knopf Doubleday Publishing Group.

Bucur, Maria (2002). *Eugenics and Modernization in Interwar Romania*. Pittsburgh: University of Pittsburgh Press.

Burch, Guy Irving (1932). "Birth Control vs. Class Suicide." *Survey Graphic*, Issue 4, April 1932, pp. 64.

Burgdörfer, Friedrich (1930a). "Eugenische Tagung in Rom." *Allgemeines Statistisches Archiv* 20:442–444.

——— (1930b). *Familie und Volk*. Berlin: Deutscher Schriftenverlag.

——— (1933). "Die kinderreichen Familien, ihre volksbiologische Bedeutung und ihre statistische Erfassung." In Corrado Gini, ed. *Atti del congresso internazionale per gli studi sulla popolazione. Rome, 7–10 septembre 1931. Vol. VIII.* Rome: Istituto poligrafico dello Stato. pp. 343–354.

——— (1936). "Bevölkerungsentwicklung im abendländischen Kulturkreis mit besonderer Berücksichtigung Deutschlands." In Hans Harmsen and Franz Lohse, eds. *Bevölkerungsfragen. Bericht des Internationalen Kongresses für Bevölkerungswissenschaft, Berlin 8/26–9/1/1935.* Munich: J. F. Lehmanns, pp. 63–92.

——— (1940). "Krieg und Bevölkerungsentwicklung." *Schulungsbrief,* Septermber 07, 1940, pp. 109–116.

——— (1942). *Kinder des Vertrauens. Bevölkerungspolitische Erfolge und Aufgaben im Großdeutschen Reich.* Berlin: Zentralverlag der NSDAP, Franz Eher Nachf. GmbH.

Burleigh, Michael (1994). *Death and Deliverance: "Euthanasia" in Germany, c. 1900 to 1945.* Cambridge: CUP Archive.

Campbell, Chloe (2007). *Race and Empire: Eugenics in Colonial Kenya.* Manchester; New York: Manchester University Press.

——— (2010). "Eugenics in Colonial Kenya." In Alison Bashford and Philippa Levine, eds. *The Oxford Handbook of the History of Eugenics.* New York; Oxford: Oxford University Press, pp. 289–300.

Campbell, Clarence G. (1936a). "The Biological Postulates of Population Study." In Hans Harmsen and Franz Lohse, eds. *Bevölkerungsfragen. Bericht des Internationalen Kongresses für Bevölkerungswissenschaft, Berlin 26. August – 1. September 1935* Munich: J. F. Lehmanns Verlag, pp. 601–611.

——— (1936b). "Die deutsche Rassenpolitik." *Der Erbarzt* 3:140–142.

——— (1936c). "The German Racial Policy." *Eugenical News* 21:25–29.

——— (1973). "The German Racial Policy." *American Mercury,* Issue 1/1973, pp. 25–28.

Cannon, Cornelia James (1922). "American Misgivings." *Atlantic Monthly,* Issue 2/1922, pp. 145–157.

Carlson, Elof Axel (2001). *The Unfit: A History of a Bad Idea.* New York: Cold Spring Harbor.

Carnegie, Andrew (1889). "Wealth." *North American Review,* Issue 6/1889, pp. 653–664.

Carol, Anne (1995). *Histoire de l'eugénisme en France. Les médecins et la procréation XIXe–XXe siècle.* Paris: Éd. Du Seuil.

Carr-Saunders, Alexander M. (1935). "Eugenics in the Light of Population Trends." *Eugenics Review* 27:11–20.

Carrel, Alexis (1935). *L'Homme, cet inconnu.* Paris: Plon.

——— (1937). *Der Mensch, das unbekannte Wesen.* Stuttgart: Deutsche Verlags-Anstalt.

Cassata, Francesco (2006a). *Il fascismo razionale: Corrado Gini fra scienza e politica.* Rome: Carocci.

——— (2006b). *Molti, sani, e forti: L'eugenetica in Italia.* Turin: Bollati Boringhieri.

——— (2011). *Bulding the New Man: Eugenics, Racial Science and Genetics in Twentieth-Century Italy.* Budapest; New York: Central European University Press.

Castle, William E. (1930). "Race Mixture and Physical Disharmonies." *Science* 71: 603–605.

Castles, Katherine (2002). "Quiet Eugenics. Sterilization in North Carolina's Institutions for the Mentally Retarded." *Journal of Southern History* 68:849–878.

Chamberlain, Houston Stewart (*1899). Die Grundlagen des neunzehnten Jahrhunderts.* München: Verlagsanstalt F. Bruckmann.

Chase, Allen (1977). The Legacy of Malthus: The Social Costs of the New Scientific Racism. New York: Knopf.

Chickering, Roger (1984). *The Men Who Feel Most German: A Cultural Study of the Pan-German League 1886–1914.* London: HarperCollins Publishers.

Chung, Yuehtsen Juliette (2002). Struggle for National Survival: Eugenics in Sino–Japanese Contexts, 1896–1945. New York; London: Routledge.

——— (2010). "Eugenics in Chinca and Hong Kong: Nationalism and Colonialism, 1890s–1940s." In Alison Bashford and Philippa Levine, eds. *The Oxford Handbook of the History of Eugenics.* New York; Oxford: Oxford University Press, pp. 258–273.

Clark, Linda Loeb (1984). *Social Darwinism in France.* Alabama: University of Alabama Press.

Clark, (1956). The Social Use of Scientific Knowledge. Eugenics in the Career of C. C. Litte. Master Thesis. University of Maine.

Cleminson, Richard (2000). Anarchism, Science and Sex: Eugenics in Eastern Spain, 1900–1937. Bern: Peter Lang.

Close, Charles (1936). "Population Tendencies in Great Britain." In Hans-Franz Lohse Harmsen, ed. *Bevölkerungsfragen. Bericht des Internationalen Kongresses für Bevölkerungswissenschaft, Berlin 8/26/-9/1/1935.* Munich: J. F. Lehmanns, pp. 93–97.

Cogdell, Christina (2004). *Eugenic Design: Streamlining America in the 1930s.* Philadelphia: University of Pennsylvania Press.

Comas, Juan (1961). "Scientific Racism Again?" *Current Anthropology* 2: 303–314.

Committee on Publication (1923). "Preface." In Charles B. Davenport, ed. *Eugenics in Race and State.* Baltimore: Williams & Wilkins Company, pp. ix.

Conklin, Edwin G. (1928). "Some Recent Criticisms of Eugenics." *Eugenical News* 13:61–65.

Connelly, Matthew (2006). "Seeing beyond the State: The Population Control Movement and the Problem of Sovereignty." *Past & Present*, no. 193: 197–233.

Conze, Werner, and Antje Sommer (1984). "Rasse." In Otto Brunner, Werner Conze, and Reinhart Koselleck, eds. *Geschichtliche Grundbegriffe.* Band 5, Stuttgart: Klett, pp. 135–178.

Coogan, Kevin (1999). Dreamer of the Day: Francis Parker Yockey and the Postwar Fascist International. New York: Autonomedia.

Crawford, Elisabeth (1990). "The Universe of International Science, 1880–1939." In Tore Frängsmyr, ed. *Solomon's House Revisited: The Organization and Institutionalization of Science.* Canton: Science History Publications, pp. 251–269.

Crew, Francis E. A. (1919). "A Biologist in a New Environment." *Eugenics Review* 11:119–123.

Crips, Liliane (1993). "Les avatars d'une Utopie scientiste en Allemagne: Eugen Fischer (1974–1967) et l'hygiène raciale." *Le Mouvement Social* 163: 7–23.

Crook, Paul (2002). "American Eugenics and the Nazis: Recent Historiography." *The European Legacy* 7:363–381.

Dahlberg, Gunnar (1942). *Race, Reason and Rubbish.* New York: Columbia University Press.

Dalsace, Jean (1935). "Stérilisation et racisme." *L'Humanité*, August 11, 1935, p. 4.

——— (1936). "A propos de la Stérilisation." In Hans Harmsen and Franz Lohse, eds. *Bevölkerungsfragen. Bericht des Internationalen Kongresses für Bevölkerungswissenschaft, Berlin 8/26–9/1/1935.* Munich: J.F. Lehmanns, pp. 706–712.

Darwin, Charles (1890). *The Descent of Man. And Selection in Relation to Sex.* 2nd ed. London: John Murray.

Darwin, Leonard (1912). "First Steps towards Eugenic Reform." *Eugenics Review* 4:26–38.

——— (1917). "Quality Not Quantity." *Eugenics Review* 8:297–321.

———(1918). "The Need for Widespread Eugenic Reform during Reconstruction." *Eugenics Review* 10:145–162.

——— (1923a). "The Aims and Methods of Eugenics Societies." Eugenics and the Family. Scientific Papers Presented at the Second International Eugenics Congress, New York, 1921. Baltimore, pp. 5–19.

―――― (1923b) "Reports of the International Congress of Eugenics Held at New York in September 1921." *Eugenics Review* 15:409–414.
―――― (1934). "Message to the International Congress of Eugenics." In Perkins, Henry F., *A Decade of Progress in Eugenics*. Baltimore: Williams & Wilkins, pp. 23–24.
Davenport, Charles B. (1911). *Heredity in Relation to Eugenics*. New York, NY: Holt.
―――― (1917). "The Effects of Race Intermingling." *Proceedings of the American Philosophical Society* 56:364–368.
―――― (1921). "Research in Eugenics." *Science* 54:397.
―――― (1928). "Race Crossing in Man." Discours prononcés à la Conférence organisée par. The International Federation of Eugenic Organisations. Paris, pp. 9–14.
―――― (1930). "Race Mixture and Physical Disharmonies." *Science* 71:603–606.
―――― (1934). "The Development of Eugenics." In Henry F. Perkins, ed. *A Decade of Progress in Eugenics*. Baltimore: Williams & Wilkins, pp. 17–22.
Davenport, Charles B., and Morris Steggerda (1929). *Race Crossing in Jamaica*. Washington: Carnegie Institution of Washington, Publication, no. 395, pp. 516.
Dawson, Patrick (1995). "Managing Quality in the Multi-Cultural Workplace." In Adrian Wilkinson and Hugh Willmott, eds. *Making Quality Critical. New Perspectives on Organizational Change*. London; New York: Routledge, pp. 173–193.
Deichmann, Ute (1992). *Biologen unter Hitler. Vertreibung, Karrieren, Forschung*. Frankfurt a.M.; New York: Campus Verlag.
Dennis, Frances (1973). "The IPPF: 21 Years of Achievement." *Journal of Biosocial Sciences* 5:413–419.
Deverell, Colville (1968). "The IPPF—Its Role in Developing Countries." *Demography* 5:574–577.
Di Trocchio, Federico (1994). *Der große Schwindel. Betrug und Fälschung in der Wissenschaft*. Frankfurt a.M.; New York: Campus Verlag.
Dice, Lee R. (1960). "Resources of Mental Ability. How Can the Supply of Superior Ability Be Conserved and Perhaps Increased?" *Eugenics Quarterly* 7:9–22.
Dietz, Bernhard (2009). "'Sterilisation of the Unfit'. Eugenikbewegung und radikale Recht im Großbritannien der 'Lost Generation.'" In Regina Wecker, Sabine Braunschweig, Gabriela Imbodeneds. *Wie nationalsozialistisch ist die Eugenik? Internationale Debatten zur Geschichte der Eugenik im 20. Jahrhundert*. Wien; Köln; Weimar: Böhlau, pp. 187–198.
Dikötter, Frank (1989). "Eugenics in Republican China." *Republican China* 15:1–17.
―――― (1998). "Race Culture: Recent Perspectives on the History of Eugenics." *American Historical Review* 103:467–478.
Dobzhansky, Theodosius (1963). "Genetics and Race Equality." *Eugenics Quarterly* 13:151–160.
―――― (1964). *Heredity and the Nature of Man*. New York: Harcourt, Brace & World.
Doelecke, Werner (1975). "Alfred Ploetz (1860–1940), Sozialdarwinist und Gesellschafts biologe." Dissertation, Frankfurt a.M.: Universität Frankfurt.
Doménech, Rosa Medina (2004). "Eugenesia y formas de hacer historia. Cuestiones para el debate." *Dynamis* 24:291–305.
Donaldson, Peter J. (1990). *Nature Against Us: The United States and the World Population Crisis, 1965–1980*. Chapel Hill: University of North Carolina.
Doron, Joachim (1980). "Rassenbewußtsein und naturwissenschaftliches Denken im Deutschen Zionismus während der wilhelminischen Ära." In Walter Grab, ed. *Jahrbuch des Instituts für deutsche Geschichte 1980*. Tel Aviv: Nateev-Printing and Publishing Enterprises, pp. 389–427.
Dorr, Gregory Michael (2008). *Segregation's Science: Eugenics and Society in Virgina*. Charlottesville: University of Virginia.
Dörner, Klaus (1967). "Nationalsozialismus und Lebensvernichtung." *Vierteljahreshefte für Zeitgeschichte* 15:121–152.

Dörner, Klaus, Christiane Haerlin, and Veronika Rau . (1980). *Der Krieg gegen die psychisch Kranken.* Rehburg-Loccum: Psychiatrischer Verlag.
Drouard, Alain (1992a). "Aux origines de l'eugenisme en France: le neo-malthusianisme (1996–1914)." *Population* 47:435–460.
——— (1992b). "La creation de 1TNED." *Population* 47:1453–1466.
——— (1992c). *Une inconnue des sciences sociales: La Foundation Alexis Carrel, 1941–1945.* Paris: Editions de la Maison des sciences de l'homme.
——— (1999). "Concerning Eugenics in Scandinavia. An Evaluation of Recent Research and Publications." *Population* 11: 261–270.
Drysdale, C. V. (1912). *Neo-Malthusianism and Eugenics.* London: W. Bell.
Dunn, Leslie Clarence (1962a). "Cross Currents in the History of Human Genetics." *The Eugenics Review*, 54, 72–82.
Dunn, Leslie Clarence (1962b). "Cross Currents in the History of Human Genetics." *Americal Journal of Human Genetics*, 14, 1–13.
Duranton-Crabol, Anne-Marie (1988). *Visages de la nouvelle droite: Le Grece et son histoire.* Paris: Presses de la Fondation nationale des sciences politiques.
Duster, Troy (2003). *Backdoor to Eugenics.* New York; London: Psychology Press.
Ebbinghaus, Angelika (1984). *Arbeiter und Arbeitswissenschaft. Zur Entstehung der "wissenschaftlichen Betriebsführung".* Opladen: Westdeutscher Verlag.
Edin, Karl Avid (1927). "Fertility in Marriage and Infantile Mortality in the Different Social Classes in Stockholm from 1919 to 1922." In Margaret Sanger, ed. *Proceedings of the World Population Conference.* London: E. Arnold, pp. 205–207.
Ehmann, Annegret (1993). "From Colonial Racism to Nazi Population Policy. The Role of the So-Called Mischlinge." Unpublished manuscript, Washington, DC.
Ehrenfels, U. R. (1962). "Critical Paragraphs Deleted." *Current Anthropology* 3:154–155.
Ehrlich, Paul R., and Shirley Feldman (1977). *The Race Bomb: Skin Color, Prejudice, and Intelligence.* New York: Quadrangle / New York Times Book Co.
Ekberg, Marryn (2007). "The Old Eugenics and the New Genetics Compared." In *Social History of Medicine* 20:581–593.
Ellinger, Tage U. H. (1942). "On the Breeding of Aryans and Other Genetic Problems of War-time Germany." *Journal of Heredity* 33:141–143.
Ellis, Havelock (1919). *The Philosophy of Conflic,t and Other Essays in War-Time.* London: Ayer Co. Pub.
Engs, Ruth Clifford (2005). *The Eugenics Movement: An Encyclopedia.* Westport; London: Greenwood.
Epps, Edgar C. (1973). "Race, Intelligence, and Learning. Some Consequences of the Misuse of Test Results." *Phylon* 34:153–173.
Etzemüller, Thomas (2007). *Ein ewigwährender Untergang: Der apokalyptische Bevölkerungsdiskurs im 20. Jahrhundert.* Bielefeld: transcript.
——— (2010). *Die Romantik der Rationalität. Alva & Gunnar Myrdal—Social Engineering in Schweden.* Bielefeld: Transcript.
——— (2012). "Die Angst vor dem Abstieg—Malthus, Burgdörfer, Sarrazin: Eine Ahnenreihe mit immer derselben Botschaft." In Michael Haller and Martin Niggeschmidt, eds. *Der Mythos vom Niedergang der Intelligenz. Von Galton zu Sarrazin: Die Denkmuster und Denkfehler der Eugenik.* Wiesbaden: VS Verlag für Sozialwissenschaften, pp. 157–184.
Eugenics Education Society (1911). The First International Congress, London. London.
——— (1912). First International Eugenics Congress. Catalogue of the Exhibition. London.
——— (1913). *Problems in Eugenics.* Report of Proceedings of the First International Eugenics Congress. London: The Eugenics Education Society.
Eysenck, Hans J. (1971). *Race, Intelligence and Education.* London: Temple Smith.
——— (1975). *The Inequality of Man.* San Diego: EDiTS.
Eysenck, Hans J. (1976). *Die Ungleichheit der Menschen.* Munich: List

——— (1982). "The Inheritance of Intelligence, and Its Critics: Some Myths Reconsidered." *New Education* 4:1–8.
Fahlbeck, Pontus (1912). "Der Neo-Malthusianismus in seinen Beziehungen zur Rassenbiologie und Rassenhygiene." *Archiv für Rassen—und Gesellschaftsbiologie* 9:30–48.
Fairchild, Henry P. (1934). "Organization for Research in Population." *Population* 1:79–84.
Falk, Raphael (2006). "Zionism, Race and Eugenics." In Geoffrey Cantor and Marc Swelitz, eds. *Jewish Tradition and the Challenge of Darwinism*. Chicago: University of Chicago Press, pp. 137–165.
——— (2010). "Eugenics and the Jews." In Alison Bashford and Philippa Levine, eds. *The Oxford Handbook of the History of Eugenics*. New York: Oxford University Press, pp. 462–476.
Fantham, H. B. (1927). "Some Thoughts on Biology and the Race." *South African Journal of Science* 24:1–20.
Fédération Internationale Latine des Sociétés d'eugénique (1937). *1er Congrès Latin d'eugénique*. Paris: Masson.
Fischer, Eugen (1913). *Die Rehobother Bastards und das Bastardisierungsproblem beim Menschen*. Jena: G. Fischer.
——— (1926). "Günther, Hans, Rassenkunde des deutschen Volkes." *Zeitschrift für Morphologie und Anthropologie* 25:160–163.
——— (1930). "Aus der Geschichte der Deutschen Gesellschaft für Rassenhygiene." *Archiv für Rassen—und Gesellschafsbiologie* 24:1–5.
——— (1933). "Eugenik." In Rudolf Dittler, ed. *Handwörterbuch der Naturwissenschaften*. Jena: Fischer, pp. 898–901.
——— (1935a). "Das Recht jedes Einzelnen auf positive und negative Rassenhygiene." *Kongress-Korrespondenz*, July 20, 1935.
——— (1935b). "Rassenhygiene als gemeinsame Aufgabe aller Kulturvölker." *Der Öffentliche Gesundheitsdienst* 1:690–715.
——— (1936a). "Einleitender Vortrag." In Hans Harmsen and Franz Lohse, eds. *Bevölkerungsfragen. Bericht des Internationalen Kongresses für Bevölkerungswissenschaft, Berlin 8/26–9/1/1935*. Munich: J. F. Lehmanns, pp. 39–62.
——— (1936b). "Kaiser-Wilhelm-Institut für Anthropologie, menschliche Erblehre und Eugenik." In Max Planck, ed., *25 Jahre Kaiser-Wilhelm-Gesellschaft zur Förderung der Wissenschaften*, 2. Vol. Berlin: Springer, pp. 348–356.
——— (1936c). "Schlußansprache." In Hans Harmsen and Franz Lohse, eds. *Bevölkerungsfragen. Bericht des Internationalen Kongresses für Bevölkerungswissenschaft, Berlin 8/26–9/1/1935*. Munich: J. F. Lehmanns, pp. 927–931.
——— (1942). "Le problème de la race et la législation raciale allemand." In Karl Epting, ed. *État et santé*. Paris: F. Sorlot, pp. 81–110.
Fisher, Irving (1915). "Eugenics—Foremost Plan of Human Redemption." Official Proceedings of Second National Conference on Race Betterment, Aug. 4–8, 1915, Battle Creek, pp. 63–66.
——— (1919). *The Necessity of a League of Nations*. Unpublished Leaflet, Boston.
——— (1921). "Impending Problems of Eugenics." *Scientific Monthly* 13:219–221.
Fleming, Rachel M. (1930). "Human Hybrids. Racial Crosses in Various Parts of the World." *Eugenics Review*, 21, 257–263.
Fleure, H. J. (1936). "Race and Politics." *Eugenics Review* 27:319–320.
Fogarty, Richard P., and Michael A. Osborne (2010). "Eugenics in France and the Colonies." In Alison Bashford and Philippa Levine, ed. *The Oxford Handbook of the History of Eugenics*. New York: Oxford University Press, pp. 332–346.
Foley, Dolores (1988). "Non-governmental Organizations as Catalysts of Policy Reform and Social Change: A Case Study of the International Planned Parenthood Federation." Ph. D. thesis, University of Southern California, Los Angeles.

Franks, Angela (2005). Margaret Sanger's Eugenic Legacy: The Control of Female Fertility. Jefferson: McFarland.
Fredrickson, George M. (2004). "Rassismus. Ein historischer Abriß." Unpublished manuscript, Hamburg.
Freeden, Michael (1979). "Eugenics and Progressive Thought: A Study in Ideological Affinity." *Historical Journal* 22:645–671.
Frercks, Rudolf (1937). *Deutsche Rassenpolitik*. Leipzig: Reclam.
Frey, Marc (2007). "Experten, Stiftungen und Politik: Zur Genese des globalen Diskurses über Bevölkerung seit 1945." *Zeithistorische Forschungen/Studies in Contemporary History, Online–Ausgabe* 4:137–159.
Frick, Wilhelm (1934). "German Population and Race Politics." *Eugenical News* 19:33–38.
——— (1936). "Eröffnungsansprache." In Hans Harmsen and Franz Lohse, eds. *Bevölkerungsfragen. Bericht des Internationalen Kongresses für Bevölkerungswissenschaft, Berlin 8/26–9/1/1935*. Munich: J. F. Lehmanns, pp. 6–33.
Friedlander, Henry (1997). *Der Weg zum NS–Genozid. Von der Euthanasie zur Endlösung*. Berlin: Berlin-Verlag.
Friese, Gerhard (1936). "Internationaler Kongreß für Bevölkerungswissenschaft." *Der Öffentliche Gesundheitsdienst* 1:641–747.
Fromm, Rainer, and Barbara Kernbach (1994). *Und morgen die ganze Welt? Rechtsextreme Publizistik in Westeuropa*. Marburg; Berlin: Schüren.
Gallagher, Nancy L. (1999). Breeding Better Vermonters: The Eugenics Project in the Green Mountain State. Hanover, NH: University Press of New England.
Galton, Francis (1865). "Hereditary Talent and Character." *Macmillan's Magazine*, vol. 12, pp. 157–166, 318–327.
——— (1869). *Hereditary Genius*. London: Macmillan and C.
——— (1908). *Memories of My Life*. London: Methuen.
Garrett, Henry E. (1947). "Negro-White Differences in Mental Ability in the United States." *Scientific Monthly* 65:329–333.
——— (1961a). "Comment on Gordon Allen's Review of Putnam's Race and Reason." *Eugenics Quarterly* 8:218–220.
——— (1961b). "The Equalitarian Dogma." *Perspectives in Biology and Medicine* 4:480–484.
——— (1961c). "Klineberg's Chapter on Race and Psychology." *Mankind Quarterly* 2:15–23.
Gasman, Daniel (1971). *The Scientific Origins of National Socialism: Social Darwinism in Ernst Haeckel and the German Monist League*. London: Macdonald.
Gates, R. Ruggles (1934). "Racial and Social Problems in the Light of Heredity." *Population* 1:25–36.
Gaudilliere, Jean-Paul (1994). "Le Cancer entre infection et heredite: Genes, virus et souris au National Cancer Institute (1937–1977)." *Revue Histoire des Sciences* 47:57–89.
Gaudilliere, Jean-Paul, and Ilana Löwy (1994). "Disciplining Cancer. Mice and the Practice of Genetic Purity." Unpublished manuscript, Paris.
Gayre, Robert (1944). *Teuton and Slav on the Polish Frontier*. London: Eyre and Spottiswoode.
——— (1972). *Miscellaneous Racial Studies 1943–1972*. Vol. 2. Edinburgh: The Armorial.
Gesellschaft für biologische Anthropologie, Eugenik und Verhaltensforschung (1987). *Ungewöhnliche Fragen*. Hamburg: Neue Anthropologie.
Geyer, Horst (1937). "Der internationale Kongreß für Bevölkerungswissenschaft 1937 in Paris." *Der Erbarzt* 4:124–126.
Ghent, William J. (1902). *Our Benevolent Feudalism*. New York: Macmillan.
Gilham, Nicholas W. (2001). *A Life of Sir Francis Galton: From African Explorer to the Birth of Eugenics*. New York; Oxford: Oxford University Press.
Gillette, Aaron (2007). *Eugenics and the Nature-Nurture Debate in the Twentieth Century*. New York: Palgrave Macmillan.

Gilman, Robbins (1914). "Better Babies." In Race Betterment Foundation, ed. *Proceedings of the First National Conference on Race Betterment*. Battle Creek: Gage Printing Company, pp. 272–278.

Gini, Corrado (1933). "Eröffnungsrede." In Corrado Gini, ed. *Proceedings of the International Congress for Studies on Population Rome 1931*, Vol. 1. Rome, pp. xlvii–lvii.

——— (1934a) *Report of the Committee for the Study of the Eugenic and Dysgenic Effects of War*. In Henry F. Perkins, *A Decade of Progress in Eugenics. Scientific Papers of the Third International Congress of Eugenics*. Baltimore: Williams & Wilkins, pp. 231–243.

——— (1934b). "Response to the Presidential Address." In Henry F. Perkins, ed. *A Decade of Progress in Eugenics. Scientific Papers of the Third International Congress of Eugenics*. Baltimore: Williams & Wilkins, pp. 25–28.

——— (1936). "Lette alla Riunione delle Società di Eugenica dell'America Latina tenutasi a Messico il 12 Ottobre 1935." *Genius* 2:77–81.

——— (1967). "Race and Sociology." In Robert E. Kuttner, ed. *Race and Modern Science*. New York: Social Science Press, pp. 261–276.

Girard, Alain (1986). *L'Institut national d'études démographiques*. Paris: Editions de l'Institut National d'Etudes Démographiques.

——— (1987). "L'Institut National d'Etudes Demographiques." *Population* 42:715–718.

Glad, John (2006). *Future Human Evolution: Eugenics in the Twenty-First Century*. Schuykill Haven: Hermitage.

——— (2011). *Jewish Eugenics*. Washington, DC; London; Tel Aviv: Wooden Shore.

Glass, Bentley (1955). Genetics in the Service of Man. Annual Report of the Board of the Smithsonian Institution. Washington, pp. 299–315.

——— (1986). "Geneticists Embattled: Their Stand against Rampant Eugenics and Racism in America During the 1920s and 1930s." *Proceedings of the American Philosophical Society* 130:130–154.

Glass, David V. (1935). "The Berlin Population Congress and Recent Population Movements in Germany." *Eugenics Review* 27:207–212.

Goddard, Henry H. (1919). *Psychology of the Normal and Subnormal*. New York: Dodd, Mead and Company.

Goldberg, David Theo (1993). *Racist Culture: Philosophy and the Politics of Meaning*. Cambridge; Oxford. Wiley-Blackwell.

Goldschmidt, Richard (1960). *In and Out of the Ivory Tower*. Seattle: University of Washington Press.

González, Armando García, Raquel Álvarez Peláez and Consuelo Naranjo Orovio (1999). *En busca de la raza perfecta: Eugenesia e higiene en Cuba (1898–1958)*. Madrid: Consejo Superior de Investigaciones Científicas.

Gordon, Linda (1974). "The Politics of Population: Birth Control and the Eugenics Movement." *Radical America*, vol. 8, pp. 61–97.

Gordon, Robert A. (1975). "Crime and Cognition: An Evolutionary Perspective." In International Center for Biological and Medico-Forensic Criminology, ed. *Proceedings of the II International Symposium on Criminology*, Vol. 4. Sao Paulo, pp. 7–55.

——— (1980). "Research on IQ, Race, and Delinquency: Taboo or Not Taboo?" In Edward Sagarin, ed. *Taboos in Criminology*. Beverly Hills: Sage, pp. 37–66.

——— (1986). "IQ-Commensurability of Black-White Differences in Crime and Delinquency." Unpublished manuscript, Baltimore.

——— (1987). "SES Versus IQ in the Race-IQ-Delinquency Model." *International Journal of Sociology and Social Policy* 7:30–96.

Gossett, Thomas F. (1997). *Race: The History of an Idea in America*. New York: Oxford University Press.

Gould, Stephen Jay (1979). "Racist Arguments and IQ." In Stephen Jay Gould, ed. *Ever Since Darwin: Reflections in Natural History*. New York: Norton, pp. 243–247.

Graham, Loren R. (1977). "Science and Values: The Eugenics Movement in Germany and Russia in the 1920s." *American Historical Review* 82:1135–1164.
——— (1981). *Between Science and Values*. New York, NY: Columbia University Press.
Grant, Madison (1916). *Passing of the Great Race: Or, the Racial Basis of European History*. New York: Scribner's Sons.
Graßmann, Paul (1936). "Friedenspreis für einen Deutschen?" *Berliner Illustrierte Nachtausgabe*, October 21, 1936.
Grebenik, E. (1959). "The Development of Demography in Great Britain." In Philip M. Hauser and Otis Dudley Duncan, eds. *The Study of Population*. Chicago: University Press, pp. 190–202.
——— (1991). "Demographic Research in Britain 1936–1986." *Population Studies* 45:3–29.
Gregor, A. James (1960). "Report on the Nineteenth International Congress of the Institute of Sociology in Mexico City." *Mankind Quarterly* 1:128.
——— (1982). "The Development of National Socialist Racial Theories, 1919–1945." *The National Socialist*, no. 3: 14–23.
Griffiths, Richard M. (1980). *Fellow Travellers of the Right: British Enthusiasts for Nazi Germany 1933–1939*. Oxford: Constable.
Grimm, Christian (2011). *Netzwerke der Forschung. Die historische Eugenikbewegung und die moderne Humangenomik im Vergleich*. Berlin: Logos Verlag.
Grobig, Hermann Ernst (1943). "Warum Rassenhygiene im Krieg?" *Archiv für Rassen-und Gesellschaftsbiologie* 37:175–179.
Grosch-Obenauer, Dagmar (1986). "Hermann Muckermann und die Eugenik." Mainz: Dissertation Johannes Gutenberg-Universität
Groß, Walter (1933). "Rassenkunde und Erbgesundheitslehre." *Der Jungdrogist*, December 15, 1933.
——— (1935a). "Die Bevölkerungs-und Rassenpolitik des neuen Deutschlands." *Rassenpolitische Auslands-Korrespondenz*, Issue 3/1935, pp. 1–6.
——— (1935b). "Was bedeutet die Rassen-und Vererbungswissenschaft für den Nationalsozialismus." *N. S. Beamten-Zeitung*, October 11, 1935.
——— (1936). "Der geistige Kampf um die Rassenpflege." Hans Harmsen and Franz Lohse, eds. *Bevölkerungsfragen. Bericht des Internationalen Kongresses für Bevölkerungswissenschaft, Berlin 8/26–9/1/1935*. Munich: J.F. Lehmanns, pp. 660–663.
——— (1939). *Der deutsche Rassegedanke und die Welt*. Berlin: Junker und Dünnhaupt.
——— (1940). "Rassen-und Bevölkerungspolitik im Kriege." *Nachrichtendienst der Reichsfrauenführung*, Issue 8/1940, pp. 13–16.
Grotjahn, Alfred (1925). "Leitsätze zur sozialen und degenerativen Hygiene." Unpublished Manuscript, Berlin, Karlsruhe.
Gruber, Max (1909). "Vererbung, Auslese und Hygiene." *Deutsche Medizinische Wochenschrift* 35:1193–1196.
Gruber, Max von, and Ernst Rüdin, eds. (1911). *Fortpflanzung, Vererbung, Rassenhygiene. Katalog der Gruppe Rassenhygiene der Internationalen Hygiene-Ausstellung 1911 in Dresden*. Munich: J. F. Lehmanns.
Günther, Hans F. K. (1922). *Rassenkunde des deutschen Volkes*. Munich: J. F. Lehmanns.
Gütt, Arthur (1936a). "Bevölkerungspolitik als Aufgabe des Staates." In Hans Harmsen and Franz Lohse, eds. *Bevölkerungsfragen. Bericht des Internationalen Kongresses für Bevölkerungswissenschaft, Berlin 8/26–9/1/1935*. Munich: J. F. Lehmanns, pp. 745–754.
——— (1936b). "Europas rassische Bedrohung durch den Bolschewismus." *Völkischer Beobachter*, March 28, 1936.
Gütt, Arthur, Ernst Rüdin, and Falk Ruttke (1936). *Gesetz zur Verhütung erbkranken Nachwuchses vom 14. Juli 1933 nebst Ausführungsverordnungen*. 2nd ed. Munich: J. F. Lehmanns.
Haeckel, Ernst (1914). "Weltkrieg und Naturgeschichte." *Nord und Süd* 38:140–147.

——— Haeckel, Ernst, and Thomas Seltzer (1916). *Eternity: World-War Thoughts on Life and Death, Religion, and the Theory of Evolution.* New York: The Truth Seeker Co..
Haldane, John B. S. (1936). "A New Approach to Eugenics. Out of the Night by H. J. Muller." *Listener,* July 8, 1936.
Haller, Mark H. (1963). *Eugenics: Hereditarian Attitudes in American Thought.* New Brunswick, N.J.: Rutgers University Press.
Haller, Michael (1981). "Europa erwache! Frankreichs Neue Rechte und die Internationale des soziobiologischen Rassismus." In Paul Lersch, ed. *Die verkannte Gefahr. Rechtsradikalismus in der Bundesrepublik.* Reinbek: Rowohlt, pp. 83–104.
Hall, Lesley A. (1990). "The Eugenics Society Archives in the Contemporary Medical Archives Centre." *Medical History* 34:327–333.
Hall, Prescott F. (1919). "Immigration Restriction and World Eugenics." *Journal of Heredity* 10:125–127.
Hankins, Frank H. (1937). "German Policies for Increasing Births." *The American Journal of Sociology* 42:630–652.
Hansen, Bent Sigurd (1996). "Something Rotten in the State of Denmark: Eugenics and the Ascent of the Welfare State." In Gunnar Broberg and Nils Roll-Hansen, eds. *Eugenics and the Welfare State: Sterilization Policy in Denmark, Sweden, Norway, and Finland.* East Lensing: Michigan State University Press, pp. 9–76.
Hansen, Randall, and Desmond King (2001). "Eugenics Ideas, Political Interests, and Policy Variance. Immigration and Sterilization Policy in Britain and the U.S." *World Politics* 53:237–263.
Harmsen, Hans (1935a). "Berichte. Der internationale Kongreß für Bevölkerungswissenschaft und der Internationale Strafrechtskongreß." *Archiv für Bevölkerungswissenschaft und Bevölkerungspolitik* 5:355–368.
——— (1935b). "Ergebnisse des Berliner Strafrechts-und Gefängniskongresses 1935." *Soziale Praxis* 44:1205–1212.
——— (1936). "Die Durchführung des Gesetzes zur Verhütung erbkranken Nachwuchses in den Anstalten der Inneren Mission." *Zeitschrift für psychische Hygiene* 9:20–22.
Harmsen, Hans, and Franz Lohse (1936). "Vorwort." In Hans Harmsen and Franz Lohse, eds. *Bevölkerungsfragen. Bericht des Internationalen Kongresses für Bevölkerungswissenschaft, Berlin 8/26–9/1/1935.* Munich: J. F. Lehmanns, pp. vi–viii.
Harris, Geoffrey (1994). *The Dark Side of Europe: The Extreme Right Today.* Edinburgh: Edinburgh University Press
Hart, Bradley W. (2012). "Watching the 'Eugenic Experiment' Unfold: The Mixed Views of British Eugenicists toward Nazi." *Journal of the History of Biology* 45:33–63.
Harten, Hans-Christian, Uwe Neirich, and Matthias Schwerendt (2006). *Rassenhygiene als Erziehungsideologie des Dritten Reiches.* Berlin: Akad. Verlag.
Hartmann, Betsy (1987). *Reproductive Rights and Wrongs.* New York, NY: Harper & Row.
Hassencahl, Frances Janet (1970). "Harry H. Laughlin, 'Expert Eugenics Agent' for the House Committee on Immigration and Naturalization, 1921–1931." PhD Diss., Case Western Reserve University.
Hauser, Philip M. (1962). "Our Population Crisis is Here and Now." *The Reader's Digest,* 2, 147–148.
Heidenreich, Gert, and Julian Wetzel (1989)." Die organisierte Verwirrung. Nationale und internationale Verbindungen im rechtsextremistischen Spektrum." In Wolfgang Benz, ed. *Rechtsextremismus in der Bundesrepublik.* Frankfurt a.M.: Fischer, pp. 151–168.
Heim, Susanne, and Ulrike Schaz (1993). *Bevölkerungsexplosion. Marketing einer Ideologie.* Hamburg: Finrage.
——— (1994). "Das Revolutionärste, was die Vereinigten Staaten je gemacht haben. Vom Aufstieg des Überbevölkerungsdogmas." In Christa Wichterich, ed. *Menschen nach Maß. Bevölkerungspolitik in Nord und Süd.* Göttingen: Larnuv, pp. 129–150.

Heim, Susanne, and Ulrike Schaz (1996). *Berechnung und Beschwörung. Überbevölkerung— Kritik einer Debatte*. Berlin: Schwarze Risse Buchhandlung und Verlag.
Herren, Madeleine (2009). *Internationale Organisationen seit 1865. Eine Globalgeschichte der internationalen Ordnung*. Darmstadt: WBG.
Herrnstein Richard J. (1971). "I. Q." *The Atlantic Monthly*, Issue 9/1971, pp. 43–64.
Herrnstein, Richard J., and Charles Murray (1994). *The Bell Curve: Intelligence and Class Structure in American Life*. New York: Free Press.
Herwerden, Mari Anne van (1926). *Erfelijkheid bij den Mensch en Eugenetiek*. Amsterdam: Wereldnbibliotheek.
——— (1933). "Eugenics Abroad. VII. in Holland." *Eugenics Review* 25:175–177.
Hesse, Hans (2001). *Augen aus Auschwitz. Ein Lehrstück über nationalsozialistischen Rassenwahn und medizinische Forschung. Der Fall Dr. Karin Magnussen*. Essen: Klartext-Verlag.
Heuer, Bernd (1989). "Eugenik/Rassenhygiene in USA und Deutschland—ein Vergleich anhand des 'Journal of Heredity' und des 'Archiv für Rassen-und Gesellschaftsbiologie' zwischen 1910 bzw. 1904 und 1939 bzw. 1933." Diss. an der Universität Bonn, Bonn.
Hietala, Marjatta (1996). "From Race Hygiene to Sterilization: The Eugenics Movement in Finland." In Gunnar Broberg and Nils Roll-Hansen, eds. *Eugenics and the Welfare State: Sterilization Policy in Denmark, Sweden, Norway, and Finland*. East Lansing: Michigan State University Press, pp. 195–258.
Hilpert, Paul (1933). "Grundsätzliches über Rassenhygiene." *Heimat und Arbeit*, vol. 6, pp. 97–115.
Hitler, Adolf (1974). *Mein Kampf*. London: Hutchinson.
Hodgson, Dennis (1991). "The Ideological Origins of the Population Association of America." *Population and Development Review* 17:1–34.
Hodson, Cora (1935). "Eugenics in Norway." *Eugenics Review* 27:41–44.
——— (1936). "International Federation of Eugenic Organization. Report of the 1936 Conference." *Eugenics Review* 28:217–219.
Hoffmann, Frederick L. (1912). "Maternity Statistics of the State of Rhode Island." In Eugencis Education Society, ed. *Problems in Eugenics*. London: The Eugenics Education Society, pp. 334–342.
Hoffmann, Geza von (1916): *Krieg und Rassenhygiene. Die bevölkerungspolitische Aufgabe nach dem Kriege*. München: J. F. Lehmanns.
Hofstadter, Richard (1955). *Social Darwinism in American Thought*. Boston: Beacon Press.
Höhn, Charlotte (1987). "Grundsatzfragen in der Entstehungsgeschichte der Internationalen Union für Bevölkerungswissenschaft (IUSSP/IUSIPP). Beitrag zur Tagung 'Bevölkerungsentwicklung und Bevölkerungstheorie in Geschichte und Gegenwart', 17–20.3.1987 in Berlin." Unpublished manuscript, Wiesbaden.
——— (1989). "Grundsatzfragen in der Entstehungsgeschichte der Internationalen Union für Bevölkerungswissenschaft (IUSSP/IUSIPP)." In Rainer Mackensen, Lydia Thilt Thoret, and Ulrich Stark, eds. *Bevölkerungsentwicklung und Bevölkerungstheorie in Geschichte und Gegenwart*. Frankfurt a.M.; New York: Campus, pp. 233–254.
Hogben, Lancelot (1931). "The Genetic Analysis of Family Traits." *Journal of Genetics*, 25, 97–112, 211–240, 293–314.
Hunt, Harrison Randall (1930). *Some Biological Aspects of War*. New York; Galton Publishing Co.
——— (1934). *Is War Dysgenic? A Decade of Progress in Eugenics*. Baltimore, pp. 244–248.
Huxley, Julian. (1927). "Too Many People Seen as Real World Danger." *The New York Times*, October 16, 1927, p. 6.
——— (1936). "Eugenics and Society." *Eugenics Review* 28:11–31.
——— (1946). *UNESCO: Its Purpose and Its Philosophy*. Paris: UNESCO.
——— (1963). "Eugenics in Evolutionary Perspective." *Perspective in Biology and Medicine* 6:155–187.

―――― (1981). *Ein Leben für die Zukunft. Erinnerungen.* Munich: Dt. Taschenbuch Verlag.
Huxley, Julian pp., A. C. Haddon, and Alexander Carr-Saunders (1935). *We Europeans.* London: CAPE.
IFEO (1930). *International Federation of Eugenic Organizations.* Report of the Ninth Conference. London.
―――― (1934). Bericht über die 11. "Versammlung der Internationalen Föderation eugenischer Organisationen. Konferenzsitzungen vom 18. bis 21. Juli 1934 im Waldhaus Dolden, Zürich." Zürich.
―――― (1936). "Bericht der 12. Versammlung der Internationalen Föderation Eugenischer Organisationen. Konferenzsitzungen vom 15. bis 20. Juli 1936 Scheveningen, Holland." Den Haag.
Inge, W. R. (1914). "Depopulation." *Eugenics Review* 6:261.
Itzkoff, Seymour W. (2006). "Preface." In John Glad, ed. *Future Human Evolution: Eugenics in the Twenty–First Century.* Schuykill Haven: Hermitage Publishers, pp. 7–12.
IUSSP (1993). *Directory of Members.* Lüttich: International Union for the Scientific Study of Population.
Jackson, John P. (2005). *Science for Segregation: Race, Law, and the Case against Brown v. Board of Education.* New York; NYU Press.
Jaschke, Hans-Gerd (1990). "Frankreich." In Franz Gress, Hans-Gerd Jaschke, and Klaus Schönekäs, eds. *Neue Rechte und Rechtsextremismus in Europa.* Opladen: Westdeutscher Verlag, pp. 17–103.
Jennings, Herbert. (1942). "Raymond Pearl, 1879–1940." In *Biographical Memoirs of the National Academy of Science* 22:295–310.
Jensen, Arthur R. (1969). "How Much Can We Boost IQ and Scholastic Achieve-ment?" *Harvard Educational Review* 33:1–123.
―――― (1972). *Genetics and Education.* London: Methuen.
―――― (1973). *Educability and Group Differences.* London: Methuan.
―――― (1975). "Rasse und Begabung (Interview)." *Nation Europa,* Issue 9/1975, pp. 19–28.
―――― (1985). "The Nature of the Black-White Difference on Various Psy-chometric Tests: Spearman's Hypothesis." *Behavioral and Brain Sciences* 8:193–263.
―――― (1992). "Preface." In Roger Pearson, ed. *Shockley on Eugenics and Race.* Washington: Scott Townsend Pub., pp. 1–14.
Johnson, Roswell H. (1914). "Eugenics and So-Called Eugenics." *American Journal of Sociology* 20:98.
―――― (1915). "Natural Selection in War." *Journal of Heredity* 6:546–548.
―――― (1934a). "Differences in the Evolutionary Situation in Various Nations—An Aspect of International Eugenics." *University of Pittsburgh Bulletin* 31:179–187.
―――― (1934b). "International Eugenics." Unpublished dissertation. University of Pittsburgh, Pittsburgh.
Johnson, Roswell H., and Paul Popenoe (1918). *Applied Eugenics.* New York: The Macmillan Company.
―――― (1933). *Applied Eugenics. 3. Aufl.* New York: Macmillan.
Jones, Greta (1980). Social Darwinism and English Thought: The Interaction between Biological and Social Theory. Sussex: Harvester Press
―――― (1986). Social Hygiene in Twentieth Century Britain. London: Taylor & Francis.
―――― (1988). *Science, Politics and the Cold War.* London; New York: Routledge.
Jordan, David Starr (1910). "War and Manhood." *Eugenics Review* 2:95–109.
―――― (1913a). "The Eugenics of War." *Eugenics Review* 5:197–213.
―――― (1913b). "The Eugenics of War." *Journal of Heredity* 4:140–147.
Joseph, Jay (2004). *The Gene Illusion: Genetic Research in Psychiatry and Psychology Under the Microscope.* New York: Algora Publishing.
―――― (2006). "The 1942 'Euthanasia' Debate in the American Journal of Psychiatry." *History of Psychiatry* 16:171–179.

Jürgens, Hans Wilhelm (1982). "Die Weltbevölkerung im Jahre 2000." *Der Überblick*, Issue 3/1982, pp. 54–57.
Kaminsky, Uwe (2008). "Die NS-'Euthanasie'. Ein Forschungsüberblick." In Klaus-Dietmar Henke, ed. *Tödliche Medizin im Nationalsozialismus. Von der Rassenyhgiene zum Massenmord*. Köln; Weimar; Vienna: Böhlau, pp. 269–290.
Kankeleit, Otto (1929). *Die Unfruchtbarmachung aus rassenhygienischen und sozialen Gründen*. Munich: J. F. Lehmanns.
——— (1931). "Die Ausschaltung geistig Minderwertiger von der Fortpflanzung." *Volk und Rasse* 6:174–179.
Kaplan, Arnold R. (1963). "Biology, Politics and Race." *Eugenics Quarterly* 10:188–190.
Kater, Michael H. (1990). *Doctors Under Hitler*. Chapel Hill; London;: The University of North Carolina Press.
Kaufmann, Doris (2003). "'Rasse und Kultur'. Die amerikanische Kulturanthropologie um Franz Boas (1858–1942) in der ersten Hälfte des 20. Jahrhunderts—ein Gegenentwurf zur Rassenforschung in Deutschland." In Hans–Walter Schmuhl, ed. *Rassenforschung an Kaiser–Wilhelm–Instituten vor und nach 1933*. Göttingen: Wallstein, pp. 309–327.
Kaup, Ignaz (1922). *Volkshygiene oder selektive Rassenhygiene*. Leipzig: S. Hirzel.
Keiter, Friedrich (1936). "Fortpflanzungsunterschiede innerhalb des Standes und ihre rassenhygienische Bedeutung." In Hans Harmsen and Franz Lohse, eds. *Bevölkerungsfragen. Bericht des Internationalen Kongresses für Bevölkerungswissenschaft, Berlin 8/26–9/1/1935*. Munich: J. F. Lehmanns, pp. 582–587.
——— (1941). "Krieg und Gegenauslese." *Das Reich*, December 7, 1941.
Kellogg, Vernon L. (1912). "Eugenics and Militarism." In Eugenics Education Society, ed. *Problems in Eugenics: Papers Communicated to the First International Eugenics Congress 1912*. London: The Eugenics Education Society, pp. 220–231.
——— (1913). "Eugenics and Militarism." *Atlantic Monthly* 112:99–108.
——— (1914). "The Bionomics of War: Race Modification by Military Selection." *Social Hygiene* 1:44–52.
Kemp, Tage (1932). "The Significance of Blood-Grouping in Anthropology." In George H. L. F. Pitt-Rivers, ed. *Problems of Population; being the Report of the Proceedings of the Second General Assembly of the IUSIPP, London 1931*. London: George Allen & Unwin, pp. 311–314.
Kemp, Tage, Mogens Hauge, and Bent Harvald (1956). *The First International Congress of Human Genetics*. Basel; New York: Karger.
Kemper, Andreas (2011). *Plagiat bei Sarrazin?* Münster: Blog von Andreas Kemper, November 30, 2011.
——— (2012). "Sarrazins deutschsprachige Quellen." In Michael Haller and Martin Niggeschmidt, eds. *Der Mythos vom Niedergang der Intelligenz. Von Galton zu Sarrazin: Die Denkmuster und Denkfehler der Eugenik*. Wiesbaden: VS Verlag für Sozialwissenschaften, pp. 49–70.
Kennedy, David M. (1970). *Birth Control in America: The Career of Margaret Sanger*. New Haven; London: Yale University Press.
Kenny, Michael G. (2002). "Toward a Racial Abyss: Eugenics, Wickliffe Draper, and the Origins of the Pioneer Fund." *Journal of History of the Behavioral Sciences* 38: 259–283.
Kerr, Anne, and Tom Shakespear (2002). *Genetic Politics: From Eugenics to Genome*. Cheltenham: New Clarion Press.
Kershaw, Ian (1995). "The Extinction of Human Rights in Nazi Germany." In Olwen Hufton, ed. *Historical Change and Human Rights: The Oxford Amnesty Lectures 1994*. New York: Basic Books, pp. 217–246.
Kevles, Daniel J. (1985). *In the Name of Eugenics: Genetics and the Use of Human Heredity*. Berkeley; Los Angeles: University of California Press.

―― (1992). "Out of Eugenics: The Historical Politics of the Human Genom." In Daniel J. Kevles and Leroy Hood, eds. *The Code of Codes. Scientific and Social Issues in the Human Genom Project*. Boston: Harvard University Press, pp. 3–36.
―― (2004). "International Eugenics." In Dieter Kuntz, ed. *Deadly Medicine: Creating the Master Race*. Washington, DC: University of North Carolina Press, pp. 41–59.
Kiesel, Hermann (1971). "Race, the 'Nation of Europe' and Ideology: A Critique." *Mankind Quarterly* 12:111–113.
Kilpatrick, James Jackson (1962). *The Southern Case for School Segregation*. New York: Crowell-Collier.
Kiminus, Manuela (2002). "Ernst Rodenwaldt. Leben und Werk." Heidelberg: Dissertation an der Universität Heidelberg.
King, Desmond (1995). *Separate and Unequal: Black Americans and the US Federal Government*. New York; Oxford: Oxford University Press.
―― (1999). *In the Name of Liberalism: Illiberal Social Policy in the USA and Britain*. Oxford; New York: Oxford University Press.
Kingsland, Sharon (1984). "Raymond Pearl: On the Frontier of the 1920's." *Human Biology* 56:1–18.
―― (1988). "Evolution and Debates over Human Progress from Darwin to Sociobiology." In Michael Teitelbaum and Jay M. Winter, eds. *Population and Resources in Western Intellectual Traditions*. New York: Population Council, pp. 167–198.
Kiser, Clyde V. (1953). "The Population Association Comes of Age." *Eugenical News* 38:107–111.
―― (1981). "The Role of the Milbank Memorial Fund in the Early History of the Association." *Population Index* 47:490–494.
Klee, Ernst (1983). *'Euthanasie' im NS-Staat. Die 'Vernichtung lebensunwerten Lebens.'* Frankfurt a.M.: Fischer.
Klineberg, Otto (1931). *"A Study of Psychological Differences Between Racial and National Groups in Europe."* New York: Archives of Psychology.
Koch, Gerhard (1993). *Humangenetik und Neuro-Psychiatrie in meiner Zeit (1932 bis 1978). Jahre der Entscheidung*. Erlangen; Jena: Palm & Enke.
Koch, Lene (1994). "On Danish Human Genetics and German Racial Hygiene in the 1930s and 1940s." Unpublished manuscript, Kopenhagen.
―― (1996). *Racehygiejne i Danmark 1920–56*. Kopenhagen: Gyldendal.
―― (2006). "Eugenic Sterilisation in Scandinavia." *The European Legacy* 11:372–392.
Kohn, Marek (1996). *The Race Gallery: The Return of Racial Science*. London: Vintage.
Koller, Siegfried (1936). "Zur Frage der Erb-und Umweltbedingtheit beruflicher Fruchtbarkeitsunterschiede." In Hans Harmsen and Franz Lohse, eds. *Bevölkerungsfragen. Bericht des Internationalen Kongresses für Bevölkerungswissenschaft, Berlin 8/26–9/1/1935*. Munich: J. F. Lehmanns, pp. 576–580.
Köbsell, Swantje (1994). "Die Guten ins Töpfchen, die Schlechten...? Alte und neue Eugenik in Deutschland." In Christa Wichterich, ed. *Menschen nach Maß. Bevölkerungspolitik in Nord und Süd*. Göttingen: Larnuv, pp. 85–106.
Köhn-Behrens, Charlotte (1934). *Was ist Rasse? Gespräche mit den größten deutschen Forschern der Gegenwart*. Munich: Zentralverlag der NSDAP, Frz. Eher Nachf.
Kranz, Harald (1984). "Rassenhygiene/Eugenik in Deutschland. Institutionalisierung und Politisierung einer Wissenschaft (1927–1945)." Unpublished master thesis, Bielefeld: Diplomarbeit an der Universität Bielefeld.
Krementsov, Nikolai (2010). "Eugenics in Russia and the Soviet Union." In Alison Bashford and Philippa Levine, eds. *The Oxford Handbook of the History of Eugenics*. New York; Oxford: Oxford University Press, pp. 413–429.
Kroll, Jürgen (1983). "Zur Entstehung und Institutionalisierung einer naturwissenschaftlichen und sozialpolitischen Bewegung: Die Entwicklung der Eugenik/Rassenhygiene

bis zum Jahre 1933." Unpublished dissertation. Tübingen: Diss. an der Universität Tübingen.

Kroll, Jürgen, and Peter Weingart (1989). "BevölkerungsWissenschaft und Rassenhygiene vor 1930 in Deutschland." In Rainer Mackensen, Lydia Thilt Thoret, and Ulrich Stark, eds. *Bevölkerungsentwicklung und Bevölkerungstheorie in Geschichte und Gegenwart.* Frankfurt a.M.; New York: Campus Verlag, pp. 215–232.

Kröner, Hans-Peter (1980). "Die Eugenik in Deutschland von 1891–1934." Unpublished dissertation, Münster: Diss. an der Universität Münster.

——— (1998). *Von der Rassenhygiene zur Humangenetik. Das Kaiser–Wilhelm–Institut für Anthropologie, menschliche Erblehre und Eugenik nach dem Kriege.* Stuttgart: g. Fischer.

Krüger, Arnd (1998). "A Horse Breeder's Perspective: Scientific Racism in Germany, 1870–1933." In Norbert Finzsch and Dietmar Schirmer, eds. *Identity and Intolerance: Nationalism, Racism, and Xenophobia in Germany and The United States.* Cambridge: Cambridge University Press, pp. 371–395.

Kunz, Gabriele (1989). "Geburtenkontrolle in Puerto Rico. Die Forschungsförderung des Population Council." In: *1999 – Zeitschrift für Sozialgeschichte des 20. und 21. Jahrhunderts,* 4:35–51. 77

Kuper, Leo, ed. (1975). *"Race, Science and Society".* Paris: The Unesco Press.

Kuttner, Robert E., ed. (1967). *Race and Modern Science.* New York: Social Science Press.

——— (1963). "Eugenics Aspects of Preventive Therapy for Mental Retardation." *Genius,* 19, 1–9.

——— (1960). "The Herd Instinct in Modern Sociology." *Mankind Quarterly* 1:105–109.

——— (1962). "Biopolitik." *Nation Europa,* Issue 12/1962, pp. 39–42.

Kühl, Stefan (1994). *The Nazi Connection. Eugenics, American Racism, and German National Socialism.* New York; Oxford: Oxford University Press.

——— (1998). "The Cooperation of German Racial Hygienists and American Eugenicists before and after 1933." In Michael Berenbaum and Abraham J. Peck, eds. *The Holocaust and History: The Known, the Unknown, the Disputed, and the Reexamined.* Bloomington: Indiana University Press, pp. 134–152.

——— (1999). "Die soziale Konstruktion von Wissenschaftlichkeit und Unwissenschaftlichkeit in der internationalen eugenischen Bewegung." In Heidrun Kaupen–Haas and Christian Saller, eds. *Wissenschaftlicher Rassismus. Analysen einer Kontinuität in den Human– und Naturwissenschaften.* Frankfurt a.M.; New York: Campus Verlag, pp. 111–121.

——— (2001). "The Relationship between Eugenics and the so–called 'Euthanasia Action' in Nazi Germany: A Eugenically Motivated Peace Policy and the Killing of the Mentally Handicapped during the Second World War." In Margit Szöllösi-Janze, ed. *Science in the Third Reich.* Oxford; New York: Berg, pp. 185–210.

Kürten, Heinz (1934). "Der Kongreß der Internationalen Föderation Eugenischer Organisationen in Zürich." *Ziel und Weg* 4:598–599.

Labisch, Alfons (1990). "Kritisches Essay." *Medizinhistorisches Journal* 25:336–349.

Ladd – Taylor, Molly (2001). "Eugenics, Sterilisation and Modern Marriage in the USA. The Strange Career of Paul Popenoe." *Gender and History* 13:298–327.

——— (2004). "The 'Sociological Advantages' of Sterilization. Fiscal Policies and Feeble–Minded Women in Interwar Minnesota." In Steven Noll and James W. Trent, eds. *Mental Retardation in America.* New York: NYU Press, pp. 281–299.

Landman, Jacob H. (1932). *Human Sterilization: The History of the Sexual Sterilization Movement.* New York: Macmillan.

——— (1936). "Sterilization and Social Betterment." *Survey Graphic* 25:162–163, 190.

Lane, Charles (1994). "The Tainted Sources of 'The Bell Curve.'" *New York Review of Books,* January 12, 1994, pp. 14–19.

Lange, Astrid (1993). *Was die Rechten lesen. Fünfzig rechtsextreme Zeitschriften.* Munich: Beck.

Lantzer, Jason pp. (2011). "The Indiana Way of Eugenics: Sterilization Laws, 1907–74." In Paul A. Lombardo, ed. *A Cenutry of Eugenics in America: From the Indiana Experiment to the Human Genome Era.* Bloomington: Indiana University Press, pp. 11–25.

Largent, Mark A. (2002). "'The Greatest Curse of the Race'. Eugenic Sterilization in Oregon, 1909–1983." *Oregon Historical Quarterly* 103:188–209.

——— (2008). *Breeding Contempt: The History of Coerced Sterilization in the United States.* New Brunswick; London: Rutgers University Press.

Laughlin, Harry H. (1934). "Historical Background of the Third International Congress of Eugenics." In Henry F. Perkins, ed. *A Decade of Progress in Eugenics.* Baltimore: Williams & Wilkins, pp. 1–14.

Lebrun, Marc (1985). *The IUSSIP in History. From Margaret Sanger to Mercedes Concepcion.* Liege: IUSSIP.

Lenz, Fritz (1914). "Bund zur Erhaltung und Mehrung der deutschen Volkskraft." *Archiv für Rassen- und Gesellschaftsbiologie* 11:555–557.

——— (1916). "Aus der Gesellschaft für Rassenhygiene." In *Archiv für Rassen-und Gesellschaftsbiologie* 12:403–410.

——— (1923). *Menschliche Auslese und Rassenhygiene.* Munich: J. F. Lehmanns.

——— (1924). "Eugenics in Germany." *Journal of Heredity* 15:223–231.

——— (1931a). "Die Stellung des Nationalsozialismus zur Rassenhygiene." *Archiv für Rassen-und Gesellschaftsbiologie* 25:300–308.

——— (1931b). *Menschliche Auslese und Rassenhygiene.* 3rd ed. Munich: J. F. Lehmanns.

——— (1956). "Über die Grenzen praktischer Eugenik." *Acta genetica* 6:13–24.

Leon, Sharon M. (2004). "'Hopelessly Entangled in Nordic–Pre-suppositions': Catholic Participation in the American Eugenics Society in the 1920s." *Journal of the History of Medicine and Allied Sciences* 59:3–49.

Leonard, Jacques (1983). "Le premier Congres international d'Eugenique et ses conséquences francaises." *Histoire des sciences medicales* 17:141–146.

Lerch, Eugen (1950). "Der Rassenwahn: Von Gobineau zur UNESCO-Erklärung." *Der Monat*, 3, 157–174.

Levine, Philippa, and Alison Bashford (2010). "Introduction: Eugenics and the Modern World." In Alison Bashford and Levine Philippa, eds. *The Oxford Handbook of the History of Eugenics.* New York, Oxford: Oxford University Press, pp. 3–26.

Lifton, Robert Jay (1986). *The Nazi Doctors: Medical Killing and the Psychology of Genocide.* New York: Basic Books.

Lilienthal, Georg (1987). "Anthropologie und Nationalsozialismus: Das erb-und rassenkundliche Abstammungsgutachten." *Jahrbuch des Instituts für Geschichte der Medizin der Robert Bosch Stiftung* 6:71–91.

Linder, Arthur (1936). "Der bereinigte Geburtenüberschuß der schweizerischen Bevölkerung." In Hans Harmsen and Franz Lohse, eds. *Bevölkerungsfragen. Bericht des Internationalen Kongresses für Bevölkerungswissenschaft, Berlin 26.8. bis 1.9.1935.* Munich: J. F. Lehmanns, pp. 106–107.

Lindsay, J. A. (1918). "The Eugenic and Social Influence of the War." *Eugenics Review* 10:133–144.

Lipphardt, Veronika (2009), "'Jüdische Eugenik'? Deutsche Biowissenschaftler mit jüdischem Hintergrund und ihre Vorstellungen von Eugenik (1900–1935)." In Regina Wecker et al., Sabine Braunschweig und Gabriela Imboden eds. *Wie nationalsozialistisch ist die Eugenik? Internationale Debatten zur Geschichte der Eugenik im 20. Jahrhundert.* Wein; Köln; Weimar: Böhlau, pp. 151–164.

Little, Clarence C. (1922). "The Second International Congress of Eugenics." *Eugenics Review* 13:511–524.

Lombardo, Paul A. (2002). "'The American Breed': Nazi Eugenics and the Origins of the Pioneer Fund." *Albany Law Review* 65:743–830.

Lorimer, E. O. (1932). "The Vinderen Biological Laboratory." *Eugenics Review* 23:336–337.
Lorimer, Frank (1949). "Meeting of the International Population Union, Geneva, 1949." *Eugenical News* 34:46–47.
——— (1971). "The Role of the International Union for the Scientific Study of Population." *Milbank Memorial Fund Quarterly* 49:86–96.
——— (1981). "How the Demographers Saved the Association." *Population Index* 47:488–490.
Lorimer, Frank, and Frederick Osborn (1934). *Dynamics of Population: Social and Biological Significance of Changing Birth-Rates in the United States*. New York: The Macmillan Company.
Lösch, Niels C. (1990). "Das Kaiser-Wilhelm-Institut für Anthropologie, menschliche Erblehre und Eugenik." Master thesis, Berlin: Magisterarbeit an der Freien Universität.
——— (1997). *Rasse als Konstrukt. Leben und Werk Eugen Fischers*. Frankfurt a.M.: Lang.
Löscher, Monika (2009). "Zur katholischen Eugenik in Österreich." In Regina Wecker, Sabine Braunschweig and Gabriela Imboden, eds. *Wie nationalsozialistisch ist die Eugenik? Internationale Debatten zur Geschichte der Eugenik im 20. Jahrhundert*. Wein; Köln; Weimar: Böhlau, pp. 233–246.
Louçã, Franciso (2008). "Emancipation Through Interaction. How Eugenics and Statistics Converged and Diverged." *Journal of the History of Biology* 42:649–684.
Ludmerer, Kenneth M. (1969). "American Geneticists and the Eugenics Movement 1905–1935." *Journal of the History of Biology* 2:337–362.
——— (1972). *Genetics and American Society*. Baltimore; Johns Hopkins University Press.
Luhmann, Niklas (1972). *Rechtssoziologie*. Reinbek: Rowohlt.
——— (1991). *Soziologie des Risikos*. Berlin; New York: de Gruyter.
——— (1992). *Die Wissenschaft der Gesellschaft*. Frankfurt a.M.: suhrkamp taschenbuch Wissenschaft.
——— (1997). *Die Gesellschaft der Gesellschaft*. Frankfurt a.M.: Suhrkamp.
——— (2002). *Die Politik der Gesellschaft*. Frankfurt a.M.: Suhrkamp.
——— (2005a). "Der medizinische Code." In *Soziologische Aufklärung 1. Aufsätze zur Theorie sozialer Systeme. 7. Aufl*. Wiesbaden: VS Verlag für Sozialwissenschaften, pp. 176–188.
——— (2005b). "Selbststeuerung der Wissenschaft." In Niklas Luhmann, ed. *Soziologische Aufklärung 1. Aufsätze zur Theorie sozialer Systeme. 7. Aufl*. Wiesbaden: VS Verlag für Sozialwissenschaften, pp. 291–316.
——— (2005c). "Soziologie des politischen Systems." In Niklas Luhmann, ed. *Soziologische Aufklärung 1. Aufsätze zur Theorie sozialer Systeme*. 7. Aufl. Wiesbaden: VS Verlag für Sozialwissenschaften, pp. 194–223.
Lundborg, Herman (1926). "Die drohende Entartung gewisser Kulturvölker." *Zeitschrift für Volksaufartung und Erbkunde* 1:3–7.
——— (1931). "Die Rassenmischung beim Menschen." *Bibliographia Genetica* 8:1–221.
Lunde, Andres S. (1981). "The Beginning of the Population Association of America." *Population Index*, 47, 479–484.
Lutzhöft, Hans Jürgen (1971). *Der Nordische Gedanke in Deutschland 1920 bis 1940*. Stuttgart: Klett.
Lünen, Alexander von (2009). "'The Perfect Astronaut Would Be a Human Without Legs.' JBS Haldane and 'Positive Eugenics.'" In Regina Wecker, Sabine Braunschweig, and Gabriela Imbodeneds. *Wie nationalsozialistisch ist die Eugenik? Internationale Debatten zur Geschichte der Eugenik im 20. Jahrhundert*. Wein; Köln; Weimar: Böhlau, pp. 127–138.
Luxenburger, Hans (1931). "Möglichkeiten und Notwendigkeiten für die psychiatrisch-eugenische Praxis." *Münchener Medizinische Wochenschrift*, 78, 753–758.
Lynn, Richard (1974). "A New Morality from Science: Beyondism, By. R. B. Cattell." *Irish Journal of Psychology* 2:205–209.
——— (2001a). *Eugenics: A Reassessment*. Westport; London: Praeger Publishers.

―――― (2001b). *The Science of Human Diversity: A History of the Pioneer Fund*. Lanham, MD: University Press of America.
―――― (2008). *The Global Bell Curve: Race, IQ, and Inequality Worldwide*. Augusta, GA: Washington Summit Publishers.
Lyons, F. pp. L. (1963). *Internationalism in Europe*. Leyden: Sythoff.
Maas, Bonnie (1976). *Population Target: The Political Economy of Population Control in Latin America*. Ontario: Charters Publishing.
Macerakis, Kristie (1993). *Surviving the Swastika: Scientific Research in Nazi Germany*. New York; Oxford: Oxford University Press.
Macfie, Roland Campbell (1917b). "The Selective Effects of War." *The New Statesman*, vol. 8, pp. 441–442.
MacKenzie, Donald A. (1981). *"Statistics in Britain 1865–1930: The Social Construction of Scientific Knowledge."* Edinburgh: Edinburgh University Press.
Macura, Milos (1986). "The Significance oft he United Nations International Population Conference." *Population Bulletin of the United Nations*, 19 und 20, S. 14–26.
Maiocchi, Roberto (1999). *Scienza italiana e razzismo fascista*. Florenz: La Nuova Italia.
Mann, Gunter (1973). "Rassenhygiene Sozialdarwinismus." In Gunter Mann, ed. Biologismus im 19. Jahrhundert. Stuttgart: Enke, pp. 73–96.
―――― (1978). "Neue Wissenschaft im Rezeptionsbereich des Darwinismus. Eugenik, Rassenhygiene." *Berichte zur Wissenschaftsgeschichte* 1:101–111.
―――― (1980). "The Third International Eugenics Congress 1932." *Medizinhistorisches Journal* 15:337–339.
Mantovani, Clara (2004). *Rigenerare la società. L'eugenetica in Italia dalle origini ottocentesche agli anni Trenta*. Soveria Mannelli: Rubbettino Editore.
Maoláin, Ciarán Ó (1987). *The Radical Right: A World Directory*. Burnt Mill: Longman Group.
Maretzki, Thomas W. (1989). "The Documentation of Nazi Medicine by German Medical Sociologists: A Review Article." *Social Science and Medicine* 29:1319–1330.
Marten, Heinz-Georg (1983). *Sozialbiologismus. Biologische Grundpositionen der politischen Ideengeschichte*. Frankfurt a.M.; New York: Campus.
Martyn, Edith How (1933). "Contribution à l'histoire du Birth Control." *Le Problème Sexuel*, Issue 2/1933, pp. 29–33.
Massin, Benoit (1993). "De l'anthropologie physique libérale à la biologie raciale eugénico-nordiciste en Allemagne (1970–1914) Virchow—Luschan—Fischer." *Revue d'Allmagne* 25:387–404.
―――― (1996). "Enseignement universitaire de l'hygiene raciale et de la raciologie (1930–1945)." Unpublished manuscript, Paris.
―――― (2003). "Rasse und Vererbung als Beruf die Hauptforschungsrichtungen am Kaiser–Wilhelm–Institut für Anthropologie, Menschliche Erblehre und Eugenik im Nationalsozialismus." In Hans–Walter Schmuhl, ed. *Rassenforschung an Kaiser–Wilhelm–Instituten vor und nach 1933*. Göttingen: Wallstein, pp. 190–244.
Matiegka, J. (1935). "Die Geschichte des Problems von der Ungleichheit oder Gleichwertigkeit der europäischen Rassen." In Karel Weigner, ed. *Die Gleichwertigkeit der europäischen Rassen und die Wege zu ihrer Vervollkommnung*. Prag: Verl. d. Tschechischen Akademie d. Wissenschaften u. Kuenste : Verl. Orbis, pp. 13–27.
May, Ronald W. (1960). "Genetics and Subversion." *The Nation*, May 14, 1960, pp. 420–422.
Mayer, Thomas (2004). "Akademische Netzwerke um die Wiener Gesellschaft für Rassenpflege (Rassenhygiene) von 1924 bis 1948." Unpublished master thesis, Vienna: Diplomarbeit Universität Wien.
―――― (2007). "Familie, Rasse und Genetik. Deutschnationale Eugeniken im Österreich der Zwischenkriegszeit." In Gerhard Baader, Veronika Hofer, and Thomas Mayer,

eds. *Eugenik in Österreich. Biopolitische Strukturen von 1900–1945.* Vienna: Czernin, pp. 162–183.

Mazumdar, Pauline (1992). *Eugenics, Human Genetics and Human Failings: The Eugenics Society, its Sources and Critics in Britain.* New York: Routledge.

McDougall, William (1921). *"Is America Safe for Democracy?"* New York: C. Scribner's Sons.

McGurk, Frank C. J. (1956). "A Scientist's Report on Race Differences." *U.S. News and World Report,* September 21, 1956.

McKusick, Victor A. (1975). "The Growth and Development of Human Genetics as a Clinical Discipline." *American Journal of Human Genetics* 27:261–273.

Meehan, Mary (1993a). "Beijing & PP Affiliate Colluding." *National Catholic Register,* March 21, 1993.

——— (1993b). "Discredited by Nazis, Eugenics Quietly Lives On." *National Catholic Register,* April 25, 1993.

———(1993c). "Eugenics: Abortion's Precursor." *National Catholic Register,* May 16, 1993.

——— (1993d). "IPPF Links to Eugenics Uncovered." *National Catholic Register,* March 14, 1993.

——— (1993e). "Nazis Were Admired by Eugenicists." *National Catholic Register,* April 4, 1993.

———(1993f). "Rescuers Go Global for the Unborn." *National Catholic Register,* February 28, 1993.

———(1993g). "Since the 60's, Eugenics Makes Steady Progress." *National Catholic Register,* March 28, 1993.

——— (1994). "Pop Council Spins Its Web." *National Catholic Register,* January 9, 1994.

Mehler, Barry (1984). "Eugenics: Racist Ideology Makes." *Guardian,* August 22, 1984.

——— (1987). "Eliminating the Inferior." *Science for the People,* Issue 6/1987, pp. 14–18.

——— (1988). "A History of the American Eugenics Society, 1921–1940." Urbana-Champaign: Diss. an der University of Illinois.

——— (1989). "Foundation for Fascism: The New Eugenics Movement in the United States." *Patterns of Prejudice,* Issue 4/1989, pp. 17–25.

——— (1997). "Beyondism: Raymond B. Cattell and the New Eugenics." *Genetica* 99:153–163.

Melville, Colonel C. H. (1910). "Eugenics and Military Service." *Eugenics Review* 2:53–60.

Mendel, Gregor (1865). "Versuche über Pflanzenhybriden." *Verhandlungen des Naturforschenden Vereins Brunn* 4:3–47.

Merton, Robert K. (1973). "The Normative Structure of Science." In Robert, K. Merton, ed. *The Sociology of Science.* Chicago:University of Chicago Press, pp. 267–414.

Methorst, H. W. (1927). "Results of Differential Birth Rate in the Netherlands." In Margaret Sanger, ed. *Proceedings of the World Population Conference.* London: E. Arnold, pp. 169–190.

Metraux, Alfred (1950). "Unesco and the Racial Problem." *International Social Science Bulletin* 2:384–390.

Meyer, David pp. (2004). "Protest and Political Opportunities." *Annual Review of Sociology* 30:125–145.

Miles, Robert (1989). "Bedeutungskonstitution und der Begriff des Rassismus." *Das Argument,* Issue 175/1989, pp. 353–367.

Miller, Adam (1994). "Professors of Hate." *Rolling Stone,* October 20, 1994, pp. 107–114.

Mintz, Frank P. (1985). *The Liberty Lobby and the American Right: Race, Conspiracy, and Culture.* Westport; London: Greenwood Press.

Mitchell, Michael (1985). "The Origins of Eugenics and the American Eugenics Movement." In Robert D. Barnes and James D. Pickering, eds. *Nature versus Nur-ture.* Lanham: University Press of America, pp. 153–172.

Mitscherlich, Alexander, and Fred Mielke (1948). *Das Diktat der Menschenverachtung*. Heidelberg: L. Schneider.
Mjöen, Jon Alfred (1914). *Racehygiene*. Kristiania: Jacob Dybwads Forlag.
——— (1923). "Harmonie and Disharmonie Race Crossings." In Charles B. Davenport, ed. *Eugenics in Race and State. Vol. II. Scientific Papers of the Second International Conference of Eugenics*. Baltimore: Williams and Wilkins, pp. 41–61.
——— (1929). "Rassenkreuzung beim Menschen." *Volk und Rasse* 4:72–77.
——— (1931). "Race-Crossing and Glands. Some Human Hybrids and their Parent Stocks." *Eugenics Review* 23:31–39.
——— (1933a). "Biologische und biochemische Untersuchungen bei Rassenmischungen." In Corrado Gini, ed. *Atti del congresso internazionale per gli studi sulla popolazione. Rome 1931*, Vol. 3. Rome: Istituto poligrafico dello Stato, pp. 199–202.
——— (1933b). "Der nordische Sippenkult und die biologische Lebensanschauung im neuen Deutschen Reich." *Völkischer Beobachter*, December 30, 1933.
——— (1934a). *Die Vererbung der musikalischen Begabung*. Berlin: Alfred Metzner Verlag.
——— (1934b). "Freiheitsrausch oder Verantwortung." *Deutsche Zeitung*, November 16, 1934.
——— (1934c). "Health Declaration Before Marriage." In Henry F. Perkins, ed. *A Decade of Progress in Eugenics: Scientific Papers of the Third International Congress of Eugenics*. Baltimore: Williams & Wilkins, pp. 222–230.
——— (1935a). "Der nordische Sippenkult und die deutsche Rassenfrage." *Völkischer Beobachter*, March 23, 1935.
——— (1935b). "Nordischer Sippenkult und Rassenfrage." *Frankfurter Zeitung*, January 23, 1935.
Montagu, Ashley (1942). *Man's Most Dangerous Myth: The Fallacy of Race*. New York: Columbia University Press.
——— (1972). *Statement on Race*. New York: Oxford University Press.
Moreau, Patrick (1983). "Die neue Religion der Rasse. Der Biologismus und die kollektive Ethik der Neuen Rechten in Frankreich und Deutschland." In Iring Fetscher, ed. *Neokonservative und "Neue Rechte."* Munich: Beck, pp. 122–162.
Mosse, George L. (1964). *The Crisis of German Ideology: Intellectual Origins of the Third Reich*. New York: Grosset & Dunlap.
——— (1978). *Toward the Final Solution: A History of European Racism*. London: Howard Fertig.
Muckermann, Hermann (1930a). "Alfred Ploetz und sein Werk." *Eugenik, Erblehre, Erbpflege* 1:261–265.
——— (1930b). "Neue Forschungen über das Problem der differenzierten Volksvermehrung." *Zeitschrift für induktive Abstammungs-und Vererbungslehre* 54:287–295.
Muller, Hermann J. (1923). "Mutation." In Charles B. Davenport, ed. *Eugenics, Genetics and the Family*. Baltimore: Williams and Wilkins, pp. 106–112.
——— (1928). "The Problem of Genic Modification." In Hans Nachtsheim, ed. *Verhandlungen des V. Internationalen Kongresses für Vererbungswissenschaft Berlin 1927, Supplementband I der Zeitschrift für induktive Abstammungs- und Vererbungslehre*. Leipzig: Borntraeger pp. 234–260.
——— (1934). "The Dominance of Economics Over Eugenics": Henry F. Perkins, *A Decade of Progress in Eugenics*. Scientific Papers of the Third International Congress of Eugenics. Baltimore: Williams & Wilkins, pp. 138–144.
——— (1936). *Out of the Night: A Biologist's View oft the Future*. New York: Vanguard Press.
Mühlen, Patrik von (1977). *Rassenideologien: Geschichte und Hintergründe*. Berlin; Bad Godesberg: Dietz Verlag J. H. W. Nachf..
Müller-Hill, Benno (1984). *Tödliche Wissenschaft. Die Aussonderung von Juden, Zigeunern und Geisteskranken*. Reinbek: Rowohlt.

Müller- Hill, Benno (2000). "Das Blut von Auschwitz und das Schweigen der Gelehrten." In Doris Kaufmann, ed. *Geschichte der Kaiser-Wilhelm-Gesellschaft im Nationalsozialismus. Bestandsaufnahme und Perspektiven der Forschung*. Göttingen: Wallstein, pp. 189–227.
Myerson, Abraham (1923). "Inheritance of Mental Disease." In Charles B. Davenport *Eugenics, Genetics and the Family*. Baltimore: Williams & Wilkins, pp. 218–225.
Myerson, Abraham, James B. Ayer, Tracy J. Putnam, Clyde E. Keeler, and Leo Alexander (1936). *Eugenical Sterilization*. New York: MacMillan.
Myrdal, Alva (1939). "The Swedish Approach to Population Policies." *Eugenical News* 24:3–8.
Nachtsheim, Hans (1927). "Der V. Internationale Kongress für Vererbungswissenschaft." *Die Naturwissenschaften* 15:988–995.
——— (1933). "Tierzüchterische Ergebnisse des VI. Internationalen Kongresses für Vererbungswissenschaft, Ithaca 1932." *Züchtungskunde* 8:193–213.
——— (1968). "Geburtenkontrolle. Ihre internationale Notwendigkeit, ihre Wege." *Die Berliner Ärztekammer*, Issue 4/1968, pp. 1–16.
——— (1963b). "Warum Eugenik?" *Fortschritte der Medizin. Internationale Halbmonatsschrift für die gesamte Heilkunde*, 81, 711–713.
——— (1963a). "Die Biologie und die Gesellschaft." In Fritz Borinski, ed. *Die Wissenschaften und die Gesellschaft*. Berlin: Duncker & Humblot, 105–121.
Naureckas, Jim (1995). "Racism Resurgent: how media let *The Bell Curve's* pseudo–science define the Agenda on race." *Extra!*, January/February 1995 URL: http://www.fair.org/extra/9501/bell.html.
Neel, James V., and William J. Schull (1954). *Human Heredity*. Chicago: University of Chicago Press.
Neidhardt, Friedhelm (1985). "Einige Ideen zu einer allgemeinen Theorie sozialer Bewegungen." In Stefan Hradil, ed. *Sozialstruktur im Umbruch*. Opladen: Leske + Budrich, pp. 193–204.
Neidhardt, Friedhelm, and Dieter Rucht (2001). "Soziale Bewegung und kollektive Aktionen." In Hans Joas, ed. *Lehrbuch der Soziologie*. Frankfurt a.M.; New York: Campus, pp. 533–556.
Niemann – Findeisen, Sören (2004): *Weeding the Garden. Die Eugenik–Rezeption der frühen Fabian Society*. Münster: Westfälisches Dampfboot.
Nilsson-Ehle, Hermann (1928). "Rassenkreuzung aus allgemein biologischen Gesichtspunkten." Institut International d'Anthropologie, *Discours prononcés à la Conférence organisée par "The International Federation of eugenic Organisations"*. Paris: E. Nourry, pp. 3–9.
Noble, David F. (1977). *America by Design: Science, Technology, and the Rise of Corporate Capitalism*. New York: Knopf.
Noordman, Jan M. A. (1989*)*. *Om de kwaliteit van het nageslacht. Eugenetica in Nederland 1900–1950*. Nijmegen: Sun.
Notestein, Frank W. (1968). "The Population Council and the Demographic Crisis of the Less-developed World." *Demography* 5:553–560.
——— (1971). "Reminiscences: The Role of Foundations, of the Population Association of America, Princeton University and the United States in Fo-stering American Interest in Population Problems." *Milbank Memorial Fund Quarterly*, Issue 2/1971, pp. 67–84.
——— (1979). "John D. Rockefeller 3d: A Personal Appreciation." *Population and Development Review* 5:501–508.
——— (1981). "Memories of the Early Years of the Association." *Population Index* 47:484–488.
Nowak, Kurt (1988). "Sterilisation und 'Euthanasie' im Dritten Reich. Tatsachen und Deutungen." *Geschichte in Wissenschaft und Unterricht* 39:327–341.
——— (1990). "Das Faktum und seine Deutung. Interpretationsmodelle zu Sterilisation und Euthanasie im Dritten Reich." In *Theologische Literaturzeitung* 115:243–254.

Nyiszli, Miklos (1960). *Auschwitz. A Doctor's Eyewitness Account*. New York: Frederick Fell Inc. Publishers.
Olson, Richard G. (2008). *Science and Scientism in Nineteenth–Century Europe*. Champaign: University of Illinois Press.
Ordover, Nancy (2003). *American Eugenics: Race, Queer Anatomy, and the Science of Nationalism*. Minneapolis; London: University of Minnesota Press.
Osborn, Frederick (1937a). "Development of Eugenic Philosophy." *American So-ciological Review* 2:389–397.
——— (1937b). "Implications of the New Studies in Population and Psy-chology for the Development of Eugenic Philosophy." *Eugenical News* 22:104–106.
——— (1938). "The Application of Measures of Quality." In *Congrès International de la Population, Paris 1937*. Vol. 8, Paris: Hermann et Cie.,Éditeurs, pp. 117–122.
——— (1939a). "The American Concept of Eugenics." *Eugenical News* 24:2.
——— (1939b). "Social Implications of the Eugenic Program." *Child Study* 16:95–97.
——— (1940). *Preface to Eugenics*. New York: Harper.
——— (1951a). "The Eugenic Hypothesis. Part I: Positive Eugenics." *Eugenical News* 36:19–21.
——— (1951b). *Preface to Eugenics*. 2nd ed. New York: Harper.
——— (1952). "The Eugenic Hypothesis. Part II: Negative Eugenics." *Eugenical News* 37:6–9.
——— (1955). "Eugenics." *Around the World News of Population and Birth Control*, Issue 39/1955, p. 1.
——— (1956). "Galton and Mid Century Eugenics." *Eugenics Review* 48:15–22.
——— (1963). "Eugenics and the Races of Man." *Eugenics Quarterly* 10:103–109.
——— (1964). "The Human Future." *Science News Letter* 86:54–55.
——— (1966). "Eugenics (Encyclopaedia Britannica)." *Eugenics Quarterly* 13:155–164.
——— (1968). *The Future of Human Heredity*. New York: Weybright and Talley.
——— (1974). "History of the American Eugenics Society." *Social Biology* 21:115–126.
——— (1983). "Eugenics." In *The New Encyclopaedia Britannica*, London: Encyclopaedia Britannica, Vol. 6, pp. 1023–1026.
Osborn, Henry Fairfield (1921). "The Second International Congress of Eugenics. Address of Welcome." *Science* 54:311–313.
——— (1932). "Birth Selection versus Birth Control." *Science* 76:173–179.
——— (1934). "Birth Selection versus Birth Control." In Henry F. Perkins and Harry H. Laughlin, eds. *A Decade of Progress in Eugenics. Scientific Papers of the Third International Congress of Eugenics*. Baltimore: Williams & Wilkins, pp. 29–41.
Otsubo, Sumiko (2005). "Between Two Worlds: Vamanouchi Shigeo and Eugenics in Early Twentieth–Century Japan." *Annals of Science* 62:205–231.
O'Keefe, Kathy (1993). "English Eugenics Society. Membership." Unpublished manuscript, N.p.
——— (1994). "American Eugenics Society. Membership." Unpublished manuscript, N.p.
——— (1995). "Democracy or Eugenics. The True Meaning of the Abortion Choice." Unpublished manuscript, N.p.
Paul, Diane (1984). "Eugenics and the Left." *Journal of the History of Ideas* 45:567–590.
——— (1989). "From Eugenics to Clinical Genetics." Unpublished manuscript, Boston.
——— (1995). *Controlling Human Heredity: 1865 to the Present*. Boston: Humanity Books.
Pearl, Raymond (1912a). "The First International Eugenics Congress." *Science* 36:395–396.
——— (1912b). "The Inheritance of Fecundity." In Eugenics Education Society, ed. *Problems in Eugenics*. Reprinted by New York, London: Garland Publishing, pp. 47–57.
——— (1925): The Biology of Population Growth. New York.
——— (1927). "The Biology of Superiority." *American Mercury* 12:257–266.
——— (1928a). "Eugenics." In Hans Nachtsheim, ed. *Verhandlungen des V. Internationalen Kongresses für Vererbungswissenschaften. Berlin 1927*. Leipzig: Borntraeger, pp. 261–282.

Pearl Raymond (1928b). *The Present Status of Eugenics.* Hanover, N.H., Minneapolis: The Sociological Press.
——— (1929). "The International Union for the Scientific Investigation of Population Problems." *Eugenical News* 14:18–21.
Pearson, Karl (1905). *National Life from the Standpoint of Science.* London: A. and C. Black.
Pearson, Roger (1959). "Pan-Nordicism as a Modern Policy." *Northern World*, Issue 3/1959, pp. 4–7.
——— (1991). *Race, Intelligence and Bias in Academe* Washington: Scott-Townsend Publishers.
——— (1992). "Introduction." In Roger Pearson, ed. *Shockley on Eugenics and Race.* Washington: Scott-Townsend Publishers, pp. 15–50.
Peltier, Roger (1949). "LTnstitut National d'Etudes Demographiques." *Population* 4:9–38.
Penrose, Lionel (1951). "Eugenics." *Scientific Worker* 6:7–8.
——— (1963). "Limitations of Eugenics." *Proceedings of the Royal Institution* 39:506–519.
Penselin, Ulla, and Ingrid Strobl (1988). *Anschlag auf die Schere im Kopf.* Hamburg: Bundesgerichtshof Hamburg.
Pernick, Martin (1996). *The Black Stork: Eugenics and The Death of "Defective" Babies in American Medicine and Motion Pictures since 1915.* New York; Oxford: Oxford University Press.
Pettigrew, Thomas F. (1964a). *A Profile of the Negro American.* Princeton: D. Van Norstand.
——— (1964b). "Race, Mental Illness, and Intelligence: A Social Psychological View." *Eugenics Quarterly* 11:189–215.
Pfeil, Elisabeth (1937). "Der internationale Bevölkerungskongreß in Paris 7/28–8/1/1937."*Archiv für Bevölkerungswissenschaft und Bevölkerungspolitik* 7:288–301.
——— (1940). "Die volksbiologische Wiedergeburt der Ostmark." *Volk und Rasse* 15:25–26.
Pinn, Irmgard, and Michael Nebelung (1990). "Kontinuität durch Verdrängung. Die 'anthropologisch-soziologischen Konferenzen' 1949–1954 als ein vergessenes Kapitel der deutschen Soziologiegeschichte." In Heinz-Jürgen Dahme Carsten Klingemann, Michael Neumann, Karl-Siegbert Rehberg, Ilja Srubar eds. *Jahrbuch für Soziologiegeschichte 1990.* Opladen: Leske + Budrich, pp. 177–218.
——— (1992). *Vom "klassischen" zum aktuellen Rassismus in Deutschland. Das Menschenbild der Bevölkerungstheorie und Bevölkerungspolitik.* 2nd ed. Duisburg.: DISS.
Pitt-Rivers, George H. L. F., ed. (1932). *Problems of Population. Being the Report of the Proceedings of the Second General Assemby of the International Union for the Scientific Investigation of Population Problems.* London: Allen & Unwin.
——— (1936). "Science of Population, Interim Report on Scientific Organization and Classification." *Population* 2:23–29.
——— (1938). *The Czech Conspiracy.* London: Boswell Pub. Co.
Pius XI. (1930). "Encyclical Letter. On Christian Marriage." In *Sixteen Encyclicals of Pope Pius XI.* Washington: National Catholic Welfare Conference.
Platen-Hallermund, Alice (1948). *Die Tötung Geisteskranker in Deutschland.* Frankfurt a.M.: Verlag der Frankfurter Hefte.
Ploetz, Alfred (1895). *Die Tüchtigkeit unserer Rasse und der Schutz der Schwachen.* Berlin: Fischer.
——— (1909). "Gesellschaften mit rassenhygienischen Zwecken." *Archiv für Rassen-und Gesellschaftsbiologie* 6:277–281.
——— (1911a). "Die Begriffe der Rasse und Gesellschaft." In Deutsche Gesellschaft für Soziologie, ed. *Verhandlungen des Ersten Deutschen Soziologentages vom 19.-22. Oktober 1910 in Frankfurt a.M.* Tübingen: Mohr, pp. 111–166.
——— (1911b). "Ziele und Aufgaben der Rassenhygiene." *Öffentliche Gesundheitspflege* 43:164–192.
——— (1913/14). "Neomalthusianismus und Rassenhygiene." *Archiv für Rassen-und Gesellschaftsbiologie* 10:166–172.

––––––– (1936a). "Rassenhygiene als Basis der Friedenspolitik." *Rassenpolitische Auslands-Korrespondenz*, Issue 2/1936, pp. 3–4.

––––––– (1936b). "Rassenhygiene und Krieg." In Hans Harmsen and Franz Lohse, eds. *Bevölkerungsfragen. Bericht des Internationalen Kongresses für Bevölkerungswissenschaft, Berlin 8/26–9/1/1935*. Munich: J. F. Lehmanns, pp. 615–620.

Pols, Hans (2010). "Eugenics in the Netherlands and the Deutsch East Indies." In Alison Bashford and Philippa Levine, eds. *The Oxford Handbook of the History of Eugenics*. New York; Oxford: Oxford University Press, pp. 347–362.

Pommerin, Rainer (1979). *Sterilisierung der Rheinlandbastarde: Das Schicksal einer farbigen deutschen Minderheit 1918–1937*. Düsseldorf: Droste.

Popenoe, Paul (1918). "Is War Necessary?" *Journal of Heredity* 9:257–262.

––––––– (1923). "Eugenics in France." *Journal of Heredity* 14:275–276.

––––––– (1924). "Rassenhygiene in den Vereinigten Staaten." *Archiv für Rassen-und Gesellschaftsbiologie* 15:184–193.

––––––– (1926). *Conservation of the Family*. Baltimore: Williams & Wilkins Company.

––––––– (1930). "The Immigrant Tide." In Madison Grant and Charles Davison, eds. *The Alien in Our Midst or "Selling Our Birthright for a Mess of Pottage."* New York, NY: The Galton Publishing Co. Inc., pp. 210–213.

––––––– (1934a). "The German Sterilization Law." *Journal of Heredity* 25:257–260.

––––––– (1934b). "The Progress of Eugenic Sterilization." *Journal of Heredity* 25:19–25.

Poulton, Edward B. (1916): "Eugenics Problems After the Great War." *Eugenics Review* 8:34–49.

Proctor, Robert N. (1988). *Racial Hygiene: Medicine Under the Nazis*. Cambridge, London: Harvard University Press.

––––––– (1991). "Eugenics Among the Social Sciences. Hereditarian Thought in Germany and the United States." In JoAnne Brown and David K. van Keuren, eds. *The Estate of Social Knowledge*. Baltimore: Johns Hopkins University Press, pp. 175–208.

Promitzer, Christian, Sevasti Trubeta, and Marius Turda (2011). "Framing Issues of Health, Hygiene and Eugenics in Southeastern Europe." In Christian Promitzer, Sevasti Trubeta, and Marius Turda, eds. *Health, Hygiene and Eugenics in Southeastern Europe to 1945*. Budapest; New York: Central European University Press, pp. 1–26.

Propping, Peter, and Bernd Heuer (1991). "Vergleich des 'Archivs für Rassen-und Gesellschaftsbiologie' (1904–1933) und des 'Journal of Heredity' (1910–1939)." *Medizinhistorisches Journal* 25:78–93.

Provine, William B. (1973). "Geneticists and the Biology of Race Crossing." *Science* 182:790–796.

––––––– (1986). "Geneticists and Race." *American Zoologist* 26:857–887.

Punnet, R.C. (1912). "Genetics and Eugenics." In Eugenics Education Society, ed. *Problems in Eugenics*. London: The Eugenics Education Society, pp. 137–138.

Putnam, Carleton (1961). *Race and Reason: A Yankee View*. Washington: Public Affairs Press.

Quine, Maria Sophia (1996). *Population Politics in Twentieth-Century Europe: Fascist Dictatorships and Liberal Democracies*. London; New York: Routledge.

––––––– (2010). "The First-Wave Eugenic Revolution in Southern Europe: Science sans frontières." In Alison Bashford and Philippa Levine, eds. *The Oxford Handbook of the History of Eugenics*. New York; Oxford: Oxford University Press, pp. 377–397.

Ramsden, Edmund (2003). "Social Demography and Eugenics in the Interwar United Sates." *Population Council* 29:547–593.

Raz, Aviad E. (2009). "Eugenic Utopias/Dystopias, Reprogenetics, and Community Genetics." *Sociology of Health and Illness* 31:602–616.

Reed, Sheldon C. (1968). "Eugenics Tomorrow." In K. R. Dronamraju, ed. *Haldane and Modern Biologie*. Baltimore: Johns Hopkins Press, pp. 231–242.

––––––– (1974). "A Short History of Genetic Counseling." *Dight Institute for Human Genetics Bulletin* 14:1–10.

Reggiani, Andres H. (2007). *God's Eugenicist: Alexis Carrel and the Sociobiology of Decline*. New York; Oxford: Berghahn Books.

——— (2010). "Dépopulation, fascisme et eugénisme 'latin' dans l'Argentine des années 1930." *Le Movuement Social*, Issue 230/2010, pp. 7–26.

Reichel, Heinrich (1931): "Alfred Ploetz und die rassenhygienische Bewegung der Gegenwart." Wiener klinische Wochenschrift, 44, 284–287.

Reilly, Philip R. (1991). *"The Surgical Solution. A History of Involuntary Sterilization in the United States."* Baltimore: The Johns Hopkins University Press.

Repp, Günther (1966). "Entwicklungshilfe—eine Notwendigkeit." *Erbe und Verantwortung*, Issue 2/1966, pp. 13–15.

——— (1967). "Die große Wende in der Eugenik." *Erbe und Verantwortung*, Issue 1/1967, pp. 10–13.

Rexilius, Günter (1980). "Die 'Neue Anthropologie'—das theoretische Organ der Rechtsradikalen in der Bundesrepublik." *Psychologie und Gesellschaftskritik* 4:104–143.

Reyer, Jürgen (1991). *Alte Eugenik und Wohlfahrtspflege. Entwertung und Funktionalisierung der Fürsorge vom Ende des 19. Jahrhunderts bis zur Gegenwart*. Freiburg im Breisgau: Lambertus.

Richter, Ingrid (2001). *"Katholizismus und Eugenik in der Weimarer Republik und im Dritten Reich. Zwischen Sittlichkeitsreform und Rassenhygiene*. Paderborn: Schöningh.

Rieck, Jörg (alias Jürgen Rieger) (1981). "Zur Debatte der Vererblichkeit der Intelligenz." In Pierre Krebs, ed. *Das Unvergängliche Erbe. Alternativen zum Prinzip der Gleichheit*. Tübingen: Grabert, pp. 315–371.

Rieger, Jürgen (1969). *Rasse—ein Problem auch für uns*. Hamburg: Selbstverlag.

Robertson, Jennifer (2005). "Blood Talks: Eugenics, Modernity and the Creation of New Japanese." *History and Anthropology* 13:191–216.

——— (2010). "Eugencis in Japan: Sanguinous Repair." In Alison Bashford and Philippa Levine, eds. *The Oxford Handbook of the History of Eugenics*. New York; Oxford: Oxford University Press, pp. 430–448.

Rockefeller, John D. (1979). "Population Growth: The Role of the Developed World." *Population and Development Review* 5:509–516.

Rodenwaldt, Ernst (1922). *Die Mestizen auf Kisar*. Den Haag: Mededeelingen van den Dienst der Volksgezondheid in Nederlandsch-Indie

——— (1934). "Vom Seelenkonflikt des Mischlings." *Zeitschrift für Morphologie und Anthropologie* 34:364–375.

——— (1939). "Nationalsozialistische Rassenerkenntnis als Grundlage für die koloniale Betätigung des neuen Europas." *Deutscher Kolonialdienst* 7:180–185.

Roelcke, Volker (2000). "Psychiatrische Wissenschaft im Kontext nationalsozialistischer Politik und 'Euthanasie'. Zur Rolle von Ernst Rüdin und der Deutschen Forschungsanstalt für Psychiatrie/Kaiser–Wilhelm–Institut." In Doris Kaufmann, ed. *Geschichte der Kaiser–Wilhelm–Gesellschaft im Nationalsozialismus. Bestandsaufnahme und Perspektiven der Forschung*, Vol. 1. Göttingen: Wallstein, pp. 112–150.

——— (2003). "Programm und Praxis der psychiatrischen Genetik an der Deutschen Forschungsanstalt für Psychiatrie unter Ernst Rüdin. Zum Verhältnis von Wissenschaft, Politik und Rasse–Begriff vor und nach 1933." In Hans–Walter Schmuhl, ed. *Rassenforschung an Kaiser–Wilhelm–Instituten vor und nach 1933*. Göttingen: Wallstein, pp. 38–67.

Roger, Jacques (1989). "L'eugénisme, 1850–1950." In Claude Benichou, ed. *L'ordre des caractères. Aspects de l'hérédité dans l'histoire des sciences de l'homme*. Paris: J. Vrin, pp. 119–145.

Roll-Hansen, Nils (1980). "Eugenics Before World War II. The Case of Norway." *History and Philosophy of Life Sciences* 2:269–298.

——— (1988). "The Progress of Eugenics. Growth of Knowledge and Change in Ideology." *History of Science* 26:293–331.

―――― (1989a). "Eugenic Sterilization: A Preliminary Comparison of the Scandinavian Experience to that of Germany." *Genome* 31:890–895.
―――― (1989b). "Geneticists and the Eugenics Movement in Scandinavia." *British Journal for the History of Science* 22:335–346.
―――― (1996a). "Conclusion. Scandinavian Eugenics in the International Context." In Gunnar Broberg and Nils Roll-Hansen, eds. *Eugenics and the Welfare State: Sterilization Policy in Denmark, Sweden, Norway, and Finland*. East Lansing: Michigan State University, pp. 307–320.
―――― (1996b). "Norwegian Eugenics: Sterilization as Part of a Program for Social Reform." In Gunnar Broberg and Nils Roll-Hansen, eds. *Eugenics and the Welfare State: Sterilization Policy in Denmark, Sweden, Norway, and Finland*. East Lansing: Michigan State University, pp. 177–230.
Rose, Steven (1976). "Scientific Racism and Ideology. The IQ Racket from Galton to Jensen." In Hilary Rose and Steven Rose, eds. *The Political Economy of Science*. London: Macmillan, pp. 112–141.
Rosen, Christine (2004). *Preaching Eugenics: Religious Leaders and the American Eugenics Movement*. Oxford: Oxford University Press.
Rost, Detlef H. (2009). *Intelligenz. Fakten und Mythen*. Weinheim: Beltz.
Roth, Karl Heinz (1986). "Schöner neuer Mensch. Der Paradigmenwechsel der klassischen Genetik und seine Auswirkungen auf die Bevölkerungsbiologie des 'Dritten Reichs.'" In Heidrun Kaupen-Haas, ed. *Der Griff nach der Bevölkerung*. Nördlingen: Greno, pp. 11–63.
Rucht, Dieter (1996). "The Impact of National Context on Social Movement Structures: A Cross–Movement and Cross–National Comparison." In Doug McAdam, John D. McCarthy, and Mayer N. Zald, eds. *Comparative Perspectives on Social Movements: Political Opportunities, Mobilizing Structures, and Cultural Framings*. Cambridge: Cambridge University Press, pp. 185–204.
Runcis, Maija (1998). *Steriliseringar i folkhemmet*. Stockholm: Ordfront.
Rushton, J. Philippe (1988a). "Genetic Similarity Theory, Intelligence, and Human Mate Choice." *Ethology and Sociobiology* 1:45–58.
―――― (1988b). "Race Differences in Behaviour: A Review and Evolutionary Analysis." *Personality and Individual Differences* 9:1009–1024.
―――― (1993). "Why We Should Study Race Differences." *Psychologische Beiträge* 32:128–142.
―――― (1995). *Race, Evolution, and Behavior*. New Brunswick: Transction Publ.
Rushton, J. Philippe, and Anthony F. Bogaert (1989). "Population Differences in Susceptibility to AIDS: An Evolutionary Analysis." *Social Science and Medicine* 28:1211–1220.
Ruttke, Falk (1934). "Erbpflege in der Deutschen Gesetzgebung." *Ziel und Weg* 4:600–603.
―――― (1935). "Erb- und Rassenpflege auf internationalen Kongressen." *Völkischer Beobachter*, September 1, 1935.
Ruzicka, Vladislav (1935). "Verbesserung des Volksstandes durch Rassenhygiene oder durch Eugenik." In Karel Weigner, ed. *Die Gleichwertigkeit der europäischen Rassen und die Wege zu ihrer Vervollkommnung*. Prag: Verlag derTschechischen Akademie der Wissenschaften undKuenste, pp. 108–129.
Rydell, Robert, Christina Cogdell, and Mark Largent (2006). "The Nazi Eugenics Exhibit in the United States, 1934–1943." In Susan Currel and Christina Cogdell, eds. *Popular Eugenics: National Efficiency and American Mass Culture in the 1930s*. Athens: Ohio University Press, pp. 359–384.
Rüdin, Ernst (1934a). "Aufgaben und Ziele der Deutschen Gesellschaft für Rassenhygiene." *Archiv für Rassen-und Gesellschaftsbiologie* 28:228–233.
―――― (1934b). "Aufgaben und Ziele der Deutschen Gesellschaft für Rassenhygiene." *Völkischer Beobachter*, July 8, 1934.

Rüdin Ernst (1934c). *Erblehre und Rassenhygiene im völkischen Staat. Tatsachen und Richtlinien.* Munich: J. F. Lehmanns.
——— (1937). "Bedingungen und Rolle der Eugenik in der Prophylaxis der Geistesstörung." Comptes rendues duIIe Congrès International d'Hygiène mentale. Paris, 19–23 juillet 1937. Paris, Vol. 1, pp. 103–115.
——— (1939). "Der uns aufgezwungene Krieg und die Rassenhygiene." *Archiv für Rassen- und Gesellschaftsbiologie* 33:443–445.
——— (1940). "Alfred Ploetz zum Gedenken." *Archiv für Rassen-und Gesellschaftsbiologie* 34:1–4.
Samaan, A.E. (2013): *From a "Race of Masters" to a "Master Race"*. Seattle: CreateSpace.
Saleeby, C. W. (1917). "Imperial Health and the Dysgenic of War." *Journal of State Medicine* 25, 3: 307–316.
Sanders, Jacob (1936). "The Twelfth Meeting of the International Federation of Eugenic Organizations." *Eugenical News* 21:106–108.
Sanger, Margaret, ed. (1927). *Proceedings of the World Population Conference.* London: E. Arnold.
——— (1919a). "Birth Control and Racial Betterment." *Birth Control Review,* Issue 2/1911, pp. 11–12.
——— (1919b). "Why Not Birth Control Clinics in America." *Birth Control Review,* Issue 5/1911, pp. 10–11.
——— (1926). "The Function of Sterilization." *Birth Control Review,* Issue 10/1926, pp. 299.
Sarrazin, Thilo (2010a). *Deutschland schafft sich ab. Wie wir unser Land aufs Spiel setzen.* Munich: Deutsche Verlags-Anstalt.
——— (2010b). "Die große Zustimmung beunruhigt mich." *FAZ,* January 10, 2010.
Sauvy, Alfred (1959). "Development and Perspectives of Demographic Research in France." In Philip M. Hauser, and Otis Dudley Duncan, eds. *The Study of Population.* Chicago: Chicago University Press, pp. 180–189.
Schade, Heinrich (1935). "Der Internationale Kongreß für Bevölkerungswissenschaften in Berlin." *Der Erbarzt* 1:140–142.
——— (1936). "Ausländische Stimmen zur deutschen Erb- und Rassenpflege." *Rassenpolitische Auslands-Korrespondenz,* Issue 5/1936, pp. 3–4.
Schaffer, Gavin (2008). *Racial Science and British Society, 1930–62.* Basingstocke: Palgrave Macmillan.
Schallmayer, Wilhelm (1903). *Vererbung und Auslese im Lebenslauf der Völker. Eine staatswissenschaftliche Studie auf Grund der neueren Biologie.* Jena: Fischer.
——— (1908). "Der Krieg als Züchter." *Archiv für Rassen-und Gesellschaftsbiologie* 5:364–400.
Scheinfeld, Amram (1958). "Changing Attitudes toward Human Genetics and Eugenics." *Eugenics Quarterly* 5:145–153.
Schenk, Faith, and A. S. Parkes (1968). "The Activities of the Eugenics Society." *Eugenics Review* 60:142–161.
Schimank, Uwe (1995). *Theorien gesellschaftlicher Differenzierung.* Opladen: Leske & Budrich.
Schlaginhaufen, Otto (1921a). "Bastardisierung und Qualitätsänderung." *Natur und Mensch* 1:53–38.
——— (1921b). "Rasse, Rassenmischung und Konstitution." *Natur und Mensch,* 1:398–411.
Schlebusch, Cornelia (1993). "Bevölkerungspolitik als Entwicklungspolitik. Eine Untersuchung zur Integration des Überbevölkerungsparadigmas in die entwicklungspolitische Theorie und Praxis." Unpublished master thesis, Aachen: Magisterarbeit an der Technischen Hochschule Aachen.
Schleiff, Hartmut (2009). "Der Streit um den Begriff der Rasse in der frühen Deutschen Gesellschaft für Soziologie als ein Kristallisationspunkt für ihrer methologischen Konstitution." *Leviathan* 37:367–388.

Schmuhl, Hans-Walter (1987). *Rassenhygiene, Nationalsozialismus, Euthanasie.* Göttingen: Vandenhoeck & Ruprecht.
—— (1991). "Max Weber und das Rassenproblem." In Manfred Hettling, Claudia Huerkamp and, Paul Nolte, eds. *Was ist Gesellschaftsgeschichte? Positionen, Themen, Analysen.* Munich: Beck, pp. 331–342.
—— (1997). "'Euthanasie'—zwei Paar Schuhe? Eine Antwort auf Michael Schwartz." *Westfälische Forschungen* 47:757–762.
—— (2003). "Rasse, Rassenforschung, Rassenpolitik. Annäherung an das Thema." In Hans–Walter Schmuhl, ed. *Rassenforschung an Kaiser-Wilhelm-Instituten vor und nach 1933.* Göttingen: Wallstein, pp. 7–37.
—— (2005). Grenzüberschreitungen. *Das Kaiser-Wilhelm-Institut für Anthropologie, menschliche Erblehre und Eugenik 1927–1945.* Göttingen: Wallstein.
Schneider, William H. (1982). "Toward the Improvement of the Human Race: The History of Eugenics in France." *Journal of Modern History* 54:268–291.
—— (1990a). "The Eugenics Movement in France, 1890–1940." In Mark B. Adams, ed. *The Wellborn Science: Eugenics in Germany, France, Brazil, and Russia.* New York; Oxford: Oxford University Press, pp. 69–109.
—— (1990b). *Quality and Quantity: The Quest for Biological Regeneration in Twentieth-Century France.* Cambridge: Cambridge University Press.
Schneidewind, H. H. (1933). *Wirtschaft und Wirtschaftspolitik der Vereinigten Staaten von Amerika.* Würzburg: Triltsch.
Schnitzler, Sonja (2009). "Fallbeispiel für rekursive Kopplung von Wissenschaft und Politik: Das 'Archiv für Bevölkerungswissenschaft und Bevölkerungspolitik' (1934–1944)." In Rainer Mackensen Jürgen Reulecke und Josef Ehmer, eds. *Ursprünge, Arten und Folgen des Konstrukts "Bevölkerung" vor, im und nach dem "Dritten Reich".* Wiesbaden: VS Verlag für Sozialwissenschaften, pp. 321–344.
Schock, Kurt (1996). "A Conjunctural Model of Political Conflict: The Impact of Political Opportunities on the Relationship between Economic Inequality and Violent Political Conflict." *Journal of Conflict Resolution* 40:98–133.
Schoen, Johanna (2011). "Reassessing Eugenic Sterilization: The Case of North Carolina." In Paul A. Lombardo, ed. *A Century of Eugenics in America: From the Indiana Experiment to the Human Genome Era.* Bloomington, Indiana: Indiana University Press, pp. 141–160.
Schouwenburg, J. Ch. van (1935). "Overzicht van de Ontwikkeling dee Eugenese in Duitsland." *Erfelijkheid bij de Mens* 1:301–310.
Schreiber, Georges (1932). "Commentaires sur les principes directeurs de l'eugénique, proposés par le comité norvegien pour l'hygiène de la race." *Revue Anthropologique* 42:318–320.
—— (1933). "Le IIIe Congrès international d'eugénique." *Revue Anthropologique* 43:479–484.
—— (1934). "Le national-socialisme et la Stérilisation eugénique en Allemagne." *Le Siècle Médical*, Issue 188/1934, pp. 1–9.
—— (1935a). "Compte rendu de la XIe assemblée de la Fédération Internationale des Organisations Eugéniques." *Revue Anthropologique* 45:78–82.
—— (1935b). "La Stérilisation eugénique en Allemagne, une année d'application de la loi du 14 juillet 1933 pour l'enraiement de l'hérédité morbide." *Revue Anthropologique* 45:84–91.
—— (1936). "Actual Aspect of the Problem of Eugenical Sterilization in France." *Eugenical News* 21:104–105.
Schrijver, Franz (1936). "Het Internationale Congres voor Bevolkingswetenschap, Berlijn 1935." *Erfelijkheid bij de Mens* 2:67–72.
Schröder-Gudehus, Brigitte (1966). "Deutsche Wissenschaft und internationale Zusammenarbeit 1914–1928. Ein Beitrag zum Studium kultureller Beziehungen in politischen Krisenzeiten." Unpblished dissertation, Geneva.

Schubnell, Hermann (1967). "Roderich von Ungern-Sternberg (1885–1965)." *Le Demograph* 14:95–96.
Schulz, Ulrike (1992). *Gene mene muh raus mußt du. Eugenik—von der Rassenhygiene zu den Gen-und Reproduktionstechnologien.* Munich: AG-SPAK-Publ.
Schuster, Edgar (1912a). *Eugenics.* London: Collins' Clear Type Press.
——— (1912b). "The First International Eugenics Congress." *Eugenics Review* 4:223–256.
Schwartz, Michael (1989). "Sozialismus und Eugenik, zur fälligen Revision eines Geschichtsbildes." *Internationale wissenschaftliche Korrespondenz zur Geschichte der deutschen Arbeiterbewegung* 4:465–489.
——— (1994). "'Proletarier' und 'Lumpen'. Sozialistische Ursprünge eugenischen Denkens." *Vierteljahreshefte für Zeitgeschichte* 42:437–470.
——— (1995a). "Konfessionelle Milieus und Weimarer Eugenik." *Historische Zeitschrift* 261:403–448.
——— (1995b). *Sozialistische Eugenik. Eugenische Sozialtechnologien in Debatten und Politik der deutschen Sozialdemokratie 1890–1933.* Bonn: Dietz.
——— (1996). "'Rassenhygiene, Nationalsozialismus, Euthanasie'? Kritische Anfragen an eine These Hans–Walter Schmuhls." *Westfälische Forschungen* 46:604–622.
——— (1998). "'Euthanasie'–Debatten in Deutschland (1895–1945).' *Vierteljahrshefte für Zeitgeschichte* 46:617–665.
——— (2008). "Eugenik und 'Euthanasie': Die internationale Debatte und Praxis bis 1933/45. In Klaus–Dietmar Henke, ed. *Tödliche Medizin im Nationalsozialismus. Von der Rassenyhgiene zum Massenmord.* Köln; Weimar; Vienna: Böhlau, pp. 65–83.
Schweisheimer, Waldemar (1918). "Bevölkerungspolitische Bilanz des Krieges 1914/19." *Archiv für Rassen- und Gesellschaftsbiologie*, 13, 10–15.
Searchlight (1984a). "The Northern League." *Searchlight*, Issue 6/1984, p. 9.
——— (1984b). "Reagan Praises Leading Fascist." *Searchlight*, Issue 111/1984, pp. 2–4.
Searle, Geoffrey R. (1971). *The Quest for National Efficiency: A Study in British Politics and British Political Thought 1899–1914.* Berkeley; Los Angeles: University of California Press.
——— (1976). *Eugenics and Politics in Britain, 1900–1914.* Leyden: Noordhoff Internat. Publ.
——— (1979). "Eugenics and Politics in Britain in the 1930s." *Annals of Science* 36:159–169.
Seidler, Horst, and Andreas Rett (1988). *Rassenhygiene. Ein Weg in den Nationalsozialismus.* Vienna; Munich: Jugend und Volk.
Semmel, Bernard (1958). "Karl Pearson: Socialist and Darwinist." *British Journal of Sociology* 9:111–125.
Sergi, Giuseppe (1917). *La Guerra e La Preservazione della Nostra Stirpe.* Rome: Direzione della Nuova antologia.
Sesin, Claus – Peter (1996). "Sind Weiße klüger als Schwarze? Der rassistische Streit um den Intelligenzquotienten in den USA." *GEO*, Issue 8/1996, pp. 46–60.
——— (2012). "Sarrazins dubiose US–Quellen." In Michael Haller and Martin Niggeschmidt, eds. *Der Mythos vom Niedergang der Intelligenz. Von Galton zu Sarrazin: Die Denkmuster und Denkfehler der Eugenik.* Wiesbaden: VS Verlag für Sozialwissenschaften, pp. 27–48.
Shapiro, Harry L. (1959). "Eugenics and Future Society." *Eugenics Quarterly* 6:3–7.
Shapiro, Thomas M. (1985). *Population Control Politics: Women, Sterilization and Reproductive Choice.* Philadelphia: Temple University Press.
Sharpe, Catherine (1937). "Shadow over Czechoslovakia." *Anglo-German Review*, 1, 202–203.
Shockley, William (1972). "Dysgenics, Geneticity, Raceology. A Challenge to the In-tellectual Responsibility of Educators." *Phi Delta Kappan*, vol. 53, pp. 297–307.
——— (1974). "Sterilization—A Thinking Exercise." *The Stanford Daily*, December 4, 1974.

——— (1992a). "An Analysis Leading to a Recommendation Concerning Inquiry into Eugenics Legislation. Press Release 4/28/1969." In Roger Pearson, ed. *Shockley on Eugenics and Race*. Washington: Scott-Townsend Publishers, pp. 127–129.

——— (1992b). "Playboy Interview with William Shockley, August 1980." In Roger Pearson, ed. *Shockley on Eugenics and Race*. Washington: Scott-Townsend Publishers, pp. 234–274.

Shuey, Audrey M. (1966). *The Testing of Negro Intelligence*. 2nd ed.. New York: Social Science Press.

Sieferle, Rolf Peter (1989). *Die Krise der menschlichen Natur*. Frankfurt a.M.: Suhrkamp.

Siegfried, Andre (1926). "American Emigration and Eugenics." *Eugenics Review* 18:217–222.

——— (1927). *Die Vereinigten Staaten von Amerika*. Zurich/Leipzig: Orell Füssli Verlag.

Siemen, Hans-Ludwig (1987). *"Menschen blieben auf der Strecke". Psychiatrie zwischen Reform und Nationalsozialismus*. Gütersloh: Jakob van Hoddis.

——— (1993). "Die Reformpsychiatrie der Weimarer Republik: Subjektive Ansprüche und die Macht des Faktischen." In Franz-Werner Kersting, Karl Teppe, and Bernd Walter, eds. *Nach Hadamar. Zum Verhältnis von Psychiatrie und Gesellschaft im 20. Jahrhundert*. Paderborn: Schöningh, pp. 98–108.

Simonnot, Anne – Laure (1999). *Hygiénisme et eugénisme au XXe siècle à travers la psychiatrie française*. Paris: Éditions Seli Arslan.

Skerlj, Bozo (1960). "The Mankind Quarterly." *Man* 60:172–173.

Smith, Frank O. (1914). *Eugenic Peace*. Washington: Unknown.

Smith, Samuel G. (1912). "Eugenics and the New Social Consciousness." In Eugenics Education Society, ed. *Problems in Eugenics*. London: The Eugenics Education Society, pp. 480–486.

Soloway, Richard Allen (1979). "Neo-Malthusians, Eugenicists and the Declining Birthrate in England, 1900–1918." *Albion* 10:264–286.

——— (1982). *Birth Control and the Population Question in England, 1877–1930*. Chapel Hill: University of North Carolina Press.

——— (1990). *Demography and Degeneration. Eugenics and the Declining Birthrate in Twentieth-Century Britain*. Chapel Hill; London: University of North Carolina Press.

Spektorowski, Albert, and Elisabet Mizrachi (2004). "Eugenics and the Welfare State in Sweden. The Politics of Social Margins and the Idea of a Productive Society." *Journal of Contemporary History* 39:333–352.

Spiro, Jonathan Peter (2008). *Defending the Master Race: Conservation, Eugenics, and the Legacy of Madison Grant*. Burlington VT: University of Vermont Press.

Stähle, Egon (1935). "Ohne Titel." *Ärzteblatt für Württemberg und Baden*, Issue 7/ 1935, p. 1.

Steinbach, Peter, and Johannes Tuchel (1984). "Die Ermordung psychisch Kranker—von der Sterilisation zur Mordaktion." In Johannes Tuchel, ed. *"Kein Recht auf Leben." Beiträge und Dokumente zur Entrechtung und Vernichtung "lebensunwerten Lebens" im Nationalsozialismus*. Berlin: WAV, pp. 11–32.

Steinwallner, Bruno (1937). "Rassenhygienische Gesetzgebung und Maßnahmen ausmerzender Art." *Fortschritte der Erbpathologie, Rassenhygiene und ihrer Grenzgebiete*, 1:193–260.

Stepan, Nancy L. (1982). *The Idea of Race in Science: Great Britain 1800–1960*. Hamden, Conn.: Archon Books.

——— (1987). "'Nature's Pruning Hook': War, Race and Evolution, 1914 bis 1918." In J. M. W. Bean, ed. *Political Culture in Modern Britain*. London: Hamish Hamilton Ltd. 129–148.

——— (1991). *The Hour of Eugenics: Race, Gender, and Nation in Latin America*. Ithaca: Cornell University Press.

Stern, Alexandra Minna (2002). "Making Better Babies: Public Health and Race Betterment in Indiana, 1920–1935." *American Journal of Public Health* 92:742–752.

Stern, Alexandra Minna (2005). *Eugenic Nationa: Faults and Frontiers of Better Breeding in Modern America*. Berkeley; Los Angeles; London: University of California Press.
Stichweh, Rudolf (1984). *Zur Entstehung des modernen Ssytems wissenschaftlicher Disziplinen. Physik in Deutschland 1740–1890*. Frankfurt a.M.: Suhrkamp.
——— (2003). "Genese des globalen Wissenschaftssystems." *Soziale Systeme* 9:3–26.
——— (2005). "Die Universalität wissenschaftlichen Wissens." Unpublished manuscript, Luzerne.
Stoddard, Lothrop (1940). *Into the Darkness: Nazi Germany Today*. New York:. Duell, Sloan & Pearce, Inc.
Stölting, Erhard (1987). "Die anthroposoziologische Schule. Gestalt und Zusammenhänge eines wissenschaftlichen Institutionalisierungsversuchs." In Carsten Klingemann, ed. *Rassenmythos und Sozialwissenschaften in Deutschland. Ein verdrängtes Kapitel sozialwissenschaftlicher Wirkungsgeschichte*. Opladen: Westdeutscher Verlag, pp. 130–171.
Stroothenke, Wolfgang (1940). *Erbpflege und Christentum. Fragen der Sterilisation, Aufnordung, Euthanasie, Ehe*. Leipzig: Klotz.
Sturtevant, Alfred H. (1965). *A History of Genetics*. New York: Harper & Row.
Suitters, Beryl (1973). *Be Brave and Angry. Chronicles of the International Planned Parenthood Federation*. London: International Planned Parenthood Federation.
Sutter, Jean (1946). "Le facteur 'qualité' en démographie." *Population* 2:299–316.
——— (1950). L'Eugénique. Problèmes, méthodes, résultats. Paris: Puf.
——— (1963). "The Relationship between Human Population Genetics and Demography." In Elisabeth Goldschmidt, ed. *The Genetics of Migrant and Isolate Populations. Proceedings of a Conference on Human Population Genetics in Israel held at the Hebrew University Jerusalem*. New York: Williams & Wilkins. pp. 160–168.
Suzuki, Zenji (1975). "Genetics and Eugenics Movement in Japan." *Japanese Studies in the History of Science* 14:157–164.
Swan, Donald A. (1973). "Rassenunterschiede-nicht nur in der Hautfarbe." *Nation Europa*, Issue 1/1973, pp. 43–46.
——— (1974). "Rassenzugehörigkeit und Intelligenz." *Nation Europa*, Issue 11/1974, pp. 37–40.
Symonds, Richard, and Michael Carder (1973). *The United Nations and the Population Question 1945–1970*. London: McGraw-Hill.
Szreter, Simon (1993). "The Idea of Demographic Transition and the Study of Fertility Change: A Critical Intellectual History." *Population and Development Studies* 19:659–701.
Thieme, Frank (1988). *Rassentheorien zwischen Mythos und Tabu. Ein Beitrag der Sozialwissenschaft zur Entstehung und Wirkung der Rassenideologie in Deutschland*. Frankfurt a.M.: Lang
Thomalla, Curt (1934). "The Sterilization Law in Germany." *Eugenical News* 19:137–140.
Thomson, J. Arthur (1915). "Eugenics and War." *Eugenics Review* 7:1–14.
Thomson, Matthew, and Paul Weindling (1993). "Sterilisationspolitik in Großbritannien und Deutschland." In Franz-Werner Kersting, Karl Teppe, and Bernd Walter, eds. *Nach Hadamar. Zum Verhältnis von Psychiatrie und Gesellschaft im 20. Jahrhundert*. Paderborn: Schöningh, pp. 137–149.
Thums, Karl (1937). "Rückblick auf den Internationalen Kongreß für Bevölkerungswissenschaft 1937 in Paris." *Ziel und Weg* 7:531–536.
Todes, Daniel (1989). *Darwin Without Malthus: The Struggle for Existence in Rus-sian Evolutionary Thought*. New York; Oxford: Oxford University Press.
Trent, James (2001). "'Who Shall Say Who Is a Useful Person?' Abraham Myerson's Opposition to the Eugenics Movement." *History of Psychiatry* 12:33–57.
Trus, Armin (2002). "Der 'heilige Krieg' der Eugeniker." In Gerhard Freiling and Günter Schärer–Pohlmann, eds. *Geschichte und Kritik. Beiträge zur Gesellschaft, Politik und Ideologie in Deutschland*. Gießen: Focus, pp. 245–286.

Tucker, William H. (1995). *The Science and Politics of Racial Research*. Urbana; Chicago: University of Illinois Press.
—— (2002). *The Funding of Scientific Racism: Wycliffe Draper and the Pioneer Fund*. Champaign: University of Illinois Press.
—— (2009). *The Cattell Controversy: Race, Science, and Ideology*. Urbana; Chicago: University of Illinois Press.
Turda, Marius (2007). "The First Debates on Eugenics in Hungary, 1910–1918." In Marius Turda and Paul J. Weindling, eds. *Blood and Homeland: Eugenics and Racial Nationalism in Central and Southeast Europe, 1900–1940*. Budapest; New York: Central European University Press, pp. 185–222.
—— (2008). "Recent Scholarship on Race and Eugenics." *The Historical Journal* 51:1115–1124.
—— (2010a). *Modernism and Eugenics*. Basingstoke: Palgrave Macmillan.
—— (2010b). "Race, Science, and Eugenics in the Twentieth Century." In Alison Bashford and Philippa Levine, eds. *The Oxford Handbook of the History of Eugenics*. New York; Oxford: Oxford University Press, pp. 62–79.
—— (2011). "Controlling the National Body: Ideas of Racial Purification in Romania, 1918–1944." In Christian Promitzer, Sevasti Trubeta, and Marius Turda, eds. *Health, Hygiene and Eugenics in Southeastern Europe to 1945*. Budapest; New York: Central European University Press, pp. 325–350.
Tydén, Matthias (2010). "The Scandinavian States: Reformed Eugenics Applied." In Alison Bashford and Philippa Levine, eds. *The Oxford Handbook of the History of Eugenics*. New York; Oxford: Oxford University Press, pp. 363–376.
UNESCO (1952). *The Race Concept: Results of an Inquiry*. Paris: Unesco.
—— (1969). *Race and Science*. New York: Columbia University Press.
Vavilov, N. I. (1928). "Geographische Genzentren unserer Kulturpflanzen." In Verhandlungen des V. Internationalen Kongresses für Vererbungswissenschaft Berlin 1927, Supplementband I der Zeitschrift für induktive Abstammungs-und Vererbungslehre, pp. 342–369.
Verschuer, Otmar Freiherr von (1931). "Internationale Organisationen auf dem Gebiet der Anthropologie, menschlichen Erblehre und Eugenik." *Forschung und Fortschritt* 7:167–168.
—— (1941). *Leitfaden der Rassenhygiene*. Leipzig: Thieme.
—— (1944). "Bevölkerungs-und Rassenfragen in Europa." *Europäischer Wissenschafts-Dienst*, Issue 1/1944, pp. 1–4.
—— (1961). "Eugenik." In Erwin von Beckerath, Carl Brinkmann und Hermann Bente, *Handwörterbuch der Sozialwissenschaften*. Bd. 3. Stuttgart: Fischer, pp. 356–357.
Voegelin, Eric (1948). "The Origins of Scientism." *Social Research* 4:462–494.
Vogel, Friedrich (1983). "Sind Rassenmischungen biologisch schädlich?" In Horst Seidler and Alois Soritsch, eds. *Rassen und Minderheiten*. Vienna: Literas-Verlag, pp. 9–20.
Wagenen, Bleecker van (1912). "Preliminary Report to the First International Eugenics Congress of the Committee of the Eugenic Section of the American Breeder's Association to Study and Report on the Best Practical Means for Cutting off the Defective Germ-plasm in the Human Population." In Eugenics Education Society, ed. *Problems in Eugenics*. London: The Eugenics Education Society, pp. 460–479.
Wagner, Gerhard (1943). *Reden und Aufrufe: 1888–1939*. Berlin; Vienna: Reichsgesundheitsverlag.
Wagner, Peter (1995). *Soziologie der Moderne*. Frankfurt a.M.; New York: Campus.
Wallace, Bruce (1962). "Race and Reason." *Eugenics Quarterly* 9:161–163.
—— (1975). "Genetics and the Great IQ Controversy." *The American Biology Teacher* 37:12–18.
Watanabe, Tsutomu (2000). "Cross–National Analysis of Social Movements: Effectiveness of the Concept of Political Opportunity Structure." *Sociological Theory and Methods* 15:135–148.

Weber, Marion, and Karin Weisemann (1989). "Wissenschaft und Verantwortung am Beispiel der Humangenetiker P. J. Waardenburg und O. Frhr. von Verschuer." *Medizinhistorisches Journal* 24:163–172.
Weber, Matthias M. (1993). *Ernst Rüdin. Eine kritische Biographie*. Berlin: Springer.
——— (1996) "Ernst Rüdin, 1974–1952. A German Psychiatrist and Geneticist." *American Journal of Medical Genetics (Neupsychiatric Genetics)* 67:323–331.
——— (2000). "Rassenhygienische und genetische Forschungen an der Deutschen Forschungsanstalt für Psychiatrie/Kaiser–Wilhelm–Institut in München vor und nach 1933." In Doris Kaufmann, ed. *Geschichte der Kaiser–Wilhelm–Gesellschaft im Nationalsozialismus. Bestandsaufnahme und Perspektiven der Forschung, Bd. 1*. Göttingen: Wallstein, pp. 95–111.
Weeks, David F. (1912). "The Inheritance of Epilepsy." In Eugenics Education Society, ed. *Problems in Eugenics*. London: The Eugenics Education Society, pp. 62–99.
Weigner, Karel (1935). "Einleitung." In Karel Weigner, ed. *Die Gleichwertigkeit der europäischen Rassen und die Wege zu ihrer Vervollkommnung*. Prag: Verlag der Tschechischen Akademie der Wissenschaften und Kuenste., pp. 5–11.
Weikart, Richard (2003). "Progress through Racial Extermination. Social Darwinism, Eugenics, and Pacifism in Germany, 1860–1918." *German Studies Review* 26:273–294.
——— (2004). *From Darwin to Hitler: Evolutionary Ethics, Eugenics, and Racism in Germany*. New York: Palgrave Macmillan.
Weindling, Paul (1987). "Die Verbreitung rassenhygienischen/eugenischen Gedankenguts in bürgerlichen und sozialistischen Kreisen der Weimarer Republik." *Medizinhistorisches Journal* 22:352–368.
——— (1989a). *Health, Race and German Politics between National Unification and Nazism, 1870–1945*. Cambridge: Cambrige University Press.
——— (1989b). "The Sonderweg of German Eugenics: Nationalism and Scientific Internationalism." *British Journal of the History of Science* 22:321–333.
——— (1993). "The Politics of International Co-ordination to Combat Sexually Transmitted Diseases, 1900–1980s." In Virginia Berridge and Philip Strong, eds. *Aids and Contemporary History*. New York: Cambridge University Press, pp. 97–107.
——— (2007). "Central Europe Confronts German Racial Hygiene: Friedrich Hertz, Hug Iltis and Iganz Zollschan as Critics of Racial Hygiene." In Marius Turda and Paul J. Weindling, eds. *Blood and Homeland: Eugenics and Racial Nationalism in Central and Southeast Europe, 1900–1940*. Budapest; New York: Central European University Press, pp. 263–282.
——— (2011a). "Critics, Commentators and Opponents of Eugenics 1880s–1950s." *East Central Europe* 38:79–96.
——— (2011b). "German Eugenic Paradigms. Racial Expertise and German Eugenic Strategies for Southeastern Europe." In Christian Promitzer, Sevasti Trubeta, and Marius Turda, eds. *Health, Hygiene and Eugenics in Southeastern Europe to 1945*. Budapest; New York: Central European University Press, pp. 27–56.
Weingart, Peter (1994). "The Thin Line Between Eugenics and Preventive Medicine." In Finzch N. and Schirmer D., eds. *Identity and Tolerance*. Washington, DC: Cambridge University Press, pp. 397–412.
——— (2012). "Ist Sarrazin Eugeniker?" In Michael Haller and Martin Niggeschmidt, eds. *Der Mythos vom Niedergang der Intelligenz. Von Galton zu Sarrazin: Die Denkmuster und Denkfehler der Eugenik*. Wiesbaden: VS Verlag für Sozialwissenschaften, pp. 19–26.
Weingart, Peter, Jürgen Kroll, and Kurt Bayertz (1988). *Rasse, Blut und Gene. Geschichte der Eugenik und Rassenhygiene in Deutschland*. Frankfurt a.M.: Suhrkamp.
Weinreich, Max (1946). *Hitler's Professors: The Part of Scholarship in Germany's Crimes Against the Jewish People*. New York: Yiddish Scientific Institute-YIVO.
Weinstein, James (1968). *The Corporate Ideal in the Liberal State: 1900–1918*. Boston: Beacon Press.

Weismann, August (1892). *Das Keimplasma. Eine Theorie der Vererbung.* Jena: Fischer.
Weissman, Steve (1973). "Die Bevölkerungsbombe ist ein Rockefeller-Baby." *Kursbuch*, Issue 33/1973, pp. 81–94.
Weiss, Sheila Faith (1987). *Race Hygiene and National Efficiency: The Eugenics of Wilhelm Schallmayer.* Berkeley; London: University of California Press.
——— (1990). "The Race Hygiene Movement in Germany 1904–1945." In Mark B. Adams, ed. *The Wellborn Science: Eugenics in Germany, France, Brazil and Russia.* New York; Oxford: Oxford University Press, pp. 8–68.
——— (2010a). "After the Fall: Political Whitewashing, Professional Posturing and Personal Refashioning in the Post–War Career of Otmar Freiherr von Verschuer." *Isis* 101:722-758.
——— (2010b). *The Nazi Symbiosis: Human Genetics and Politics in the Third Reich.* Chicago; London: University of Chicago Press.
Weiss, Volkmar (1982). *Psychogenetik. Humangenetik in Psychologie und Psychiatrie.* Jena: Gustav Fischer.
——— (2000): *Die IQ–Falle. Intelligenz, Sozialstruktur und Politik.* Graz: Stocker.
——— (2012). *Die Intelligenz und ihre Feinde. Aufstieg und Niedergang der Industriegesellschaft.* Graz: ARES Verlag.
Wendt, Georg Gerhard, und Peter E. Becker (1975). *Erbkrankheiten: Risiko und Verhütung: Bericht über die Tagung am 17. Und 18. Februar 1975 in Marburg a.d. Lahn; mit 39 Tabellen und einem Bericht über den dreijährigen Modellversuch „Genetische Beratungsstelle für Nordhessen".* Marburg: Medizinische Verlagsgesellschaft.
Weß, Ludger (1986). "Aktuelle Programme der Humangenetik. Moderne Methoden—altbekannte Ziele." *Mitteilungen der Dokumentationsstelle zur NS-Sozialpolitik*, Issue 11/1986, pp. 5–48.
——— (1989). *Die Träume der Genetik. Gentechnische Utopien von sozialem Fortschritt.* Nördlingen: Greno.
Weyer, Johannes (1984). *Westdeutsche Soziologie 1945–1960. Deutsche Kontinuitäten und nordamerikanischer Einfluß.* Berlin: Duncker & Humblot.
——— (1986). "Der 'Bürgerkrieg in der Soziologie'. Die Westdeutsche Soziologie zwischen Amerikanisierung und Restauration." In Sven Papcke, ed. *Ordnung und Theorie. Beiträge zur Geschichte der Soziologie in Deutschland.* Darmstadt: Wiss. Buchges., pp. 280–304.
Weyher, Harry F. (2001). "My Years with the Pioneer Fund." In Richard Lynn, ed. *The Science of Human Diversity: A History of the Pioneer Fund.* Lanham: University Press of America, pp. ix–lxii.
Weyl, Nathaniel (1973a). "Population Control and the Anti-eugenic Ideology." *Mankind Quarterly* 14:63–82.
——— (1973b). "Race, Nationality and Crime." *Mankind Quarterly* 14:41–48.
Wichterich, Christa (1994). "Menschen nach Maß—Bevölkerung nach Plan—Die Neue Weltordnung der Fortpflanzung." In Christa Wichterich, ed. *Menschen nach Maß. Bevölkerungspolitik in Nord und Süd.* Göttingen: Larnuv, pp. 9–39.
Wiebe, Robert (1976). *The Search for Order.* New York: American Century Series.
Wieth-Knudsen, Knud Asbjörn (1936). "Das Bevölkerungsproblem des Nordens." In Hans Harmsen and Franz Lohse, eds. *Bevölkerungsfragen. Bericht des Internationalen Kongresses für Bevölkerungswissenschaft, Berlin 8/26–9/1/1935.* Munich: J. F. Lehmanns, pp. 98–105.
Wiggam, Albert Edward (1923). *The New Decalogue of Science.* Indianapolis: The Bobbs-Merrill Company.
Winau, Rolf (1989). "Die Freigabe der Vernichtung 'lebensunwerten Lebens.'" In Johanna Bleker and Norbert Jachertz, eds. *Medizin im "Dritten Reich."* Köln: Deutscher Ärzte-Verlag, pp. 162–174.
Winkler, Wilhelm (1936). "Der Geburtenrückgang in Österreich." In Hans Harmsen and Franz Lohse, eds. *Bevölkerungsfragen. Bericht des Internationalen Kongresses für Bevölkerungswissenschaft, Berlin 8/26–9/1/1935.* Munich: J. F. Lehmanns, pp. 108–114.

Winston, Andrew (1998). "Science in the Service of the Far Right: Henry E. Garrett, the IAAEE and the Liberty Lobby." *Journal of Social Issues* 53:179–209.
Wolf, Julius (1933). "Differenzialgeburtenziffer bei den verschiedenen Gesellschaftsklassen." In Corrado Gini, ed. *Atti del congresso internazionale per gli studi sulla popolazione. Rome, 7–10 septembre 1931. Vol. VIII.* Rome: Istituto poligrafico della Stato, pp. 85–88.
Wolf, Maria A. (2008). *Eugenische Vernunft. Eingriffe in die reproduktive Kultur durch die Medizin 1900–2000.* Vienna; Köln; Weimar: Böhlau.
Wright, David (1990). "The Study of Idiocy. The Professional Middle Class and the Evolution of Social Policy on the Mentally Retarded in England, 1848–1914." Montreal: MA thesis an der McGill University.
Zmarzlik, Hans-Günther (1963). "Der Sozialdarwinismus in Deutschland als geschichtliches Problem." *Vierteljahreshefte für Zeitgeschichte* 11:246–273.
Zollschan, Ignaz (1938). "Die Bedeutung des Rassenfaktors für die Kulturgenese." In Démographie de la France-d'Outremer, ed, *Congrès International de la Population, Paris 1937.* Paris: Hermann et Cie., Éditeurs, pp.93–105.
——— and Julian Huxley (1942). *Racialism against Civilisation.* London: New Europe Publ. Co.

INDEX OF PERSONS

Abel, Wolfgang 128, 135
Abir-Am, Pnina 69
Adam, George 41
Allen, Garland 83
Allen, Gordon 157
Ammon, Otto 30, 32
Amsel, Hans Georg 175
Anderson, Hjalmar 36
Angell, Robert C. 139
Apert, Eugene 16, 25, 109
Armstrong, Clariette 165
Astel, Karl 97, 106, 108
Aznar, Severino 87

Bajema, Carl 146
Baker, John R. 160, 171f, 174ff
Balfour, Arthur James 26
Balibar, Étienne 52
Bambaren, Carlos 110
Barderleben, General Carl von 33
Barker, Lewellys F. 39
Barlow, Thomas 18
Bauman, Zygmunt 52
Baur, Erwin 44f, 48, 51, 84, 92, 165
Beck, Maximilian 113f, 116
Becker, Peter Emil 136
Bedwell, Cyril E.A. 42
Beethoven, Ludwig van 22
Bell, Alexander Graham 18
Bell, Julia 79
Bellamy, Edward 64
Benoist, Alain de 174ff
Berger, Erich 95
Bergmann, Ernst 125
Bernard, Leon 87
Bertillon, Jacques 16
Bertram, Colin 166

Binding, Karl 124
Biswas, P.C. 168
Blacker, C.P. 74f, 79, 81f, 89, 101, 130ff, 141, 146ff, 149, 152, 154, 159, 181
Bluhm, Agnes 44, 93, 124
Boas, Franz 110ff, 116, 131, 141, 169
Bodenheimer, Fritz S. 152
Bodmer, Walter F. 171, 173
Boldrini, Marcello 34
Bonnevie, Kristine 80, 108
Boudreau, Frank G. 151, 155
Bougle, Celestin 112
Bouhler, Philipp 126
Boverat, Fernand 100
Brandt, Karl 126
Brush, Dorothy 148, 154
Burgdörfer, Friedrich 84, 93, 101, 108, 127
Burkhardt, Hans 168

Cadbury, George W. 154
Campbell, Clarence G. 101, 103f, 131
Carnegie, Andrew 13
Carr-Saunders, Alexander M. 81f, 100
Carrel, Alexis 133, 179
Carter, Cedric O. 146, 150
Carto, Willis A. 168
Castex, Mariano R. 109
Castle, William 55, 74
Cattell, Raymond 170
Caty, Louis 24
Cavalli-Sforza, Luigi Luca 171, 173
Chamberlain, Houston Stewart 51, 92, 179
Chetverikov, S.S. 68
Chiarelli, Brunetto 177
Churchill, Winston 18, 96
Clauß, Ludwig F. 136
Close, Charles 88, 99ff, 104, 114f

INDEX OF PERSONS

Cole, Leon J. 117
Comas, Juan 139, 164, 168f
Conklin Edwin G. 38, 40, 140
Cook, Robert C. 75, 117, 149, 176
Coon, Carleton S. 165
Correa, Adrián 109
Cox, Ernest Sevier 166
Crew, Francis Albert Eley 34, 78, 82, 85f, 118
Cuénot, Lucien 40, 85

Dahlberg, Gunnar 69, 80, 104, 107ff, 116, 118, 130ff, 138, 140f, 145
Darlington, Cyril Dean 118, 142f, 145, 160, 164f, 175f
Darwin, Charles 13f, 30
Darwin, Charles Galton 160, 167
Darwin, Leonard 24, 33, 36f, 40, 48, 52, 65, 101, 107, 130, 160
Davenport, Charles B. 15, 20, 24, 39f, 42ff, 47, 49, 51f, 54ff, 63ff, 68, 71f, 74, 77f, 80, 82, 85ff, 98, 101, 104f, 107, 118, 122, 131f, 134, 145, 159, 181
Davis, Watson 118
Davis, Kingsley 152
Dice, Lee R. 149f
Dobshansky, Theodosius 118, 140f, 145f, 159, 168, 171
Doumer, Paul 24
Draper, Wickliffe P. 47f, 57, 96, 101, 157f, 160ff, 166f, 169, 176
Drysdale, Charles Vickery 65
Dublin, Louis 88
Duncan, Otis Dudley 174
Dunn, L.C. 69, 74, 140f, 145, 164f, 168
Dupont, A.F. 66

East, Edward M. 74, 81
Eckland, Bruce K. 174
Edin, Karl Avid 83, 114
Ehrenfels, Umar Rolf von 169
Eickstedt, Egon von 164f
Ellinger, T.U.H. 128, 160
Ellis, Havelock 37, 64
Emerson, Rollins Adams 118
Eysenck, Hans J. 171ff, 176, 178

Fairchild, David 39
Fairchild, Henry Pratt 81, 84, 86, 147
Fantham, Harold B. 53

Faure, Fernand 24
Federley, Harry 53, 55, 100, 107
Fetscher, René Rainer 97
Filipchenko, Jurius 68
Fischer, Eugen 27, 53, 55, 58, 60, 72, 84, 87, 89, 91ff, 99ff, 103, 105ff, 114, 128ff, 144, 164f
Fisher, Irving 29, 31, 34, 37, 42, 52
Fisher, Ronald A. 40, 52, 55, 79, 143
Fleure, Herbert J. 141
Forel, Auguste 18, 53, 55
Frets, G.P. 63, 80, 85, 97f, 100, 105ff, 109, 114f, 130f
Freyer, Hans 164
Frick, Wilhelm 93, 99, 108, 122f
Fürth, Henriette 64

Galton, Francis 11ff, 16ff, 22, 25, 30, 34, 37, 41, 75, 101, 142, 159, 172, 184
Gardner, Eldon John 147
Garrett, Henry E. 157ff, 162f, 165, 167ff, 173ff
Gates, Reginald Ruggles 40, 51f, 55, 59, 79, 133, 138, 157ff, 166ff, 173
Gayre, Robert 157f, 163, 166ff, 173, 175
Gedda, Luigi 165, 167, 170, 175
Gehlen, Arnold 164
Gemelli, Agostino 109
George, Wesley C. 157
Geyer, Horst 129, 136
Gilbert, Oliver 166
Gilman, Robbins 11
Gini, Corrado 23f, 53, 63, 73, 78, 82f, 85, 87f, 100, 109f, 133, 147, 162, 164f, 167f, 170
Ginsberg, Morris 139
Glass, Bentley 146
Glass, David V. 102, 104, 147
Gobineau, Joseph Arthur de 51, 101
Goebbels, Josef 99
Goethe, Charles M. 106
Goldschmidt, Richard 82, 135
Gordon, Robert A. 172, 177
Gottfredson, Linda 177
Gotto, Sybil 24
Govaerts, Albert 109
Grant, Madison 50, 61f, 101, 159, 179
Grebe, Hans 136
Gregor, A. James 163f, 165, 167

Grobig, Hermann Ernst 127
Groß, Walter 91, 95, 97, 100, 122, 125, 127, 135, 158, 163
Grotjahn, Alfred 38, 82, 84
Guangdan, Pan 54
Grüneberg, Hans 135
Günther, Hans F.K. 72, 92, 135, 158, 163, 166, 179
Gütt, Arthur 102
Guttmacher, Alan F. 154

Haag, Ernest van den 165
Haddon, Alfred C. 110f
Haeckel, Ernst 34
Haldane, J.B.S. 67f, 74, 79f, 82, 110, 116, 118, 138f, 141, 145, 168
Hall, Prescott F. 47, 62
Hankins, Frank H. 101
Hansen, Sören 23f, 53, 85
Hardin, Garrett J. 174
Harmsen, Hans 84, 101f, 147, 154
Helms, Jesse 177
Henlein, Konrad 115
Herriot, Edouard 122
Herrnstein, Richard J. 171ff, 177ff
Herskovits, Melville 55, 110
Herwerden, Marianne van 82, 85
Heß, Rudolf 94, 108
Heuyer, Georges 107
Heyde, Werner 126
Hilpert, Paul 91
Hitler, Adolf 53, 91ff, 98ff, 103, 107, 111, 121f, 125f, 134, 138, 158, 166, 174, 178
Hoche, Alfred 124
Hodson, Cora 66, 79, 86, 100, 102, 106, 108, 130f, 133, 137
Höhn, Charlotte 181
Hoffman, Frederick Ludwig 21
Hoffmann, Geza von 34
Hofmeyr, J.D.J. 165
Hogben, Lancelot 74, 80, 116
Hoover, Herbert 39, 96
Houghton, Vera 154
Houssay, Frederic 24
Hunt, Harrison R. 117
Huntington, Ellsworth 75, 84
Huxley, Julian S. 74, 81f, 101, 110f, 116ff, 130f, 139ff, 146, 160

Inge, Wiliam R. 31
Issac, Julius 147
Itzkoff, Seymour 177f

Jennings, Herbert S. 40, 43, 74
Jensen, Arthur 171, 176f
Jensen, Thit 64
Johannsen, Wilhelm 16
Johnson, Albert 62
Johnson, Roswell H. 31, 33f, 49ff, 81
Jordan, David Starr 31
Jürgens, Hans Wilhelm 174, 176

Kabir, Humayun 139
Kallmann, Franz 147
Kaplan, Arnold R. 158
Kaup, Ignaz 37
Kehl, Renato Ferraz 109
Keilhau, Wilhelm 53, 82
Keiter, Friedrich 101f, 135, 164f
Keith, Arthur 79, 101, 141, 165
Kellogg, Vernon 30ff
Kemp, Tage 80, 89, 106, 130, 133, 138, 145, 147, 149f
Keynes, John Maynard 82
Kirk, Dudley 155
Kiser, Clyde V. 84, 147, 151
Klineberg, Otto 55, 139, 164
Koch, Gerhard 136
Koller, Siegfried 101, 135
Koltzoff, N.N. 68
Komai, Taku 168
Kosiek, Rolf 175
Krohne, Otto 45
Kropotkin, Pëtr Alekseevič 22
Kuczynski, Robert 84
Kürten, Heinz 97
Kusserow, Wilhelm 166
Kuttner, Robert 163f, 167, 175

Lamarck, Jean-Baptiste de 20
Lamoine, Roger 174
Landauer, Walter 67, 116f, 118, 131
Landman, Jacob H. 241
Landry, Adolphe 100, 112ff, 147
Lapouge, Georges Vacher de 50, 66, 101, 165
Laughlin, Harry H. 49ff, 55, 62f, 74, 80, 82, 85f, 95f, 98, 101f, 104, 131f
Laugier, Henri 112

Lehmann, Wolfgang 136
Lenz, Fritz 32, 35, 45, 50ff, 60, 84, 92f, 108, 124, 126, 129, 135ff, 143, 147, 153, 164f
Leslie, Murray 21
Lévi-Strauss, Claude 139, 164
Levin, Michael A. 177
Levit, Solomon G. 117
Lewis, Aubrey Julian 146
Lidbetter, Ernest 20, 83
Lincoln, Abraham 22
Linden, Herbert 108f, 126
Linder, Arthur 101
Lindsay, James Alexander 37
Little, Clarence C. 40, 78, 81, 84, 117
Livi, Livio 100, 114, 147
Loeffler, Lothar 97, 108, 135f
Lohse, Franz 102
Lombroso, Cesare 172
Lorimer, Frank 84, 86, 130, 147, 152
Lundborg, Herman 38, 43, 53, 55f, 59, 63f, 80, 83, 95, 100, 103f, 106f, 117, 130f
Lundman, Bertil J. 175
Luther, Martin 22
Luxenburger, Hans 125
Lynn, Richard 177, 179f

MacBride, Ernst W. 52
Macfie, Ronald Campbell 30
Mahaim, Ernst 114
Maier, Hans W. 106
Mallet, Bernard 81, 84, 86
Malthus, Thomas Robert 63f
Manouvrier, Leonce 16
March, Lucien 24, 26, 36, 42, 66, 82, 85
Marin, Louis 129
Marinesco, G. 109
Masaryk, Tomás Garrigue 112
Mauco, Georges 115, 129f
McGurk, Frank J.C. 162
Melville, Colonel Charles H. 30, 32
Mendel, Gregor Johann 19, 21
Mengele, Josef 128, 135
Methorst, Henri W. 82f, 114
Métraux, Alfred 141
Michels, Roberto 18
Mjöen, Jon Alfred 16, 23f, 41, 50, 53, 59, 80, 95, 98, 100, 105f, 123

Mohr, Otto 69, 80, 104f, 108ff, 130, 138
Mombert, Paul 84
Montagu, Ashley 139ff, 145
Moore, Eldon 84
Morgan, Thomas H. 40, 43, 74
Mosley, Oswald 163
Mozart, Wolfgang Amadeus 22
Muckermann, Hermann 51, 84, 93, 105
Müller, Karl Valentin 135, 164
Muller, Hermann Joseph 40f, 67ff, 77, 80, 113, 116ff, 131, 138f, 142ff
Murray, Charles 178f
Mussolini, Benito 7, 87, 96
Myerson, Abraham 41
Myrdal, Alva 132
Myrdal, Gunnar 132

Nachtsheim, Hans 128, 135, 141, 143ff, 149, 152f, 168
Nagai, Hisomu 54
Napoleon 22
Needham, Joseph 67, 116, 118, 139f
Neel, James 150
Niceforo, Alfredo 18, 23, 82
Nilsson-Ehle, Herman 53, 55, 58, 80, 95, 100, 104, 106f, 117, 131
Nitsche, Paul 126
Notestein, Frank 130, 151, 155

Oliver, Clarence P. 147, 161, 167
Olson, Harry F. 82
Osborn, Frederick H. 74, 81, 84, 114, 130f, 155, 158, 173
Osborn, Henry Fairfield 38, 74
Osborne, R. Travis 177
Ossietzky, Carl von 124
Ottesen-Jensen, Elise 153

Panse, Friedrich 174
Paschall, Davis Y. 165
Pearl, Raymond 20, 40, 42, 74, 77, 81ff, 86ff, 99, 104
Pearson, Karl 16, 30, 142, 172
Pearson, Roger 166, 173, 175, 177
Penrose, Lionel 80, 142
Pfeil, Elisabeth 114
Pinchot, Gifford 18, 39

Pitt-Rivers, Captain George H.L.F. 53, 85, 88, 95, 100, 104, 114f, 131
Platen-Hallermund, Alice 126
Ploetz, Alfred 14ff, 22ff, 31, 35, 37, 43ff, 50, 53ff, 64ff, 74, 82, 91ff, 97, 101, 105, 121ff, 134, 137, 145, 181, 184
Pohlisch, Kurt 105, 126, 134f
Popenoe, Paul 31, 34, 36f, 39
Possony, Stefan T. 175
Poulton, Edward Bagnall 34
Punnett, Reginald 20
Putnam, Carleton 157, 165, 168, 176

Querton, Louis 24

Ramos, Arthur 139
Raphael, Nancy Rose 154
Rau, Rama 153
Reagan, Ronald 177f
Reche, Otto 165
Reed, Sheldon C. 147, 149ff, 153, 161
Reichel, Heinrich 87, 107
Reiter, Hans 108
Renner, Otto 116
Repp, Günther 153
Rhodes, E.C. 88
Rieger, Jürgen 175f, 181
Ritter, Robert 114
Rivet, Paul 112
Roberts, J.A. Fraser 146f, 150, 161
Rockefeller, John D. 13
Rockefeller III, John D. 155f
Rodenwaldt, Ernst 58, 97, 105, 113f
Rolleston, Humphrey 82
Roosevelt, Theodore 96
Ross, Edward A. 31
Roth, Karl Heinz 119
Rüdin, Ernst 15f, 35, 53ff, 61, 63f, 84, 89, 93, 97, 101f, 104ff, 112, 114, 117, 121, 124, 126, 135ff, 176
Rushton, Philippe J. 172f, 177, 181
Ruttke, Falk 97f, 102, 106, 108, 125f

Sais, Puig 109
Saleeby, Caleb W. 34
Saller, Karl 144

Sand, Rene 67, 82
Sanders, Jacob 47, 105, 131
Sanger, Margaret 64, 81ff, 153, 155
Sarrazin, Thilo 179f
Sauvy, Alfred 147
Schade, Heinrich 103, 135, 168, 174
Schallmayer, Wilhelm 31
Schaxel, Julius 116f
Scheidt, Walter 144f, 164f, 168
Schieffelin, Barbara 56
Schlaginhaufen, Otto 53, 55f, 100
Schneider, William H. 110
Schreiber, Georges 53, 97f, 108, 110
Schrijver, Franz 103
Schuster, Edgar 31
Schwidetzky, Ilse 164f, 168, 174
Scott, Ralph S. 175
Serebrovskii, A.S. 68
Sergi, Giuseppe 34
Sergi, Sergio 167
Shapiro, Harry L. 146, 164
Shapiro, Thomas M. 134
Shockley, William B. 173f, 177
Shuey, Audrey M. 167
Siegfried, Andre 51, 62, 82
Sjögren, Torsten 53, 55, 104, 106ff, 117, 130, 137, 147, 167, 170
Skerlj, Božo 169
Slater, Eliot T.O. 147, 174
Smith, Frank 31
Smith, Samuel G. 22
Snyder, Laurence H. 80, 133, 146, 161
Sovorgnan, Franco 100
Spearman, Charles 56
Spencer, Herbert 30
Stähle, Eugen 125
Steggerda, Morris 56f, 86, 131
Steinberg, Arthur G. 147
Stephenson, W.R.S. 56
Stern, Curt 140
Stoddard, Lothrop 40, 101, 128
Stöcker, Helene 64
Stone, Abraham 155
Stone, Robert 15
Stopes, Marie 64, 100, 153
Sturtevant, Alfred H. 144f
Sutter, Jean 133, 147
Swan, Donald A. 163ff, 167f, 175f

Tanner, James M. 146
Taylor, Frederick Winslow 13
Teleki, Paul 35
Thomalla, Kurt 99, 108
Thompson, Warren S. 114, 130, 152
Thomson, J. Arthur 32
Thums, Karl 114, 176
Tietze, Christopher 156
Tildesley, Miriam L. 56, 141
Tirala, Lothar 97
Trocchio, Federico di 171

Ungern-Sternberg, Roderich von 48

Valla, Jean-Claude 174
Vandervelde, Émile 96
Vavilov, N.I. 77, 118
Verschuer, Otmar Freiherr von 93, 97, 99, 103, 105, 109, 127ff, 135ff, 147, 168
Vogel, Friedrich 174
Vogt, William 155

Waardenburg, Petrus J. 80, 107, 147
Wagenen, Bleecker van 21
Wad, Gunnar 106
Wagner, Gerhard 98
Wagner, Richard 22
Wallace, Bruce 158, 173
Walter, Francis E. 161

Ward, Robert DeCourcy 62
Warden, Colonel 33
Webster, Martin 166
Weeks, David F. 20
Weinert, Hans 135, 144f, 165
Weismann, August 19ff, 24
Wendt, Georg 174
Wettstein, Fritz von 108
Weyher, Harry F. 164, 166ff
Weyl, Nathaniel 172
Whelpton, Pascal 151
White, Arnold 33
Whitney, Leon F. 95
Whyte, G.A. 154
Wieth-Knudsen, Knud Asbjörn 101
Wiggam, Albert E. 49f, 179 (Alfred)
Wilson, Edwin B. 87
Wimmer, August 53, 55, 85
Winkler, Wilhelm 101
Wirz, Franz 123f
Wolf, Julius 84
Woltmann, Ludwig 51
Woods, Adam 24

Yates, Frank 147
Yerkes, Robert M. 40

Zahn, Friedrich 84
Zollschan, Ignaz 110ff, 116, 146

Organizations, Conferences, Journals and Newspapers

American Anthropological Association 172
American Birth Control League 84
American Breeders Association 15, 21f, 24, 39
American Ethnology Smithsonian Institution 44
American Eugenics Society 42, 49, 62, 73ff, 78, 84, 86, 95, 99, 101, 117, 131, 134, 146f, 149, 151ff, 157ff, 173f; *see also* Society for the Study of Social Biology
American Genetic Association 39
American Mercury 163, 215
American Museum of Natural History 38, 146
American Psychological Association 159
American Sociological Society 172
Annals of Eugenics 142
Annals of Human Genetics 142
Archiv für Rassen- und Gesellschaftsbiologie 31, 44, 92, 107
Around the World News of Population and Birth Control 154f
Associacion Argentina de Biotipologia, Eugenesia y Medicina Social 109
Association for Voluntary Sterilization 154

Berliner Illustrierte Nachtausgabe 124
Berliner Tageblatt 18
Birth Control International Information Center 154
Birth Control Investigation Committee 154
Birth Control Review 64

British Birth Control Association 154
British National Service League 33
British Population Society 84
Bund Heimattreuer Jugend 175
Bureau of Human Heredity 78f

Central Association for Mental Welfare 17
Cold Spring Harbor Laboratory 57, 60, 159
Comissao Central Brasileira de Eugenia 109
Comitato italiano per lo studio dei problemi della popolazione 85
Council for Social and Economic Studies 177f
Current Anthropology 168f
Czechoslovak State Office for Eugenics 79

Danish Anthropological Society 23f
Deutsche Allgemeine Zeitung 103
Dight Institute 147, 153, 161

Éléments 175
Erbarzt 103
Erfelijkheid bij de Mens 103
Eugenetsche Vereeniging in Nederlandsch-Indië 54
Eugenical News 66, 73, 83, 98f, 103, 107, 131f, 147; *see also Eugenics Quarterly* and *Journal of Social Biology*
Eugenics Education Society 12, 16ff, 20, 22ff, 26, 33f, 36f, 160; *see also* Galton Institute
Eugenics Quarterly 149, 158f, 165; see also *Eugenical News* and *Journal of Social Biology*

Eugenics Record Office in Cold Spring
    Harbor 20, 26, 39, 49, 74, 79
Eugenics Registry of the Race
    Betterment Foundation 39
Eugenics Research Association 39, 56, 73,
    86, 98, 101, 165
Eugenics Review 34, 79, 84, 104, 149;
    see also *Journal of Biosocial Science*
Eugenika 169
The European 163

Fédération Internationale Latine des
    Sociétés d'Eugénique 109, 188
Ford Foundation 155
Foreign Policy Research Institute 178
Fondation française pour l'étude des
    problèmes humains 133
Foundation for Human
    Understanding 177f
Foundation for Race Betterment 154
Foundation for Research and Education
    on Eugenics and Dysgenics 177

Galton Institute 149
Galton Society 39, 73, 98
Genealogical Record Office 39
Genetics Society of America 118, 172
German Society for Genetic Science 116
German Society for Hereditary Health
    Care 153
German Society for Race
    Hygiene 44f, 93f, 99, 108, 124;
    see also International Society
    for Race Hygiene
German Statistical Society 99
Groupement de recherche et d'etudes
    pour la civilisation europeenne
    (GRECE) 174f, 178

Heritage Foundation 178
Herold 33

Immigration Restriction League 62
Imperial Bureau of Animal Genetics 78
Indian Eugenics Society 54
Institut International d'Anthropologie
    109, 129
Institut National d'Etudes
    Demographiques 133

Institute for Human Heredity
    Research 136
Institute for Research on the Gifted
    for the Ministry of Education of Lower
    Saxony 135
Institute for the Study of Man 163, 177f
International Association for the
    Advancement of Ethnology and
    Eugenics 158f, 161, 181
International Committee for Planned
    Parenthood 154; see also
    International Planned
    Parenthood Federation
International Congress for Anthropology
    and Ethnology (1934) 111
International Congress for Child
    Psychiatry, First (1937) 121
International Congress for Mental Health,
    Second (1937) 121
International Congress for Population
    Research (1931) 87ff, 99f
International Congress of Genetics,
    Fifth (1927) 76, 118
International Congress of Genetics,
    Sixth (1932) 78
International Congress of Genetics,
    Seventh (1937) 116f
International Congress of Human
    Genetics, First (1956) 146
International Eugenic Conference
    Amsterdam (1927) 45
International Eugenic Conference
    Brussels (1922) 44f
International Eugenic Conference
    Farnham (1930) 59, 78
International Eugenic Conference
    London (1919) 38
International Eugenic Conference Lund
    (1923) 45
International Eugenic Conference
    Milan (1924) 45, 67
International Eugenic Conference
    Munich (1928) 47, 57ff, 63
International Eugenic Conference Paris
    (1913) 23ff
International Eugenic Conference Paris
    (1926) 51, 62
International Eugenic Conference Rome
    (1929) 59, 87

International Eugenic Conference
  Scheveningen (1936) 105ff, 115
International Eugenic Conference Zurich
  (1934) 97ff, 115, 123, 137
International Eugenic Congress, First
  (1912) 17f, 71, 83
International Eugenic Congress, Second
  (1921) 38ff, 55, 71, 76
International Eugenic Congress, Third
  (1932) 65, 68, 71ff, 110, 113
International Eugenic Congress, Fourth
  (1940) 108ff
International Federation of Eugenic
  Organization 52, 72f, 76ff, 80f,
  85ff, 96ff, 102, 109f, 115f, 123, 126,
  129f, 133f, 137; see also Permanent
  International Eugenics Commission
  and Permanent International
  Eugenics Committee
International Group for Human
  Heredity 80, 137, 188
International Human Heredity
  Committee 109
International Hygiene Exposition
  (1911) 17
International Institute for Advanced
  Race Research 166
International Institute of Intellectual
  Cooperation 112, 140
International Institute of Sociology
  133, 164
International Labor Office 182
International Planned Parenthood
  Federation 153, 182f; see also
  International Committee for Planned
  Parenthood
International Society for Race
  Hygiene 15ff, 19, 22ff, 26, 35
International Union for the Scientific
  Investigation of Population
  Problems 63, 84, 96, 99f, 104, 107,
  114f, 129, 188; see also International
  Union for the Scientific Study of
  Population
International Union for the Scientific
  Study of Population 83, 147, 183;
  see also International Union for
  the Scientific Investigation of
  Population Problems

Japanese Society for Race Hygiene 54
Journal of Biosocial Science 149; see also
  Eugenics Review
Journal of Heredity 34, 41, 75, 117
Journal of Social Biology 149; see also
  Eugenical News and *Eugenics Quarterly*
Journal of Social Hygiene 32
Julius-Klaus-Foundation for Anthropology
  and Genetics 79

Kaiser-Wilhelm-Institute for Anthropology,
  Human Genetics, and Eugenics 60, 79,
  82, 93, 103, 128, 135f
Kaiser-Wilhelm-Institute for
  Psychiatry 79, 107, 114, 125, 127, 137
Kenya Society for the Study of Race
  Improvement 54
Ku-Kux-Klan 166f

Le Matin 18
League of Red Cross Societies 66f, 82
Liga nacional de Higiene y Profilaxia
  Social 110

Malthusian League 65
Man 168
Mankind Quarterly 166ff, 172ff
Mensch und Maß 175
Milbank Memorial Fund 87, 151f, 155
Moral Education League 17

Nation Europa 163
National Committee of Maternal
  Health 156
National Conference for Race
  Betterment 11, 36
National Council for Mental Hygiene 17
Neue Anthropologie 174ff, 178f, 181
Neues Volk 98
New Health Society 17
New Jersey State Village for Epileptics 20
New York Times 82, 140
Nieuwe Rotterdamische Courant 106
Northern League 166ff
Northlander 167, 174; see also
  Western Destiny
Norwegian eugenics consultation
  committee 23
Nouvelle Ecole 174ff, 178f

Oficina Central Panamericana de
  Eugenesia y Homicultura 54, 63, 188

Permanent International Eugenics
  Commission 43ff, 52; International
  Federation of Eugenic Organization
  and Permanent International Eugenics
  Committee
Permanent International Eugenics
  Committee 22ff, 26, 35, 38, 40, 42f;
  see also International Federation of
  Eugenic Organization and Permanent
  International Eugenics Commission
Pioneer Fund 47, 96, 158,
  160f, 163f, 166f, 172, 176ff, 181, 188
Planned Parenthood Federation of
  America 154; see also International
  Planned Parenthood Federation
Playboy 173
Population 88
Population Association of America
  84, 151
Population Council 155f, 182f
Population Investigation Committee
  84, 154
Population Reference Bureau 84
Pro Familia Deutschland 154
Professional Classes War Relief Council 36
Public Opinion 18

Race Policy Office of the NSDAP 91,
  95f, 98, 117
Revue Anthropologique 97
Revue Internationale de Sociologie 164
Revue Politique et Parlementaire 24
Right 163, 168
Rockefeller Foundation 137, 155
Royal Anthropological Institute 141
Royal Army Medical College 30
Royal College of Surgeons 56
Royal Society of Edinburgh 34

Science Service 118
Scientific Monthly 159
Scripps Memorial Foundation for
  Population Research 114, 152
Svenska Sällskapet för Rashygien 16, 22
Swedish State Institute for Race
  Biology 36, 79
Sociedad mexicana de Eugenesia 109f
Società italiana di genetica ed
  eugenitica 23f, 109
Societé francaise d'eugénique 23, 36,
  66f, 129
Society for Biological Anthropology,
  Eugenics, and Behavioral
  Research 175f, 178
Society for the Psychological Study of
  Social Issues 172
Society for the Study of Social
  Biology 149; see also American Eugenics
  Society
Society Solvay Institut 23f
South African Eugenics Society 54

UNESCO 134, 138ff, 162ff, 170, 182
United Nations Fund for Population
  Activities 183

Vinderen Laboratory 24
Vossische Zeitung 18

Western Destiny 174; see also *Northlander*
World Anti-Communist League 177
World Peace Foundation 31
World Population Congress, First (1927)
  81ff, 100f
World Population Congress, Second (1935)
  99ff, 110, 122
World Population Congress, Third (1937)
  107ff, 114ff, 130

Ziel und Weg 94f

The manufacturer's authorised representative in the EU is Springer Nature Customer Service Centre GmbH, Europaplatz 3, 69115 Heidelberg, Germany. If you have any concerns regarding our products, please contact ProductSafety@springernature.com

Printed and bound by CPI Group (UK) Ltd, Croydon, CR0 4YY
23/03/2026
02076449-0019